Neuroendocrine Perspectives

E.E. Müller and R.M. MacLeod
Series Editors

John A.H. Wass Maurice F. Scanlon
Editors

Neuroendocrine Perspectives

Volume 6

Proceedings of the Third Congress of the European Neuroendocrine Association, September 9–11, 1987, London, U.K.

With 113 Illustrations

Springer-Verlag New York Berlin Heidelberg
London Paris Tokyo

J.A.H. Wass
Department of Endocrinology
St. Bartholomew's Hospital
London EC1A 7BE, U.K.

M.F. Scanlon
Department of Medicine
University of Wales College
of Medicine
Cardiff CF4 4XN, U.K.

Series Editors
Eugenio E. Müller
Department of Pharmacology
University of Milan
School of Medicine
20129 Milan, Italy

Robert M. MacLeod
Department of Internal Medicine
University of Virginia
School of Medicine
Charlottesville, Virginia 22908,
U.S.A.

Assistant Editor
Jo Ann M. Eliason
Department of Internal Medicine
University of Virginia
School of Medicine
Charlottesville, Virginia 22908,
U.S.A.

Volumes 1–5 of *Neuroendocrine Perspectives* were published by Elsevier Science Publishers B.V., Amsterdam, The Netherlands.

ISSN: 0168–0617

Printed on acid-free paper

Softcover reprint of the hardcover 1st edition 1989

Camera-ready copy provided by the editors.

9 8 7 6 5 4 3 2 1

ISBN-13: 978-1-4612-8117-7 e-ISBN-13: 978-1-4612-3478-4
DOI: 10.1007/978-1-4612-3478-4

CONTENTS

A. DEVELOPING AREAS OF NEUROENDOCRINE RESEARCH

B. THE CONTROL OF GONADOTROPIN SECRETION

C. GROWTH HORMONE NEUROREGULATION I

D. THE CONTROL AND FUNCTION OF THE OPIATE PATHWAYS

E. GROWTH HORMONE NEUROREGULATION II

F. THE FUTURE OF NEUROENDOCRINE RESEARCH

A. DEVELOPING AREAS OF NEUROENDOCRINE RESEARCH

INTRODUCTION

In the first section of this volume, we have attempted to bring together some of the papers that reflect exciting new areas of development in relation to neuroendocrine investigation. Very recently, specific nucleotide recognition sequences for thyroid hormones, steroid hormones and the fundamental intracellular regulator, cyclic AMP, have been determined. In this section, the preliminary characterization and investigation of the nuclear binding proteins that mediate the actions of cAMP are described. Not only does this represent an important advance in cell biology, but it may represent a further level of alteration in control in normal and disease states. Intercellular pituitary "cross talk" is well described in the in vitro setting. The active molecules are clearly able to exert significant actions at the subnanomolar level. It is now important to design experiments in order to define the precise physiological relevance of these novel and potentially important observations.

Paracrine and autocrine cellular interactions are of established importance in growth control in a variety of body tissues. But it is only very recently that the investigation of normal and abnormal anterior pituitary growth has reached the forefront of neuroendocrine research. This is perhaps surprising, beause the inhibitory effects of doapmine agonism on lactotroph growth and differentiation have been appreciated for some time. The anterior pituitary gland produces numerous growth factors that exert a variety of functional effects on pituitary hormone synthesis and release. Numerous questions remain to be answered concerning the cellular origins and targets of such molecules, their effects on cell growth and differentiation, their relationship to oncogene activation and their involvement in the pathological growth of the anterior pituitary. Paracrine and autocrine interactions may also be involved in the emerging discipline of "neuroimmunology." There is a growing belief that the brain and brain peptides can influence immune responses and POMC-derived peptides have been implicated in the mediation of these events. Not only may circulating peptides exert an influence on lymphocyte function, but there is also evidence for the production of POMC-related peptides by lymphocytes themselves, which may then exert autocrine or paracrine actions.

The search for novel peptides continues. The regulation of fluid balance is complex due to the interplay of a large array of controlling factors. The control of such events by the hypothalamus has been the subject of intense investigation over many years. Several groups have devoted their energies to the isolation and characterization of a hypothalamic regulator of Na^+-K^+ATPase, which is central to fluid and electrolyte balance. Evidence is presented for the existence of such a molecule as a small peptide, although only further characterization, synthesis and biological activity studies will confirm its nature and relevance.

M. F. Scanlon

REGULATION OF EXPRESSION OF THE HUMAN PRE-PROVIP GENE

R. H. Goodman, M. Verhave, S. Kasper,
T. Tsukada, K. Walton, G. Mandel
and J. S. Fink

Division of Molecular Medicine, Department of Medicine
Tufts-New England Medical Center Hospital
Boston, Masssachusetts, U.S.A.

Many of the classical studies of endocrinology have been directed towards determining the mechanisms involved in intercellular communication. Within the last decade, a new endocrinology of intracellular communication has evolved. This new endocrinology addresses the pathways that transduce signals from the surface of the cell membrane to the nucleus. New techniques in molecular biology, coupled with recent insights into the biochemistry of signal transduction, have allowed investigators to begin to understand the mechanisms involved in the regulation of gene expression. This review will focus on the elucidation of a DNA element located upstream from the human vasoactive intestinal polypeptide (VIP) gene, which mediates the regulation of gene expression through the protein kinase A and protein kinase C pathways.

VIP is a 28-amino acid hormone found primarily in the central and peripheral nervous systems, the gastrointestinal, respiratory and urogenital tracts, the exocrine glands, and the thyroid, pituitary and adrenal glands (1). Although VIP was initially characterized as a vasodilatory hormone in the small intestine, recent studies have indicated that VIP functions as a neurotransmitter and neuroendocrine releasing factor (2). When overproduced by the pancreatic islet cell or neural crest tumors, VIP has been associated with the "pancreatic cholera" syndrome (3). The effects of

VIP are thought to be mediated through activation of adenylate cyclase (1). The subsequent elevation of cAMP results in the increased water and electrolyte secretion, hepatic glycogenolysis, dilation of systemic vessels, and hypercalcemia seen in the "pancreatic cholera" syndrome.

VIP is synthesized as part of a larger precursor, pre-proVIP. Itoh et al., using mRNA isolated from a neuroblastoma cell line, characterized a cDNA encoding the VIP precursor (4). From the nucleotide sequence of the cDNA, they deduced the amino acid sequence of a hormone precursor, which in addition to VIP, encoded a 27-amino acid peptide designated PHM. PHM is structurally related to PHI-27, an amidated peptide isolated from the porcine intestine (5). PHI-27 has several biological activities, including inhibition of VIP-binding to cellular receptors, stimulation of cAMP production, and stimulation of insulin, glucagon, prolactin and amylase secretion. The VIP precursor is therefore an example of a polyprotein hormone, like pro-opiomelanocortin, proenkephalin and pro-TRH.

The ability of normal adrenal cells and certain neural crest-derived tumors to produce large amounts of VIP in culture is a distinct advantage for studying the regulation of VIP gene expression. Equally important is the finding that VIP production can be modulated in these cells. In cultured bovine adrenal cells, for example, forskolin (a postreceptor activator of adenylate cyclase), cholera toxin, isobutylmethylxanthine, 8-Br-cAMP and dibutyryl cAMP all increase VIP levels 10- to 20-fold within 24 hr (6). Pulse-chase studies in human neuroblastoma cells have indicated that dibutyryl cAMP increases the incorporation of radioactive amino acids into proVIP (7). Several studies carried out in our lab and others have suggested that this increase in proVIP synthesis occurs, at least in part, at the transcriptional level (7).

The VIP gene also appears to be regulated through the protein kinase C pathway. TPA, a phorbol ester, causes a 40-fold increase in VIP levels in cultured bovine adrenal cells (6). Enkephalin levels within the adrenal cells are elevated by forskolin but not by TPA, suggesting that the effects of TPA are specific for VIP. Immunohistochemical studies suggest that these effects of TPA may be due to recruitment of chromaffin cells that normally do not produce VIP (6). Alternatively, it is possible that adrenal cells constitutively produce VIP, but at a level below the detection limits of the histochemical procedures employed.

To begin our analysis of VIP gene expression, we examined the effects of forskolin and TPA on VIP mRNA levels in a cultured human neuroblastoma cell line (Fig. 1). These agents test whether the level of VIP mRNA can be regulated through the protein kinase A or protein kinase C pathways, respectively. Elevation of VIP mRNA levels by either of these two agents could explain the elevations in VIP levels measured immunologically, but would not elucidate the mechanism of the regulation. RNA levels could be elevated by increasing transcription of the VIP gene or by stabilizing the mRNA within cells.

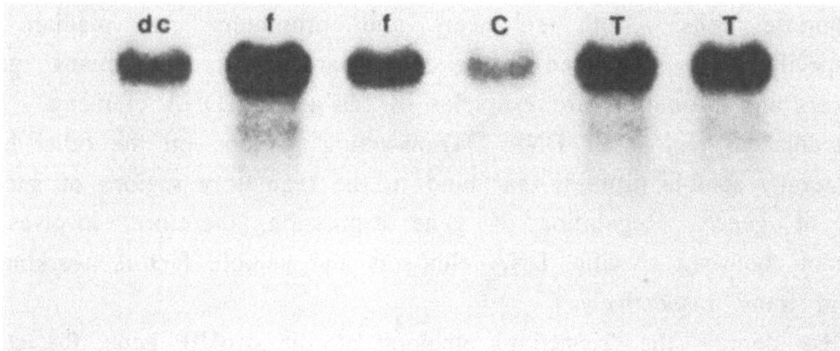

Figure 1. VIP mRNA regulation by cAMP and phorbol esters. RNA was isolated from a human VIP-producing neuroblastoma cell line treated with dibutyryl cAMP (dc), forskolin (f) or TPA (T). C represents control untreated cells. RNA was transferred to nitrocellulose and probed with a [^{32}P]-labeled VIP cDNA.

ProVIP mRNA is approximately 1800 bases in length and migrates as a single band on agarose gel electrophoresis. As shown in Figure 1, VIP mRNA levels in the neuroblastoma cells are elevated by both forskolin and TPA. These results support the findings of Ohsawa et al., who also found that the effects of forskolin and TPA on VIP mRNA levels are additive (8). According to these researchers, proVIP mRNA levels are increased 9- to 11-fold by dibutyryl cAMP, 4-fold by TPA, and 36-fold by a combination of the two agents (8). Our results, on the other hand, suggest that the effects of cAMP and TPA are nearly equal. Transcriptional run-off assays using nuclei from forskolin-treated neuroblastoma cells indicate that cAMP increases proVIP mRNA synthesis, as opposed to increasing stability of the message (7). The mechanism underlying the increase in VIP synthesis after treatment with TPA is unknown. It is possible, given the additive effects of forskolin and TPA, that TPA may increase both transcription and stability of proVIP mRNA.

The mechanisms involved in eukaryotic gene transcription are complex and poorly understood. Even the definition of a eukaryotic promoter, a relatively simple concept, has changed significantly during the past five years. Originally thought to consist of a single concensus sequence, promoters are now most commonly associated with multiple sequence motifs. Enhancer elements, which alter the basal level of transcription, can be distinguished operationally from promoter elements by their ability to function independently of orientation and position. The effects on promoter activity can be positive, as originally defined, or negative, as in the "silencer" elements that depress transcription in inappropriate cells. Both enhancers and promoters can mediate the tissue-specific and regulated expression characteristics of many genes. Enhancers and promoters are examples of "cis-acting" DNA elements - they act on adjacent regions of DNA. "Trans-acting" factors, on the other hand, are generally soluble proteins that bind to the regulatory regions of specific classes of genes. Regulation of gene expression, therefore, involves an interaction between specific DNA elements and soluble factors working in "cis" and "trans," respectively.

To identify the "cis-acting" elements of the proVIP gene, the cloned proVIP cDNA was used to screen a bacteriophage genomic library containing inserts of DNA isolated from human fetal liver. The proVIP gene, depicted in Figure 2, is approximately 9 kilobases in length and is interrupted by six introns (9). These introns divide the VIP precursor into domains that correspond very closely to the known products of proVIP. Exon 1 encodes the untranslated region, exon 2 encodes the signal region, exon 3 encodes the amino-terminal peptide, exon 4 encodes PHM, exon 5 encodes VIP, exon 6 encodes the carboxy-terminal peptide, and exon 7 encodes the 3′ untranslated region. The presence of an intron within the 3′ untranslated region is an unusual feature of the proVIP gene.

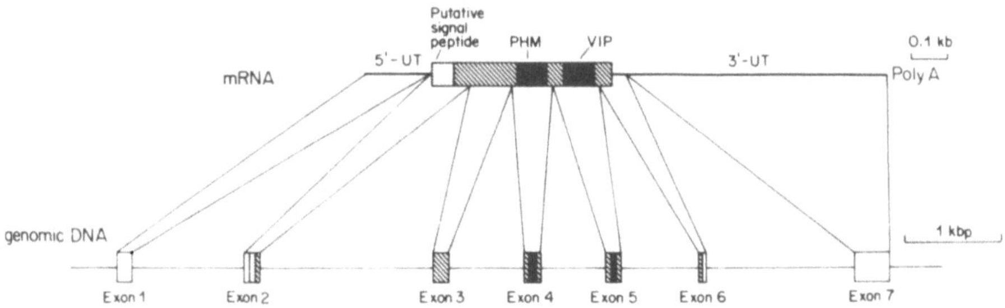

Figure 2. Structure of the human VIP mRNA (top) and gene (bottom). UT designates untranslated region. Reprinted from ref. 9, with permission.

"Cis-acting" transcriptional regulatory sequences could reside in any of the noncoding regions of the proVIP gene. By analogy with many other eukaryotic genes, we assumed that many of the important regulatory sequences would be found in the 5′ flanking region of the gene. Consequently, we focused our efforts on determining whether the 5′ flanking region contained elements that could mediate the effects of either cAMP or TPA on proVIP gene transcription (10).

A fusion gene containing 2 kilobases of proVIP 5′ flanking sequences linked to the bacterial reporter enzyme chloramphenicol acetyl transferase (CAT) was introduced into cultured PC12 cells by calcium phosphate precipitation. CAT activity in the cell lysates was determined in a transient assay system. PC12 cells, derived from a rat pheochromocytoma, do not normally produce VIP but do express fusion genes containing neuroendocrine gene promoters. The level of CAT activity, determined by the acetylation of [^{14}C]chloramphenicol, gives a measure of the activity of the particular promoter studied. By examining the promoter activity of the fusion gene in the presence and absence of forskolin or TPA, we could determine whether the 2 kilobase region of DNA contains a cAMP- or TPA-responsive element.

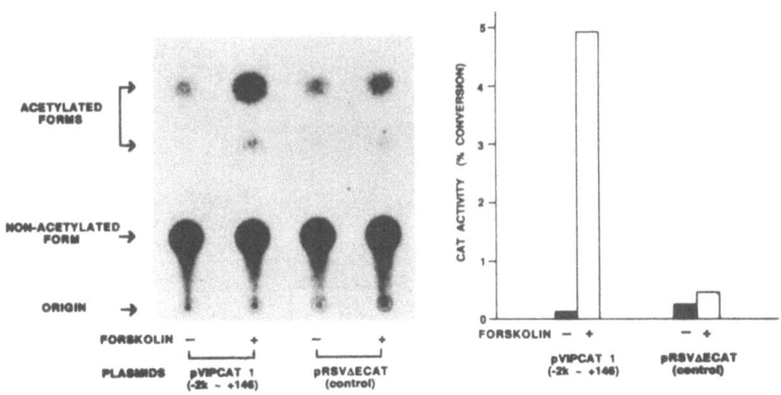

Figure 3. Activity of a 2 kilobase VIP-CAT fusion gene in transfected PC12 cells. Acetylated and nonacetylated forms of chloramphenicol are shown in the thin layer chromatogram at the left of the figure. Results are represented graphically at the right of the figure. Solid bars represent unstimulated cells, open bars represent cells treated with forskolin.

As shown in Figure 3, activity of the VIP-CAP fusion gene was increased 20- to 30-fold by treatment with forskolin. To delineate the boundaries of the cAMP-regulatory element further, deletions were made from the 5′-end of the fusion gene. Fusion genes containing 497, 340 and

94 base pairs of the 5′ flanking region responded to forskolin by increasing their CAT activity. Genes deleted further did not respond to forskolin. These studies indicated that the 5′ boundary of the cAMP-responsive element was approximately 94 base pairs upstream from the transcriptional initiation site.

To identify the 3′ boundary of the regulatory element, a more complex fusion gene was constructed. This gene contained a portion of the proVIP 5′ flanking region, between -340 and -31, linked to a Rous sarcoma virus (RSV) promoter-CAT fusion gene. In this construction, the RSV promoter was placed under the control of the proVIP regulatory elements. Cyclic AMP-responsiveness of the VIP-RSV-CAT gene was indistinguishable from the 2 kilobase VIP-CAT gene. By sequentially deleting fragments from the 3′-end of the VIP sequences, we were able to determine that the 3′ limit of the cAMP-responsive element was approximately 64 base pairs upstream from the transcriptional initiation site. Further deletional analysis indicated that the cAMP-responsive element extended only from -86 to -70 (Fig. 4). These studies also demonstrated that the cAMP-responsive element could direct transcription from a heterologous promoter.

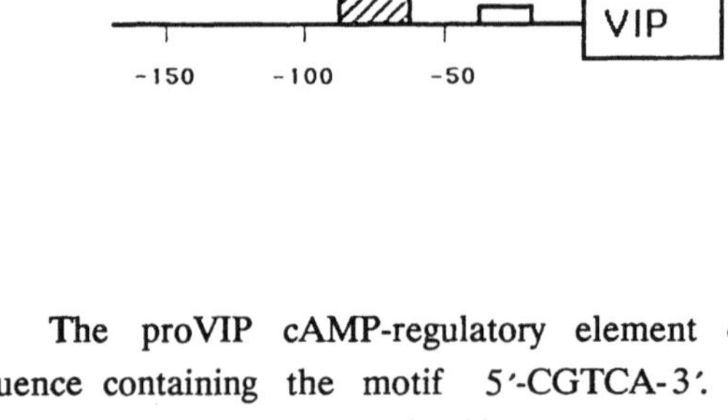

Figure 4. Location of the VIP cyclic AMP-regulated element (CRE).

The proVIP cAMP-regulatory element contains an inverted repeat sequence containing the motif 5′-CGTCA-3′. This five-base-pair motif is also found within the 5′ flanking sequences of other genes regulated by cAMP including preprosomatostatin, phosphoenolpyruvate carboxykinase and preproenkephalin (11-13). It is likely, therefore, that this motif is important for cAMP-responsiveness of many genes. A GC-rich region upstream from the CGTCA motif is also present in the preprosomatostatin, phosphoenolpyruvate carboxykinase and preproenkephalin genes. Our studies indicate that this GC-rich region is not necessary for activity of the proVIP cAMP regulatory element, however.

To confirm the importance of the five base pair motif in the function of the cAMP-regulatory element, we have examined a series of mutated proVIP gene fragments. These mutated DNA elements were examined by inserting the proVIP fragments upstream from the RSV-CAT fusion gene. As shown in Figure 5, mutations in the five base pair motifs severely decrease the responsiveness of the element. Mutations that fall between the motifs have little effect on function. Earlier studies suggested that both five base pair motifs were required for full function of the element. Whether the two motifs form a single stem-loop binding site or provide two separate binding sites is unknown. Replacement of the CGTCA sequences in the element with another inverted repeat causes a loss of function, suggesting that specific sequences, and not just a general DNA structure, are required for activity.

The experiments described above indicate tht the seventeen base pair element is sufficient to mediate cAMP-responsiveness when fused to an RSV promoter. To determine whether this element was also necessary for proVIP gene expression, we examined another mutated VIP-CAT fusion gene. As previously described, the 2 kilobase proVIP-CAT fusion gene was actively expressed in PC12 cells and was regulated by cAMP. When a four base pair deletion was made within the regulatory element, the 2 kilobase gene was no longer functional. This study suggests that the regulatory element may be required for basal as well as regulated expression.

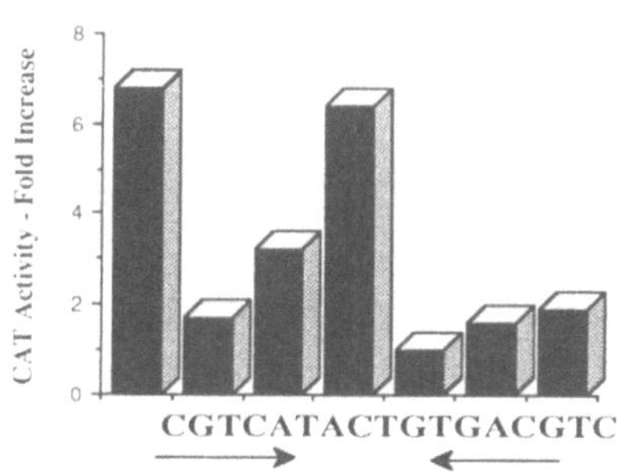

Figure 5. Deletional analysis of the VIP cyclic AMP-regulated enhancer. CGTCA motifs are denoted by arrows.

The finding that the proVIP regulatory element can regulate a heterologous viral promoter (RSV) suggested that the element might be an enhancer. Enhancers, unlike promoters, typically have the ability to regulate gene expression in an orientation- and position-independent fashion. To determine whether the cAMP-regulatory element is an enhancer, we ligated the element in both orientations upstream from the RSV-CAT fusion gene. The activity of the element was not dependent on its orientation. In addition, the element was functional when it was inserted as far as 300 base pairs upstream from the transcriptional initiation site or when it was placed at the 3′ end of the CAT-fusion gene. These studies suggested that the regulatory element was indeed an enhancer. Unlike many of the previously described enhancers, however, the activity of the proVIP element was dependent upon cAMP.

To fulfill the criteria for an enhancer, the element must "enhance" transcription from the appropriate promoter. To determine whether the proVIP element initiates transcription at the correct start site, S-1 nuclease mapping studies were performed (Fig. 6). RNA isolated from cells expressing the proVIP-RSV-CAT fusion genes was hybridized to a labeled DNA probe, digested with a nuclease specific for single-stranded DNA, and analyzed on a denaturing polyacrylamide gel. Initiation at the appropriate site should result in protection of a 107 base fragment. Figure 6 demonstrates that the expected 107 base fragment is protected whether the regulatory element is inserted in the correct or inverted orientation. Although some upstream

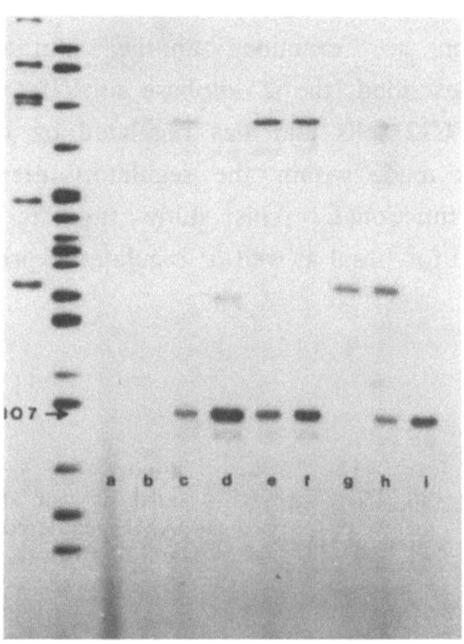

Figure 6. S-1 nuclease analysis of transfected gene products. The appropriately-initiated product is 107 nucleotides long. Lanes represent (a) nontransfected PC12 cells, (b,c) untreated and forskolin-treated PC12 cells transfected with a VIP-CAT fusion gene, (d) C6 glioma cells transfected with an RSV-CAT gene, (e,f) untreated and forskolin-treated C6 glioma cells transfected with a VIP-CAT fusion gene, (g,h) untreated and forskolin-treated C6 gioma cells transfected with a fusion gene containing an inverted copy of the VIP enhancer, and (i) a repeated experiment of C6 glioma cells transfected with RSV-CAT.

transcription is seen when the element is inserted in reverse direction, only the "appropriate" transcription appears to be regulated by cAMP.

Previously, nuclear run-off analyses had suggested that the regulation of proVIP gene expression occurred at the transcriptional level, although nothing was known about the regulation of proVIP production by phorbol esters. Comb et al. had previously suggested that the cAMP-responsive element of the proenkephalin gene also mediated responsiveness to phorbol esters, but the effect of TPA required the addition of isobutylmethylxanthine (IBMX), a phosphodiesterase inhibitor (13). This agent by itself could elevate cAMP levels and stimulate the enhancer.

TRE (from Tjian)									C/G	T	G	A	C	T	C/A	G
									|	|	|	|	|			
VIP-CRE	C	G	T	C	A	T	A	C	T	G	T	G	A	C	G	T C

Figure 7. Sequence similarity of the VIP-CRE and a consensus TPA responsive element ("TRE").

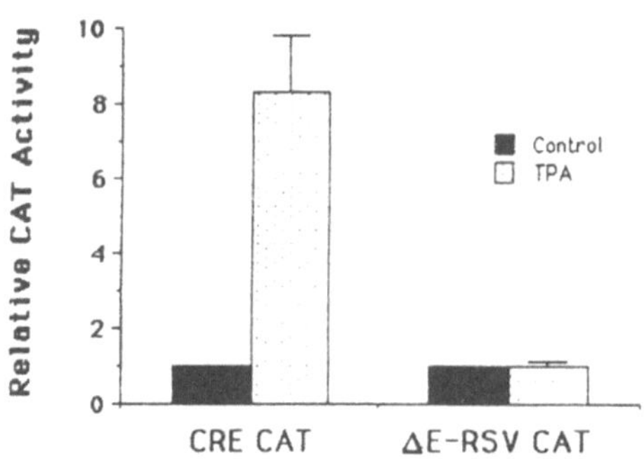

Figure 8. Response of the VIP cyclic AMP-responsive element to TPA.

Comparison of the cAMP regulatory region of the proVIP gene with the TPA-responsive regions of other genes such as metallothionein and collagenase revealed a striking similarity (14,15; Fig. 7). This similarity raised the possibility that the cAMP-responsive element might indeed mediate responsiveness to TPA as well. When the CAT fusion genes containing the cAMP-responsive element were examined in PC12 cells, no regulation by TPA was observed. Although PC12 cells respond to phorbol esters by several parameters, it is possible that these cells lack the factors required to mediate some of the transcriptional effects of TPA. Consequently, we examined the VIP-CAT fusion genes for their ability to respond to TPA in HeLa cells, which express genes known to be transcriptionally regulated by TPA (Fig. 8). In HeLa cells, the fusion genes responded to TPA as well as to cAMP. This study indicated first, that TPA can regulate proVIP gene expression at the transcriptional level, and second, that the proVIP cAMP-regulatory element can also mediate responsiveness to phorbol esters.

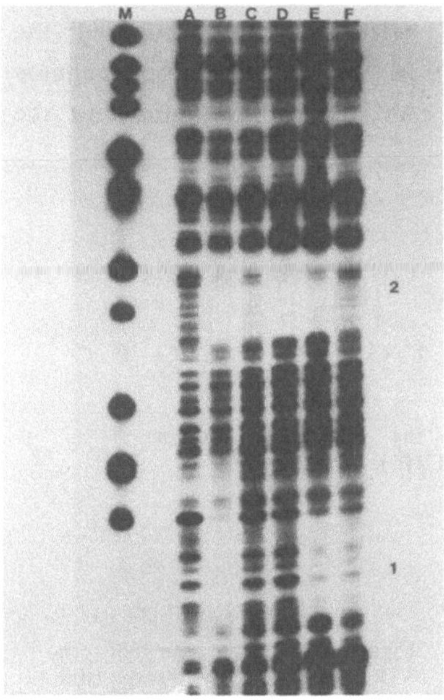

Figure 9. DNase footprinting analysis of the VIP 5′ flanking region. M denotes size markers, lane A shows fragments generated from unprotected DNA, lanes B-F show protection from different fractions of nuclear extract. Footprint 1 corresponds to the enhancer region.

The nature of the factors that regulate the proVIP gene are unknown. Angel et al. and Lee et al. have suggested that TPA-responsiveness of the metallothionein gene is mediated by the transcription factor AP-1 (14,15). It is possible tht this factor also mediates the TPA-responsiveness of the proVIP gene. Nuclear extracts from HeLa, PC12, and C6 glioma cells bind specifically to the proVIP regulatory element, as determined by DNase 1 footprinting assays (Fig. 9). These assays measure the ability of a DNA-binding protein to protect a fragment of DNA from digestion by a nuclease. Suprisingly, protection is seen in extracts that have not been treated with cAMP or TPA, suggesting that cellular proteins bind constitutively to the element. According to Montminy et al., binding characteristics of these cellular proteins may be altered by phosphorylation (16). By cross-linking a BUdR- and [^{32}P]-labeled enhancer element with nuclear extracts from HeLa, PC12 and C6 glioma cells, we have determined that the proVIP enhancer-binding protein has an approximate molecular weight of 43 kD (Fig. 10). This size is similar to the somatostatin enhancer-binding protein identified by Montminy (16).

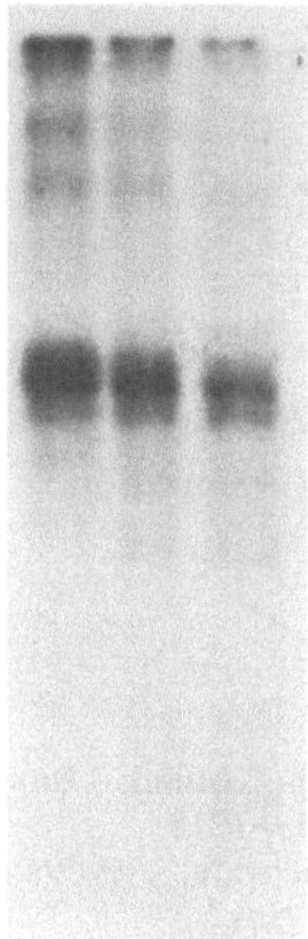

Figure 10. DNA-affinity cross-linking experiment using three dilutions of PC12 cell nuclear extract. A single band of 43 kD binds to the DNA probe.

Recent studies have indicated that sequences homologous to the proVIP enhancer are found in viral genomes, such as adenovirus and cytomegalovirus (17,18). In the viral genes, the homologous elements act as enhancers, but it is not known whether they mediate responsiveness to second messengers such as cAMP. It is possible that the related enhancer elements in specific cellular and viral genes represent a family of protein-binding domains. Each of these domains may allow binding of a unique set of transcription factors. Alternatively, binding of specific transcription factors may be somewhat promiscuous, depending on the intracellular levels of each factor. Transcription factors containing similar binding, but different activity, domains could compete with each other to regulate the expression of a given gene. Our observation that proteins isolated from cellular extracts of yeast (Saccharomyces cerevisiae) bind specifically to the proVIP regulatory element supports the hypothesis that a variety of transcription factors can bind to a single DNA sequence.

In summary, our studies have defined a DNA element required for regulated transcription of the proVIP gene. This element, or closely related variants, is found in other cellular and viral genes where it appears to function as an enhancer. Recent studies from two laboratories have suggested that this regulatory element may also mediate the high basal levels of hormone gene expression seen in some varieties of tumors (19,20). It is not known whether the same transcription factors required for regulated gene expression by second messenger systems are also involved in the pathogenesis of ectopic hormone syndromes. Further characterization of these transcription factors is necessary to elucidate the controls of hormone gene expression in both physiological and pathological states.

ACKNOWLEDGMENTS

This work was supported by grants from the National Institutes of Health (CA 37370, KO7 NS00885, P30 AM34928), the American Parkinson's Disease Association, and the Toyobo Biotechnology Foundation.

REFERENCES

1. Said SI, Mutt V (1970) Polypeptide with broad biological activity: isolation from small intestine. Science 169: 1217-1218
2. Said SI (ed) (1982) Vasoactive intestinal peptide. Advances in Peptide Hormone Research Series. New York: Raven Press
3. Verner JV, Morrison AB (1958) Islet cell tumor and a syndrome of refractory watery diarrhea and hypokalemia. Amer J Med 25: 374-380
4. Itoh N, Obata K, Yanaihara N, Okamoto H (1983) Human preprovasoactive intestinal polypeptide contains a novel PHI-27 like peptide. Nature 304: 547-549
5. Tatemoto K, Mutt V (1981) Isolation and characterization of the intestinal peptide porcine PHI (PHI-27), a new member of the glucagon secretion family. Proc Natl Acad Sci USA 781: 6603-6607
6. Pruss RM, Moskel JR, Eiden LE, Beinfeld MC (1985) Specific regulation of vasoactive intestinal polypeptide biosynthesis by phorbol ester in bovine chromaffin cells. Endocrinology 117: 1020-1026
7. Hayakawa Y, Obata K, Itoh N, Yanaihara N, Okamoto H (1984) Cyclic AMP regulation of provasoactive intestinal polypeptide/PHM-27 synthesis in human neuroblastoma cells. J Biol Chem 259: 9207-9211
8. Ohsawa K, Hayakawa Y, Nishizawa M, Yamagami T, Yamamoto H, Yanaihara N, Okamoto H (1985) Synergistic stimulation of VIP/PHM-27 gene expression by cyclic AMP and phorbol esters in human neuroblastoma cells. Biochem Biophys Res Commun 132: 885-891
9. Tsukada T, Horovitch SJ, Montminy MR, Mandel G, Goodman RH (1985) Structure of the human vasoactive intestinal polypeptide gene. DNA 4: 293-300
10. Tsukada T, Fink JS, Mandel G, Goodman RH (1986) Identification of a region in the human vasoactive intestinal polypeptide gene responsible for regulation by cyclic AMP. J Biol Chem 262: 8743-8747
11. Montminy MR, Sevarino KA, Wagner JA, Mandel G, Goodman RH (1986) Identification of a cyclic AMP-responsive element within the rat somatostatin gene. Proc Natl Acad Sci USA 83: 6682-6686
12. Short JM, Wynshaw-Boris A, Short HP, Hanson RW (1986) Characterization of the phosphoenolpyruvate carboxykinase promoter-regulatory region. J Biol Chem 261: 9721-9726
13. Comb M, Birnberg NC, Seasholtz A, Herbert E, Goodman HM (1986) Characterization of a cAMP and phorbol ester inducible DNA element. Nature 323: 353-356
14. Angel P, Imagawa M, Chiu R, Stein B, Imbra RJ, Rahmsdorf HF, Jonat C, Herrlich P, Karin M (1987) Phorbol ester-inducible genes contain a common cis element recognized by a TPA-modulated trans-acting factor. Cell 49: 729-739
15. Lee W, Mitchell P, Tjian R (1987) Purified transcription factor AP-1 interacts with TPA-inducible elements. Cell 49: 741-752

16. Montminy MR, Bilezikjian LM (1987) Binding of a nuclear protein to the cyclic AMP-responsive element of the somatostatin gene. Nature 328: 175-178
17. Lee KAW, Green MR (1987) A cellular transcription factor E4F1 interacts with an Ela-inducible enhancer and mediates constitutive enhancer function in vitro. EMBO J. 6: 1345-1353
18. Ghazal P, Lubon H, Fleckenstein B, Hennighausen L (1987) Binding of transcription factors and creation of a large nucleoprotein complex on the human cytomegalovirus enhancer. Proc Natl Acad Sci USA 84: 3658-3662
19. Andrisani OM, Hayes TE, Roos B, Dixon JE (1987) Identification of promoter sequences involved in the cell-specific expression of the rat somatostatin gene. Nuc Acids Res 15: 5715-5728
20. Delegeane AM, Ferland LH, Mellon PL (1987) Tissue-specific enhancer of the human glycoprotein hormone alpha-subunit gene: dependence on cyclic AMP-inducible elements. Mol Cell Biol 7: 3994-4002

PARACRINE INTERACTIONS IN THE ANTERIOR PITUITARY

W. Robberecht and C. Denef

Laboratory of Cell Pharmacology, Department of Pharmacology,
University of Leuven School of Medicine
Leuven, Belgium

INTRODUCTION

Paracrine interactions are defined as cell-to-cell communications within a tissue in which chemical substances are used as local messengers. To elucidate the nature of such messages, one clearly has to dissect a tissue into its composing cells, and thus most of the studies concerning paracrine interactions in the anterior pituitary gland (AP) have been done in vitro. This implies that, for most of the interactions described, the physiological importance is still uncertain, as very few in vivo studies have addressed the issue.

In this paper, two main lines of evidence for the possible involvement of paracrine interactions in AP physiology are presented: the specific cellular arrangements in the gland and their possible functional significance, and the presence within the AP of a variety of substances which may have local regulatory roles.

SPECIFIC ARRANGEMENTS OF AP CELLS AND THEIR FUNCTIONAL CORRELATES

Several investigators have shown that the topographical distribution of the various AP cell types is not random. In the rat, both corticotrophs (1-3) and thyrotrophs (4) can be found in close association with somatotrophs. In addition, extensions of corticotrophs can encircle a growth hormone (GH)-containing cell (2). Although these arrangements suggest a close interaction between these cells, it remains to be shown whether there is a functional significance.

The closest cellular association has been described between rat pituitary gonadotrophs and a certain subtype of lactotrophs. According to the classification of Nogami and Yoshimura, the type IV lactotroph is a cup-shaped cell with scarce and mostly spherical polymorphic granules (5). These type IV lactotrophs are mainly found in the vicinity of the dorsocephalic zone of the gland, a region rich in gonadotrophs (6). The cup-shaped lactotrophs closely associate with these gonadotrophs and can extend cytoplasmic processes to engulf this cell type. In addition, gonadotrophs and lactotrophs form specialized junctions between each other (7).

These associations have been shown to be dynamic in nature: the cup-shaped lactotrophs only appear at puberty and their immunostainability varies throughout the estrous cycle (6). Furthermore, they show a clear sex difference, being more numerous in males than in females (6). This situation resembles one found in teleost fish. In this species, lactotrophs are found in the rostral part of the gland and gonadotrophs in the caudal part; when the animals reach sexual maturity, gonadotrophs also appear in the lactotroph-dominated part of the gland (8).

Several in vitro studies have demonstrated that these anatomical relationships have functional correlates. It has been shown for rat (9) and human (10) pituitary cells that gonadotrophs may pass a signal to lactotrophs which mediates the prolactin (PRL)-stimulating properties of gonadotropin releasing hormone (GnRH). This effect is inhibited by estrogens (11). A factor with PRL-releasing activity is present in medium conditioned by gonadotrophs (9). These in vitro observations have been confirmed by in vivo studies: in a number of clinical reports, the PRL releasing properties of GnRH have been described, although the hormonal conditions required are still disputed (12-19).

Further possible in vivo correlates for a functional interaction between gonadotrophs and lactotrophs come from the observation that hyperprolactinemia induced in rats by treatment with haloperidol or by transplantation of the AP under the kidney capsule can be abolished by treatment of the animals with antiserum to GnRH or by desensitization of the gonadotroph by treating the rats with a GnRH analogue (20,21). These findings have recently been confirmed in vitro in our laboratory by long-term perifusion of reaggregated AP cells: when cells in such a system are treated with high doses of GnRH, a clearcut decline of PRL release is seen (Mignon and Denef, unpublished results). This clearly demonstrates that normal gonadotroph function is necessary for the lactotroph to maintain its normal hormone release.

Further evidence for a functional interaction between lactotrophs and gonadotrophs comes from studies on embryonic development: gonadotrophs can induce the differentiation of pituitary stem cells into lactotrophs using the α-subunit of luteinizing hormone (LH) as a paracrine factor (22-24). This observation might have an in vivo correlate as well, as increased number of lactotrophs have been found in the pituitary gland of human anencephalic fetuses (25), a condition where only the α-subunit of LH is secreted while almost no intact LH is present (26).

INTRAPITUITARY SUBSTANCES WITH A POSSIBLE LOCAL FUNCTION

In AP cells, a variety of substances are present which are not considered to be "classical" hormones; for some of these, the cell types in which they reside have been identified (27). In order to establish an intrapituitary paracrine role for these substances, their synthesis within the pituitary itself has to be demonstrated. This has been done for vasoactive intestinal peptide (VIP) (28), atrial natriuretic peptide (ANP) (29) and dynorphin (30). Met-enkephalin is still present in somatotrophs, even after 10 days of culture, which is equally compatible with its local synthesis (31) and propeptides of gastrin and cholecystokinin have been isolated, which may suggest that these peptides are synthesized in situ as well (32). Thyrotropin releasing hormone (TRH) and GnRH also persist in long-term AP cultures (33-34). A complete renin-angiotensin system seems to be present in various tissues including the AP: local synthesis of renin in gonadotrophs has been shown by demonstrating its mRNA (35).

Most of these substances are known to modulate basal or stimulated release of AP hormones, when administrated exogenously (27). Obviously, this does not mean that the endogenous substances have a comparable role. To evaluate this, one has to study the effect on AP hormone release caused by blocking the activity of these intrinsic substances with antisera or antagonists. Several such studies have been performed.

An autocrine or paracrine effect on PRL release has been suggested for VIP by demonstrating a decrease of PRL release from AP cells in vitro by adding VIP antiserum to the cultures (36). Both for human pituitaries (37) and for rat pituitary cells (38), an interaction between corticotrophs and gonadotrophs has been described by the demonstration of the LH-releasing activity of corticotropin releasing factor (CRF). In rat pituitary cells, this activity was blocked by naloxone and anti-β-endorphin-antiserum. It was therefore speculated that an endogenous opioid peptide (EOP) was involved as a paracrine factor (38).

Using antagonists of angiotensin II (AII) and EOP, we recently found evidence for multiple peptide interactions in the regulation of the release of GH (Fig. 1): in reaggregated AP cells from 14-day-old rats, GnRH has a negligible effect on GH release. However, if the endogenous opioid activity is blocked by adding Mr 2266, an opioid antagonist, significant GH releasing activity for GnRH

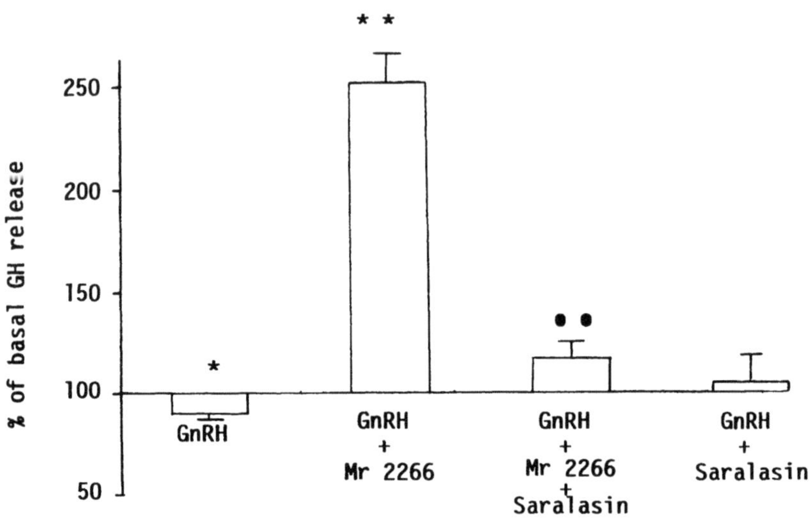

Figure 1. Effect of Mr 2266 (1 μM) and saralasin (0.1 μM) on the effect of GnRH (0.1 nM) on GH release in AP cell aggregates from 14-day-old rats, cultured in serum-free medium supplemented with 80 nM of dexamethasone and 0.05 nM T3. *P < 0.05 to basal release; **P < 0.005 to basal release; ●●P < 0.005 to GnRH + Mr 2266. (Mean ± SEM; 3 independent experiments.)

is established. This effect of GnRH can be seen in concentrations as low as 0.1 pM. This GH releasing activity of GnRH can be blocked by saralasin, suggesting that AII is the factor mediating this activity. This fits with the fact that AII is present in gonadotrophs (39) and that it has GH releasing properties in vitro, at least in immature animals (40). This suggests that GnRH is able to release a GH releasing substance, AII, and at the same time an GH release inhibiting substance, probably an EOP, blocked by Mr 2266 (Fig. 2). It can be speculated that this latter EOP is dynorphin, which is synthesized in AP gonadotrophs (30) and which can be released by GnRH (38).

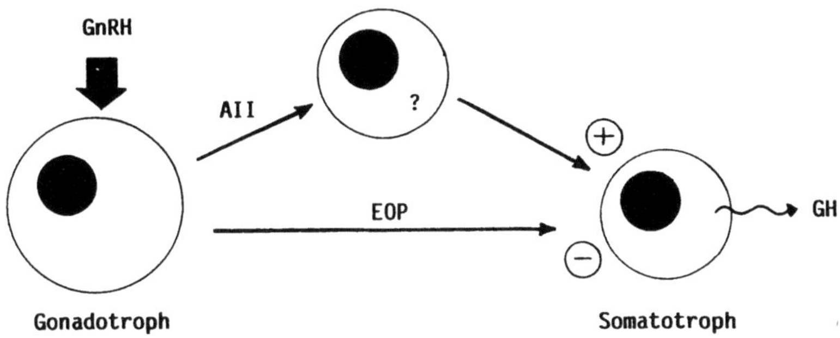

Figure 2. Hypothetical model for the effect of GnRH, AII and EOP on GH release. The action of AII is indirect as shown before (40).

REGULATORY CONTROL

That these intrapituitary substances are involved in pituitary physiology is further substantiated by the fact that most of them are under control of known regulatory factors, such as adrenal and gonadal steroids and thyroid hormone: examples are renin (42) and kallikrein (43), which are sex steroid-dependent and substance P, which is regulated by an interaction between androgens and thyroid hormone (44). We have shown that the GH response to AII strongly depends on the presence of glucocorticoids (40).

The content of intrapituitary EOP is modulated by multiple hormones. Apart from the known interactions of glucocorticoids with the corticotroph, the content of β-endorphin is dependent on thyroid hormone (45) and sex steroids (46). The pituitary enkephalin system is under similar gonadal control (47-48) and the release of dynorphin from AP is modulated by dexamethasone and 5α-dihydrotestosterone (49).

The regulation of these substances can become deficient in certain conditions, as abnormalities in their content have been found in pathological animal models. Pituitary content of met-enkephalin is two-fold higher in genetically obese diabetic mice than in controls, while leu-enkephalin is lower at 4 weeks of age in these animals, but reverts to higher levels in animals of 6-30 weeks of age (50). The pituitary content of dynorphin is decreased in the spontaneously hypertensive rat (SHR) (51), which is interesting, because treatment with EOP antagonists lowers the blood pressure of SHR and not that of controls (52).

DEVELOPMENTAL CHANGES

In addition to the influence of known regulatory factors, it has been shown that the concentrations of some of these intrapituitary substances undergo significant changes during postnatal development. A dramatic increase of the concentration of proenkephalin B derived peptides in the rat AP has been reported by Seizinger et al. (53). The same authors demonstrated the opposite evolution for the concentration of β-endorphin (54).

The pituitary concentration of motilin present in somatotrophs (55) reaches its maximum just before birth, and this declines to adult levels during the first postnatal weeks (56). Interestingly, a decrease of plasma GH levels by treatment of rats with anti-motilin-antiserum has been reported (57), suggesting the possibility that this peptide may have a development-dependent GH release modulating role within the gland.

In addition, the influence of regulatory factors on the effects of intrapituitary peptides on AP hormone release, can undergo significant changes during development: we have shown that glucocorticoids clearly enhance the GH releasing properties of AII in neonatal rats, but establish GH release inhibiting activity of this peptide in cell aggregates from adult animals, the effect on aggregates from juvenile rats being intermediate (40).

CONCLUSION

In vitro studies suggest that hormone release from the AP is modulated by a complicated network of intercellular communication systems consisting of stimulatory and inhibitory substances, the action of which is dependent on hormonal and developmental factors. By switching on and off certain cell types, these messengers might bring AP cells to a condition which will determine the final effect of hypothalamic releasing factors reaching the pituitary tissue through the portal vessel system.

Most of the evidence for the importance of these interactions inevitably comes from in vitro studies with rats as AP cell donors. Once the multiplicity of interactions and the responsible intercellular messengers have been identified, their physiological significance will have to be studied in the intact animal to evaluate their relevance in the in vivo setting. It is obvious that an understanding of the local messages of AP cells will be needed to understand the physiology and pathophysiology of this tissue.

ACKNOWLEDGMENTS

This work was supported by the "Belgian National Fonds voor Wetenschappelijk Onderzoek" and "Geconcerteerde Onderzoeksakties." The technical assistance of K. Rillaerts and the typographical assistance of M. Bareau is greatly appreciated.

REFERENCES

1. Nakane PK (1970) Classification of pituitary cell types with immunoenzyme histochemistry. J Histochem Cytochem 18: 9-20
2. Siperstein ER, Miller JK (1970) Further cytophysiologic evidence for the identity of the cells that produce adrenocorticotrophic hormone. Endocrinology 86: 451-486
3. Nagata M, Mizunaga A, Ema S, Yoshimura F (1980) Various types of the pituitary folliculo-stellate cells involving the Siperstein's corticotroph in the normal rats. Endocrinol Jpn 27: 13-22
4. Yoshimura F, Nogami H (1980) Immunohistochemical characterization of pituitary stellate cells in rats. Endocrinol Jpn 27: 43-51
5. Nogami H, Yoshimura F (1982) Fine structural criteria of prolactin cells identified immunohistochemically in the male rat. Anat Rec 202: 261-274
6. Sato S (1980) Postnatal development, sexual difference and sexual cyclic variation of prolactin cells in rats: special reference to the topographic affinity to a gonadotrophs. Endocrinol Jpn 27: 573-583
7. Horvath E, Kovacs K, Ezrin C (1977) Junctional contact between lactotrophs and gonadotrophs in the rat pituitary. IRCS Med Sci 5: 511
8. Margolis-Kazan W, Schreibman MP (1984) Sexually dimorphic, age-related changes in pituitary gonadotroph distribution mechanisms. Aging Developm 24: 325-333
9. Denef C, Andries M (1983) Evidence for paracrine interaction between gonadotrophs and lactotrophs in pituitary cell aggregates. Endocrinology 112: 813-822
10. Blackwell RE, Rodgers-Neame NT, Bradley EL Jr, Asch RH (1986) Regulation of human prolactin secretion by gonadotropin-releasing hormone in vitro. Fertility and Sterility 46: 26-31
11. Denef C (1984) Functional interrelationships between pituitary cells. In: Lamberts SWJ, Tilders FJH, Vanderveen EA, Assies J (eds) Proceedings of the 3rd European Workshop on Pituitary Adenomas. Amsterdam, Free University Press: pp 25-35
12. Giampetto O, Moggi G, Chisci R, et al. (1979) Unusual prolactin response to luteinizing hormone-releasing hormone in some anovulatory women. J Clin Endocrinol Metab 49: 141-143
13. Yen SSC, Hoff JD, Lasley BL, Casper RF, Sheehan K (1980) Induction of prolactin release by LRF and LRF agonist. Life Sci 26: 1963-1967
14. Casper RF, Yen SSC (1981) Simultaneous pulsatile release of prolactin and luteinizing hormone induced by luteinizing hormone-releasing factor agonist. J Clin Endocrinol Metab 52: 934-936
15. Cetel NS, Yen SSC (1983) Concomitant pulsatile release of prolactin and luteinizing hormone in hypogonadal women. J Clin Endocrinol Metab 56: 1313-1315
16. Braund W, Roeger DC, Judd SJ (1984) Synchronous secretion of luteinizing hormone and prolactin in the human luteal phase: neuroendocrine mechanisms. J Clin Endocrinol Metab 58: 293-297
17. Van Loon GR (1978) Bromocriptine-induced inhibition of plasma dopamine, noradrenaline and adrenaline responses to LH-RF. Nature 275: 331-332
18. Zhang Y, Yen SSC (1984) Prolactin-releasing action of LRF: a central catecholamine mediated event? Life Sci 34: 653-657

19. Barbarino A, De Marinis L, Manicini A, Farabegoli C (1982) Estrogen-dependent plasma prolactin response to gonadotropin-releasing hormone in intact and castrated men. J Clin Endocrinol Metab 55: 1212-1216
20. Debeljuk L, Torres-Aleman I, Schally AV (1985) Antiserum to LH-RH blocks haloperidol-induced hyperprolactinemia in female rats. Neuroendocrinology 40: 185-187
21. Debeljuk L, Torres-Aleman I, Schally AV (1985) D-Trp-6-LHRH inhibited hyperprolactinemia in female rats. Endocrinology 116: 2227-2231
22. Bégeot M, Hemming FJ, Dubois MP, Dubois PM (1983) Gonadotropin releasing hormone (GnRH) stimulates immunoreactive lactotrope differentiation. Endocrinology 112: 2224-2226
23. Bégeot M, Hemming FJ, Dubois PM, et al. (1984) Induction of pituitary lactotrope differentiation by luteinizing hormone α-subunit. Science 226: 566-568
24. Bégeot M, Morel G, Rivest RW, et al. (1984) Influence of gonadoliberin on the differentiation of rat gonadotrophs: an in vivo and in vitro study. Neuroendocrinology 38: 217-225
25. Bégeot M, Dubois MP, Dubois PM (1984) Evolution of lactotrophs in normal and anencephalic human fetuses. J Clin Endocrinol Metab 58: 726-730
26. Hagen C, McNeilly AS (1975) Identification of human luteinizing hormone, follicle-stimulating hormone, luteinizing hormone β-subunit and gonadotropin α-subunit in fetal and adult pituitary glands. J Endocrinol 67: 49-57
27. Denef C (1986) Paracrine interactions in the anterior pituitary. Clin Endocrinol Metab 15: 1-32
28. Arnaout MA, Garthwaite TL, Martinson DR, Hagen TC (1986) Vasoactive intestinal polypeptide is synthesized in anterior pituitary tissue. Endocrinology 119: 2052-2057
29. Gardner DG, Deschepper CF, Ganong WF, Hane S, Fiddes J, Baxter JD, Lewicki J (1986) Extra-atrial expression of the gene for atrial natriuretic factor. Proc Natl Acad Sci USA 83: 6697-6701
30. Khachaturian H, Sherman TG, Lloyd RV, Civelli O, Douglass J, Herbert E, Akil H, Watson SJ (1986) Pro-dynorphin is endogenous to the anterior pituitary and is co-localized with LH and FSH in the gonadotrophs. Endocrinology 119: 1409-1412
31. Weber E, Voigt KH, Martin R (1978) Pituitary somatotrophs contain [Met]enkephalin-like immunoreactivity. Proc Natl Acad Sci USA 75: 6134-6139
32. Rehfeld JF (1985) Accumulation of nonamidated preprogastrin and preprocholecystokinin products in porcine pituitary corticotrophs. J Biol Chem 261: 5841-5847
33. Li JY, Knapp RJ, Sternberger LA (1984) Immunocytochemistry of a "private" luteinizing-hormone-releasing hormone system in the pituitary. Cell Tissue Res 235: 263-266
34. Childs (Moriarty) GV, Cole DE, et al. (1978) Endogenous thyrotropin-releasing hormone in the anterior pituitary: sites of activity as identified by immunocytochemical staining. J Histochem Cytochem 26: 901-908
35. Deschepper CF, Mellon SH, Cumin F, Baxter JD, Ganong WF (1986) Analysis by immunocytochemistry and in situ hybridization of renin and its mRNA in kidney, testis, adrenal, and pituitary of the rat. Proc Natl Acad Sci USA 83: 7552-7556
36. Hagen TC, Arnaout MA, Scherzer WJ, Martinson DR, Garthwaite TL (1986) Antisera to vasoactive intestinal polypeptide inhibit basal prolactin release from dispersed anterior pituitary cells. Neuroendocrinology 43: 641-645
37. Blumenfeld Z, Kuhn RW, Jaffe RB (1986) Corticotropin-releasing factor can stimulate gonadotropin secretion by human fetal pituitaries in superfusion. Am J Obstet Gynecol 154: 606-612
38. Blank MS, Fabbri A, Catt KJ, Dufau ML (1986) Inhibition of luteinizing hormone release by morphine and endogenous opiates in cultured pituitary cells. Endocrinology 118: 2097-2101
39. Deschepper CF, Crumrine DA, Ganong WF (1986) Evidence that the gonadotrophs are the likely site of production of angiotensin II in the anterior pituitary of the rat. Endocrinology 119: 36-43

40. Robberecht W, Denef C (1987) Stimulation and inhibition of pituitary GH release by angiotensin II in vitro. Endocrinology, in press
41. Knepel W, Schwaninger M, Döhler KD (1985) Corelease of dynorphin-like immunoreactivity, luteinizing hormone, and follicle-stimulating hormone from rat adenohypophysis in vitro. Endocrinology 117: 481-487
42. Naruse K, Naruse M, Obana K, Demura R, Demura H, Inagami T, Shizume K (1986) Renin in the rat pituitary coexists with angiotensin II and depends on testosterone. Endocrinology 118: 2470-2476
43. Powers CA, Hatala MA (1986) Dopaminergic regulation of the estrogen-induced glandular kallikrein in the rat anterior pituitary. Neuroendocrinology 44: 462-469
44. Coslovsky R, Braverman LE, Leeman SE, Aronin N (1985) The differential effects of thyroid and gonadal hormones on substance P content in the anterior pituitary of the prepubertal rat. Endocrinology 117: 2198-2202
45. Cheng MC, Smith AI, Funder JW (1986) β-Endorphin and its congeners in rat pituitary and thyroid: effects of propylthiouracil and thyroid hormone administration. Endocrinology 119: 642-647
46. Petraglia F, Penalva A, Locatelli V, Cocchi D, Paneral AE, Genazzani AR, Müller EE (1982) Effects of gonadectomy and gonadal steroid replacement on pituitary and plasma β-endorphin levels in the rat. Endocrinology 111: 1224
47. Hong JS, Yoshikawa K, Lamartiniere CA (1982) Sex-related difference in the rat pituitary [Met^5]-enkephalin level-altered by gonadectomy. Brain Res 251: 380-383
48. Hong J-S, Yoshikawa K, Hudson PM, Uphouse LL (1982) Regulation of pituitary and brain enkephalin systems by estrogen. Life Sci 31: 2181-2184
49. Schwaninger M, Knepel W, Döhler KD, Sandow J (1987) Release of dynorphin-like immunoreactivity from rat adenohypophysis in vitro during inhibition of anterior pituitary hormone secretion from individual cell types. Endocrinology 121: 167-174
50. Timmers K, Voyles NR, Zalenski C, Wilkins S, Recant L (1986) Altered β-endorphin, met- and leu-enkephalins, and enkephalin-containing peptides in pancreas and pituitary of genetically obese diabetic (db/db) mice during development of diabetic syndrome. Diabetes 35: 1143-1152
51. Kouchich FJ, Quock RM, Tseng L-F (1984) Dynorphin-(1-13)-like immuno-reactivity in central nervous system and pituitary gland of spontaneously hypertensive rats. Clin Exper Hyper, Theory and Practice A6 (3): 699-708
52. Delbarre B, Casset-Senon D, Delbarre G, Sestillange P, Christin O (1982) Naloxone effects on blood pressure, analgesia and diuresis in spontaneously hypertensive and normotensive rats. Neurosci Lett 30: 167
53. Seizinger BR, Liebisch DC, Grimm C, Herz A (1984) Ontogenetic development of the pro-enkephalin B (= pro-dynorphin) opioid peptide system in the rat pituitary. Neuroendocrinology 39: 414-422
54. Seizinger BR, Höllt V, Herz A (1984) Postnatal development of β-endorphin-related peptides in rat anterior and intermediate pituitary lobes: evidence for contrasting development of proopiomelanocortin processing. Endocrinology 115: 136-142
55. Loftus CM, Nilaver G, Beinfeld MC, Post KD (1986) Motilin-related immunoreactivity in mammalian adenohypophysis. Neurosurgery 19: 201-204
56. Korchak DM, Nilaver G, Beinfeld MC (1984) The development of motilin-like immunoreactivity in the rat cerebellum and pituitary as determined by radioimmunoassay. Neurosci Lett 48: 267-272
57. Samson WK, Lumpkin MD, Nilaver G, McCann SM (1984) Motilin: a novel growth hormone releasing agent. Brain Res Bull 12: 57-62

PITUITARY GROWTH FACTORS

S. Melmed

Division of Endocrinology & Metabolism
Cedars-Sinai Medical Center - UCLA School of Medicine
Los Angeles, California

Growth factors act on neighboring or autologous cells to effect specific cellular responses. These include regulation of the cell cycle and cell growth, induction of cell-specific unique proteins or regulation of a differentiated cell function (1). The latter function of growth factors is mediated by a direct regulation of phenotypic gene expression. The pituitary gland has been shown to contain increasing numbers of specific growth factors, most of which are listed in Table 1. These pituitary-derived factors have been isolated to varying degrees of purity (2).

TABLE 1

PITUITARY-DERIVED GROWTH FACTORS

Epidermal Growth Factor	Chondrocyte Growth Factor
Fibroblast Growth Factor	Adipocyte Growth Factor
Insulin-like Growth Factors	Ovarian Growth Factor
Thyroid Hormone-inducible Growth Factor	Glial Growth Factor
Mammary Tumor Cell Growth Factor	

To date, however, few of these factors have been convincingly shown to regulate specific pituitary function (Table 2). EGF has been shown to be a potent selective stimulator of PRL secretion (3). FGF at nonmitogenic concentrations, stimulates TRH-induced PRL and TSH secretion by primary rat pituitary cells (4). PDGF, in contrast, appears to inhibit selectively PRL secretion by rat pituitary tumor cells (5).

For the past several years, our laboratory has been studying the effects of IGF-I on pituitary function. The data presented below will demonstrate the role of IGF-I in the regulation of specific pituitary gene expression.

TABLE 2

GROWTH FACTOR ACTION ON ANTERIOR PITUITARY FUNCTION

Factor	Pituitary Receptors	Regulation of Pituitary Function
EGF	yes	Stimulates PRL
FGF	?	Stimulates PRL and TSH
IGF-I	yes	Suppresses GH
IGF-II	yes	Suppresses GH
Thyroid Hormone-inducible Growth Factor		Stimulates cell replication
PDGF	?	Suppresses PRL
Insulin	yes	Suppresses GH and Stimulates PRL

REGULATION OF GH SECRETION

GH is under dual hypothalamic stimulatory and inhibitory control (6,7). GHRH stimulates GH gene expression and secretion, while SRIF inhibits the secretion of GH (8). Peripheral hormones, including triiodothyronine, hydrocortisone and insulin also directly regulate the transcription of the rat GH gene (9-11). Hydrocortisone enhances the stimulatory action of hypothalamic GHRH on the somatotroph (12). Most of the peripheral growth promoting actions of GH are mediated by somatomedin-C (IGF-I), a growth factor that stimulates sulfate incorporation by cartilage (13). IGF-I, which has a high homology with pro-insulin and exhibits a 50% homology with insulin, has a molecular weight of 7650. The C peptide linkage is retained, resulting in a single chained polypeptide (14,15). GH-dependent IGF-I is secreted predominantly by the liver, but the IGF-I gene is also expressed in multiple tissues of the rat including kidney, brain, lung, muscle, heart and the pituitary itself (16,17).

PITUITARY RECEPTORS

Specific binding sites for insulin, IGF-I and IGF-II, have been well characterized in rat and human anterior pituitary tissue. The existence of these abundant high affinity pituitary IGF receptors is consistent with a biological role for the somatomedin peptides in the regulation of GH secretion (18-20).

FEEDBACK REGULATION OF GH BY IGF-I

In Vivo Studies

Autoregulation of GH secretion has been demonstrated in rats, rhesus monkey and man. The stimulation of hypothalamic somatostatin secretion by high levels of GH and by IGF-I has been implicated as a cause for the autoregulation of GH secretion (21). In animals bearing GH-secreting tumors with elevated circulating GH and IGF-I levels, pituitary GH transcription rates were selectively suppressed (22). The intraventricular injection of Somatomedin-C to conscious, freely moving, chronically catheterized rats markedly suppressed the episodic pulsatile secretion of pituitary GH (23,24). The action of IGF-I may occur at the level of the hypothalamus by stimulating SRIF production or decreasing GHRH secretion. Based on in vitro evidence presented below, it would appear that this inhibition occurs directly at the level of the pituitary. Circulating IGF-I may therefore be implicated as participating in a long and short negative feedback regulation of pituitary GH gene expression.

In Vitro Studies

The recent availability of recombinant human IGF-I and pure native peptide have facilitated the in vitro study of the effects of IGF-I on rat and human GH gene expression.

Effects of IGF-I on Basal Rat GH Secretion

A purified somatomedin-C preparation was shown to inhibit dibutyryl cyclic AMP-stimulated GH secretion by rat pituitary cells (25). Somatomedin-C also stimulated the in vitro secretion of hypothalamic SRIF, further delineating the participation of this growth factor in the feedback regulation of GH secretion. Both basal and GHRH-stimulated GH secretion in vitro were suppressed by several IGF-I preparations (25-32).

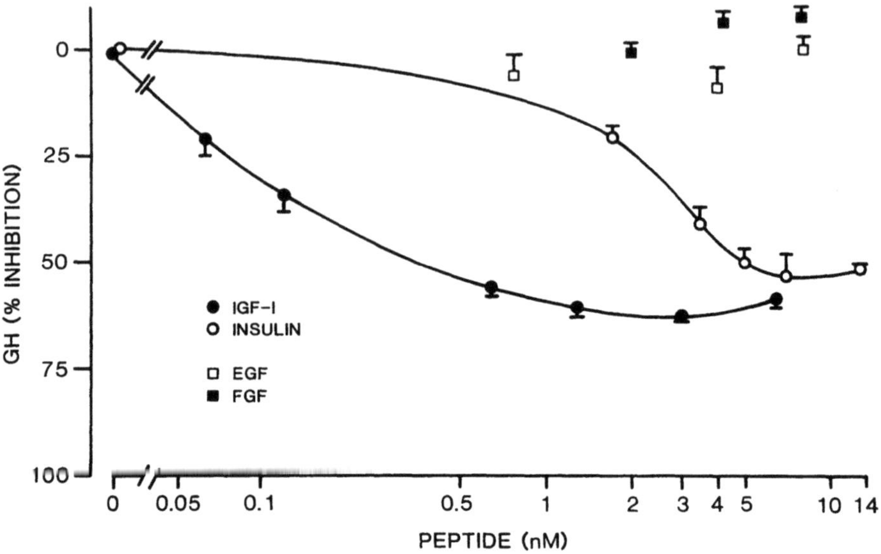

Figure 1. Dose-response effect of IGF-I on GH secretion. Pituitary cells were treated for 72 hr with the indicated doses of peptide added to serum-free medium. From ref. 28, with permission.

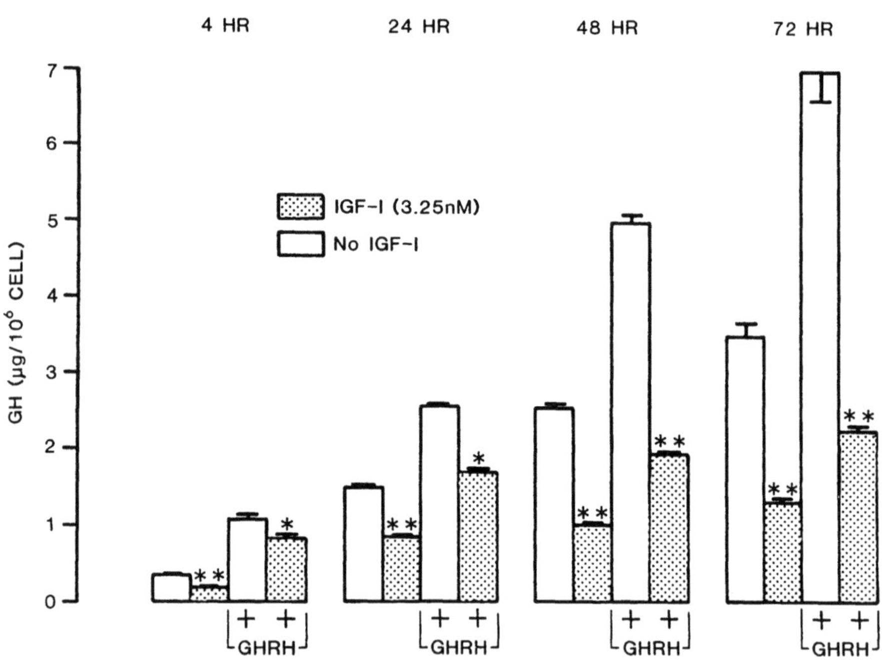

Figure 2. Effect on IGF-I on GHRH-stimulated GH secretion. Pituitary cells were grown in serum-free defined medium containing IGF-I with or without GHRH (1 nM) as indicated. *P < 0.05, **P < 0.001 (vs. respective IGF-I-free controls). From ref. 28, with permission.

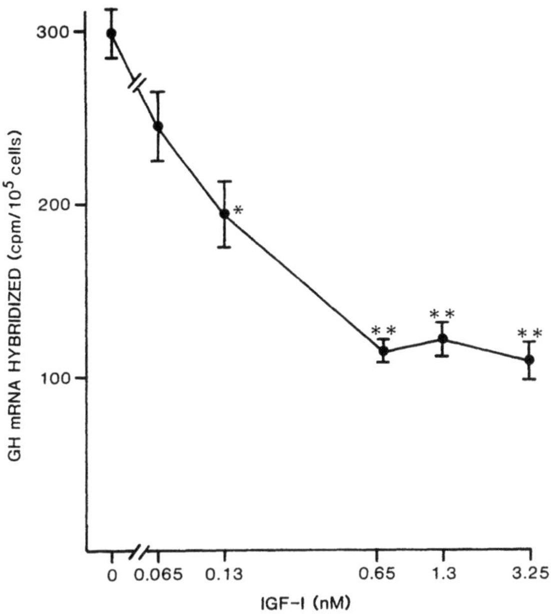

Figure 3. Dose-responsiveness of IGF-I suppression of GH mRNA levels. Primary rat pituitary cells were treated for 72 hr before mRNA extraction and measurement by dot blot hybridization. *P < 0.05, **P < 0.001. From ref. 28, with permission.

The IGF-I effect on the pituitary is specific inasmuch as other growth factors including EGF and FGF do not alter GH gene expression. IGF-I also does not alter the secretion of prolactin by pituitary cells, suggesting that the inhibition of GH induced by IGF-I is indeed selective. The effective inhibitory concentration of IGF-I correlates well with the displacement curves for IGF-I binding to its receptor sites on pituitary cells. Fifty percent displacement of [^{125}I]-IGF-I was achieved by about 0.5-2.5 nM (25).

Effects of IGF-I on Stimulated GH

GHRH has been shown to stimulate GH secretion, mRNA levels, and GH gene transcription (33,34). IGF-I inhibits the stimulation of GH mRNA levels as well as GH gene transcription induced by GHRH. The T3-induced GH mRNA levels and secretion in pituitary glands derived from hypothyroid rats was also inhibited by IGF-I (Fig. 4). In addition, IGF-I attenuates cAMP and forskolin-induced GH mRNA levels (Fig. 5).

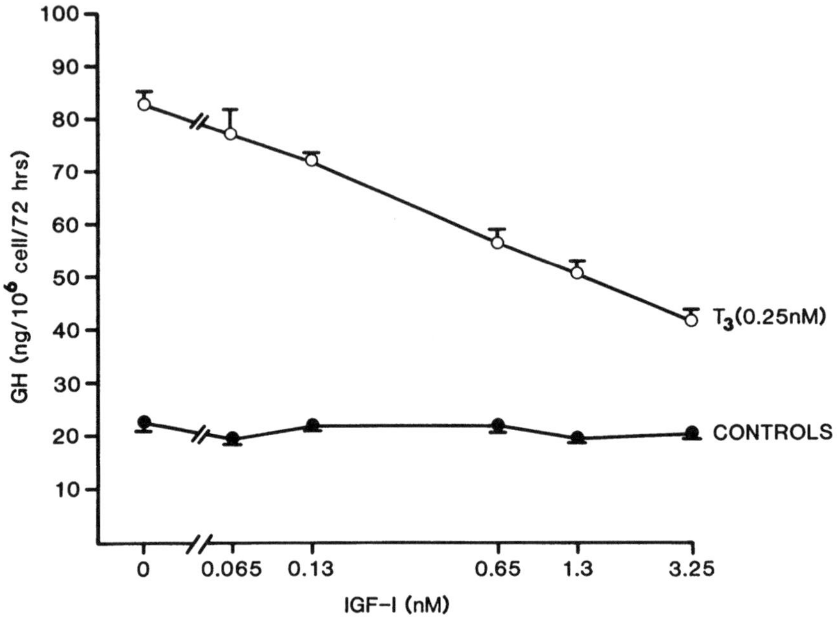

Figure 4. Effects of increasing doses of IGF-I on T_3-stimulated GH secretion. Pituitary cells derived from thyroidectomized rats were preincubated for 24 hr in medium depleted of thyroid hormone. Incubations were then continued for 72 hr in the presence of the indicated doses of IGF-I with or without added T_3. Controls received no added T_3. From ref. 30, with permission.

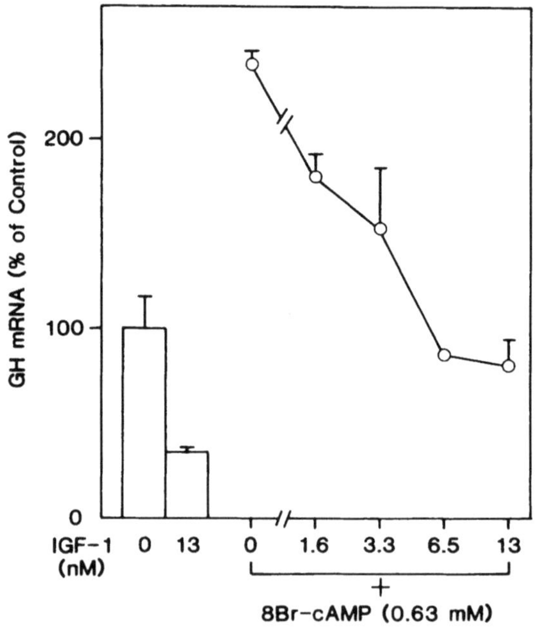

Figure 5. IGF-I suppresses GH mRNA induced by 8-Br-cAMP in rat pituitary cells. Cells were treated for 72 hr and GH mRNA levels were then measured. From ref. 29, with permission.

Effects of IGF-I on Human Pituitary Cells

IGF-I inhibits the release of GH by human pituitary tumor cell cultures (20,35). The 50% stimulation of GH secretion and mRNA levels induced by GHRH was also blocked by simultaneous exposure of the cells to IGF-I (35) (Fig. 6). The effects of IGF-I were prevented by α-IR3, a specific antibody to the IGF-I receptor. Maximal inhibition of human GH mRNA was seen after treatment with 6.5 nM IGF-I. These data demonstrate the regulation of human GH mRNA in phenotypic human somatotroph cells.

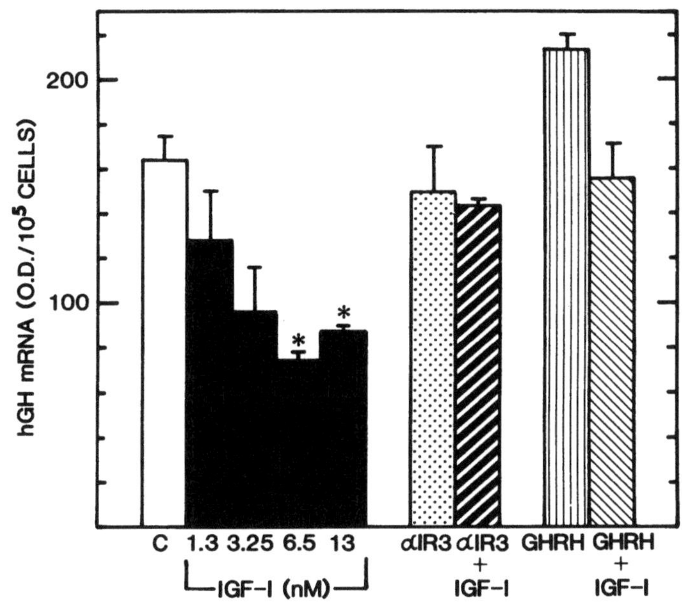

Figure 6. Effects of IGF-I, α-IR3 and GHRH on relative levels of GH mRNA sequences in cultured human somatotropinoma cells. After indicated treatments for 48 hr, extracted RNA was immobilized and hybridized with [^{32}P]cDNA for hGH. From ref. 35, with permission.

Transcriptional Effect of IGF-I on GH Gene Expression

The inhibitory action of IGF-I on GH gene expression may happen directly at the transcriptional level or may occur post-transcriptionally by altering mRNA stability or translation. In order to elucidate further the effect on GH gene expression, nuclei derived from primary cultures of rat anterior pituitary cells treated with added IGF-I were incubated in an in vitro transcriptional run-off assay (36).

IGF-I treatment suppressed the transcriptional activity of the GH gene in a dose-responsive manner. Maximal inhibition was seen with 6.5 nM IGF-I by 4 hours. After 24 hours, GH gene transcription was still suppressed by about 50% as compared to control untreated cells. PRL gene transcriptional activity was not altered by IGF-I.

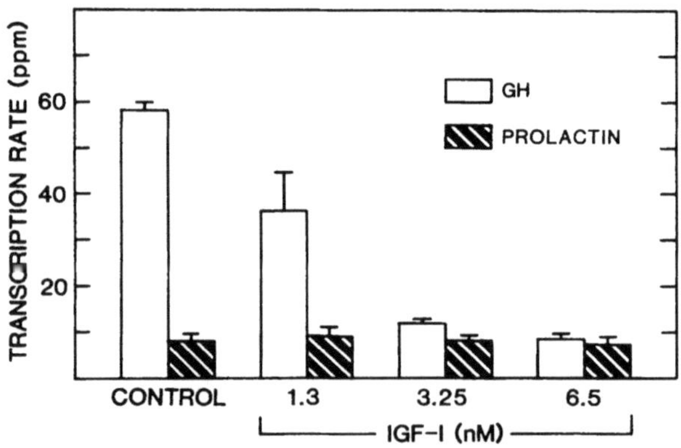

Figure 7. Dose response of IGF-I effect on GH and prolactin gene transcription. Primary cultures of rat anterior pituitary cells were incubated in serum-free defined medium with or without added IGF-I. New GH mRNA and prolactin mRNA were measured by transcriptional runoff assay with [^{32}P]-labeled GTP and hybridization against immobilized GH cDNA and prolactin cDNA. From ref. 36, with permission.

The stimulation of new GH mRNA synthesis induced by GHRH was also blocked by IGF-I. Other growth factors, including EGF and FGF, did not alter GH transcription. Relatively high doses of insulin also suppressed the GH gene transcriptional activity. This may indicate that insulin is acting via the IGF-I receptor, or alternatively, insulin may specifically inhibit GH gene transcription in these cells, as has been shown for GH_3 cell GH gene transcription (37).

Effect of IGF-I on the Transfected Human GH Gene

In order to delineate further the transcriptional mechanism of IGF-I inhibition of human GH gene expression, a human GH gene was transfected into human JEG3 cells. A 2.6 kb Eco RI fragment of the hGH genomic clone (38,39) containing the entire hGH gene along with 497 bp of the 5′-flanking region and

540 bp of the 3′-flanking region was subcloned into PUC18 without a heterologous promoter. This allowed us to examine directly the suppressive effects of IGF-I on the gene (40). A highly specific and sensitive human GH RIA was used. The RIA exhibits <0.001% cross-reaction with human placental lactogen, and has a sensitivity of 60 pg/ml.

Transfected cells secreted low amounts of human GH into the serum-free medium. The GH secretion after transfection were similar in all untreated wells, suggesting relatively uniform transfection efficiency. Treatments did not alter the final cell number in each well. IGF-I (6.5 nM) significantly suppressed the stimulation of GH induced by 8-Br-cAMP and hydrocortisone (Fig. 7).

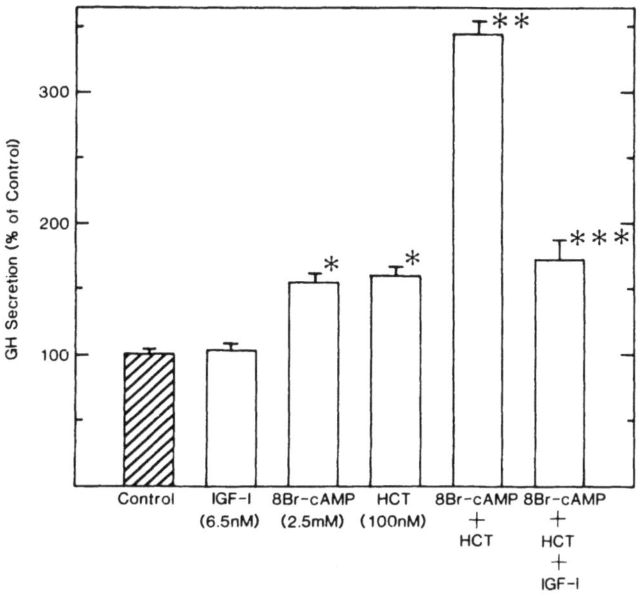

Figure 8. Regulation of human GH secretion in transfected JEG3 cells. After transfection with 2.6 kb human GH gene, cells were grown in serum-free medium with the indicated treatments for 72 hr. Nontransfected cells had undetectable GH in the medium (<0.13 ng/ml). From ref. 40, with permission.

The transfected cells were also tested for evidence of de novo GH synthesis by specific immunoprecipitation and gel analysis. Treatment of the transfected cells with hydrocortisone and 8-Br-cAMP for 72 hr in serum-free medium induced a newly synthesized protein band of 22 kDa similar to the size of authentic labeled human GH standard. IGF-I (6.5 nM) attenuated the incorporation of ^{35}S-methionine into the newly synthesized 22 kDa GH. Non-transfected JEG3 cells did not incorporate [^{35}S]methionine into this 22 kDa band. The 22 kDa band was competitively inhibited by addition of excess cold human GH.

The dose-response of the IGF-I effect on new GH synthesis in tranfected cells is depicted in Figure 9. The highest dose of IGF-I used, 13 nM, suppressed the strong induction of GH by cAMP + hydrocortisone to below control levels.

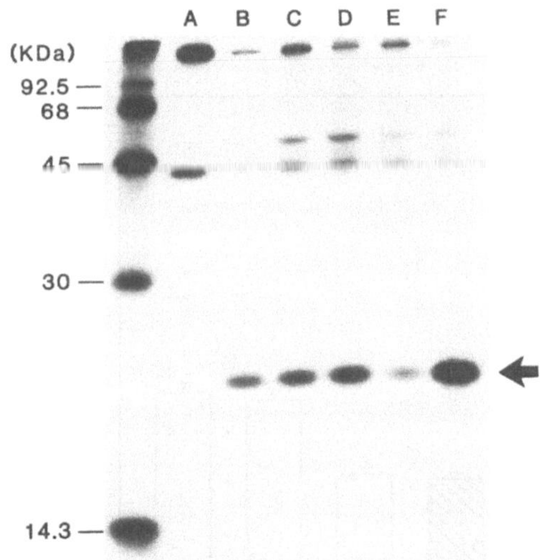

Figure 9. Immunoprecipitation and SDS PAGE analysis of human GH in transfected JEG3 cells. 72 hr after the indicated treatments in serum-free medium, cells were exposed to [^{35}S]methionine for 4 hr in methionine-free medium. Cytosol extracts were subjected to immunoprecipitation with highly specific human GH antiserum (provided by Dr. A.F. Parlow) and separated. Arrows indicate 22 kDa size and migration of [^{125}I]hGH marker. Lane A: control untreated transfected cells; B-D: cAMP + hydrocortisone + IGF-I; E: cAMP + hydrocortisone + insulin; F: cAMP + hydrocortisone. From ref. 40, with permission.

Northern analysis showed that nontransfected JEG cell RNA did not hybridize to the hGH cDNA probe. Both the mature 1.0 kb GH mRNA as well as the 2.2 kb GH mRNA precursor were suppressed by IGF-I treatment of the cells.

The above results showing regulation of the transfected human GH gene confirm the action of IGF-I at the level of transcription of the rat GH gene. It therefore appears that the 2.6 kb GH gene fragment contains regulatory elements responsive to IGF-I. Specific IGF-I responsive cis-acting elements may therefore reside in the 0.5 kb 5′-flanking region, although other locations cannot be presently ruled out.

Pituitary IGF-I

The IGF-I gene is expressed in the pituitary itself, as evidenced by IGF-I RIA, as well as by hybridization of specific mRNA with IGF-I cDNA probes (41,42). As circulating IGF-I binds to carrier proteins with an affinity constant comparable to the affinity of pituitary IGF receptors, it is possible that locally produced pituitary IGF-I may be bioactive in the pituitary. The mechanism of the autoregulation of GH secretion may therefore involve an interactive regulation of the expression of both the GH and IGF-I genes in neighboring pituitary cells (paracrine) or at a single cell level (autocrine). Pituitary GH secretion may therefore be under tonic inhibition by locally produced pituitary IGF-I.

CONCLUSION

The evidence presented above indicates a potent and selective feedback inhibition of IGF-I on rat and human GH gene transcription. This feedback loop appears to be analogous to the negative feedback of thyroid hormone and adrenal steroids on their respective pituitary trophic hormone expression (43,44).

This information may explain the discordancy between serum IGF-I and GH levels during starvation, protein-calorie malnutrition, and cachexia (45). The high GH levels in these states may be caused by a protective lowering of peripheral or local IGF-I production leading to unrestrained GH secretion. Clearly therefore, the central hypothalamic control of GH secretion is modulated by endocrine and/or paracrine signal input of IGF-I in determining the final net effect on GH secretion.

ACKNOWLEDGMENTS

The author thanks Dr. James Fagin for helpful discussions and is grateful to Joye L. Nunn for editorial assistance in the preparation of this manuscript. Work in the author's laboratory was supported by NIH grants DK 33802 and DK 34824. Figures 1-3 are reproduced from S. Yamashita, S. Melmed, Insulin-like growth factor I action on rat anterior pituitary cells: suppression of growth hormone secretion and messenger ribonucleonic acid levels. *Endocrinology* 118(1):176-182, © by The Endocrine Society, 1986. Figure 4 is reproduced from S. Melmed, S. Yamashita, Insulin-like growth factor I action on hypothyroid rat anterior pituitary cells: suppression of 3,5,3′-triiodothyronine-induced growth hormone secretion and messenger ribonucleic acid levels. *Endocrinology* 118(4): 1483-1490, © by The Endocrine Society, 1986. Figure 5 is reproduced from S.

Morita, S. Yamashita, S. Melmed, IGF-I action on rat anterior pituitary cells: Effects of intracellular messengers on GH secretion and mRNA levels. *Endocrinology* 121(6): 2000-2006, © by The Endocrine Society, 1987. Figure 6 is reproduced from S. Yamashita, M. Weiss, S. Melmed, Insulin-like growth factor I regulates growth hormone secretion and messenger RNA levels in human pituitary tumor cells. *J Clin Endocrin Metab* 63(3): 730-735, © by The Endocrine Society, 1986. Figure 7 is reproduced from *The Journal of Clinical Investigation*, 1987, 79:449-452 by copyright permission of the American Society for Clinical Investigation. Figures 8 and 9 are reproduced from S. Yamashita S, Ong J, Melmed S (1987) Regulation of human GH gene expression by IGF-I in transfected cells. J Biol Chem 262: 13254-13257

REFERENCES

1. Sporn MB, Todaro GJ (1980) Autocrine secretion and malignant transformation of cells. N Engl J Med 303: 878-880
2. Kasper S, Friesen HG (1987) Growth factor: a selected review. In: Ludecke D, Tolis G (eds) Growth Hormone, Growth Factor and Acromegaly. New York: Raven Press
3. Johnson LK, Baxter JD, Vlodovsky I, Gospadorowicz D (1980?) EGF and expression of specific genes: effects on cultured rat pituitary cells are dissociable from the mitogenic response. Proc Natl Acad Sci 77: 394-398
4. Baird A, Mormede P, Ying SY, Wehrenberg WB, Ueno N, Ling N and Guillemin R (1985) A non-mitogenic pituitary function of fibroblast growth factor: regulation of TSH and PRL secretion. Proc Natl Acad Sci USA 82: 5545-5549
5. Sullivan N, Tashjian AH (1983) PDGF selectively decreases PRL production in pituitary cells in culture. Endocrinology 113: 639-642
6. Ling N, Zeytin F, Bohlen P, Esch F, Brazeau P, Wehrenberg WB, Baird A, Guillemin R (1985) Growth hormone releasing factors. Ann Rev Biochem 54: 403-423
7. Frohman LA, Jansson JO (1986) Growth hormone-releasing hormone. Endocr Rev 7: 223-253
8. Brazeau P, Vale W, Burgus R, Ling N, Butcher M, Rivier J, Guillemin R (1973) Hypothalamic polypeptide that inhibits the secretion of immunoreactive pituitary growth hormone. Science 179: 77-79
9. Yaffe BM, Samuels HH (1984) Hormonal regulation of the growth hormone gene: relationship of the rate of transcription to the level of nuclear thyroid hormone-receptor complexes. J Biol Chem 259: 6284-6291
10. Evans RM, Birnberg NC, Rosenfeld MG (1982) Glucocorticoid and thyroid hormones transcriptionally regulate growth hormone gene expression. Proc Natl Acad Sci USA 79: 7659-7663
11. Yamashita S, Melmed S (1986) Insulin regulation of rat GH gene transcription. J Clin Invest 78: 1008-1014
12. Wehrenberg WB, Baird A, Ling N (1983) Potent interaction between glucocorticoids and growth hormone-releasing factor in vivo. Science 221: 556-558
13. Salmon WD, Daughaday W (1957) A hormonally controlled serum factor which stimulates sulfate incorporation by cartilage in vitro. J Lab Clin Med 49: 825-836
14. Rinderknecht E, Humbel RE (1978) The amino acid sequence of human insulin-like growth factor I and its structural homology with proinsulin. J Biol Chem 253: 2769-2775
15. Klapper DG, Svoboda ME, Van Wyk JJ (1983) Sequence analysis of somatomedin-C: confirmation of identity with insulin-like growth factor I. Endocrinology 112: 2215-2221

16. D'Ercole AJ, Stiles AD, Underwood LE (1984) Tissue concentrations of somatomedin C: Further evidence for multiple sites of synthesis and paracrine or autocrine mechanisms of action. Proc Natl Acad Sci USA 81: 935-939
17. Fagin JA, Pixley S, Slanina S, Ong U, Melmed S (1987) Insulin-like growth factor I gene expression in GH_3 rat pituitary cells mRNA content, immunocytochemistry and secretion. Endocrinology 120: 2037-2043
18. Rosenfeld RG, Ceda G, Wilson DM, Dollar LA, Hoffman AR (1984) Characterization of high affinity receptors for insulin-like growth factors I and II on rat anterior pituitary cells. Endocrinology 114: 1571-1575
19. Goodyer CG, Stephano LD, Lai WH, Guyda H, Posner BI (1984) Characterization of insulin-like growth factor receptors in rat anterior pituitary hypothalamus and brain. Endocrinology 114: 1187-1195
20. Ceda GP, Hoffman AR, Silverberg GD, Wilson DM, Rosenfeld RG (1985) Regulation of GH release from cultured human pituitary adenomas by somatomedins and insulin. J Clin Endocrinol Metab 60: 1204-1210
21. Tannenbaum GS (1980) Evidence for autoregulation of GH secretion via the central nervous system. Endocrinology 107: 2117-2220
22. Yamashita S, Slanina S, Kado H, Melmed S (1986) Autoregulation of pituitary growth hormone messenger RNA levels in rats bearing transplantable mammosomatotrophic pituitary tumors. Endocrinology 118: 915-918
23. Tannenbaum GS, Guyda HJ, Posner BI (1983) Insulin-like growth factors: a role in growth hormone negative feedback and body weight regulation via brain. Science 220: 77-79
24. Abe H, Molitch ME, Van Wyk JJ, Underwood LE (1983) Human growth hormone and somatomedin C suppress the spontaneous release of growth hormone in unanesthetized rats. Endocrinology 113: 1319-1324
25. Berelowitz M, Szabo M, Frohman LA, Firestone S, Chu L, Hintz RL (1981) Somatomedin-C mediates growth hormone negative feedback by effects on both the hypothalamus and the pituitary. Science 212: 1279-1281
26. Brazeau P, Guillemin R, Ling N, Van Wyk J, Humbel R (1982) Inhibition by somatomedins of growth hormone secretion stimulated by hypothalamic growth hormone releasing factor (somatocrinin, GRF) on the synthetic peptide hpGRF. C R Acad Sci [D] (Paris) 295: 651-654
27. Goodyer CC, De Stephano L, Guyda HJ, Posner BI (1984) Effects of insulin-like growth factor on adult male rat pituitary function in tissue culture. Endocrinology 115: 1568-1574
28. Yamashita S, Melmed S (1986) Insulin-like growth factor I action on rat anterior pituitary cells: suppression of growth hormone secretion and messenger ribonucleic acid levels. Endocrinology 118: 176-182
29. Morita S, Yamashita S, Melmed S (1987) IGF-I action on rat anterior pituitary cells: Effects of intracellular messengers on GH secretion and mRNA levels. Endocrinology 121: 2000-2006
30. Melmed S, Yamashita S (1986) Insulin-like growth factor I action on hypothyroid rat anterior pituitary cells: suppression of 3,5,3'-triiodothyronine-induced growth hormone secretion and messenger ribonucleic acid levels. Endocrinology 118: 1483-1490
31. Goodyer CG, Marcovitz S, Hardy J, Lefebvre Y, Guyda HJ, Posner BL (1986) Effect of insulin-like growth factors on human foetal, adult normal and tumour pituitary function in tissue culture. Acta Endocrinol 112: 49
32. Ceda GP, Davis RG, Rosenfeld RG, Hoffman AR (1987) Growth hormone releasing hormone-GH-somatomedin axis: evidence for rapid inhibition of GHRH-elicited GH release by insulin-like growth factors I and II. Endocrinology 120: 1658-1662
33. Barinaga M, Yamamoto G, Rivier C, Vale W, Evans R, Rosenfeld MG (1983) Transcriptional regulation of growth hormone gene expression by growth hormone releasing factor. Nature 306: 84-86
34. Gick GG, Zeytin FN, Brazeau P, Ling NC, Esch FS, Bancroft C (1984) Growth hormone-releasing factor regulates growth hormone mRNA in primary cultures of rat pituitary cells. Proc Natl Acad Sci USA 81: 1553-1557

35. Yamashita S, Weiss M, Melmed S (1986) Insulin-like growth factor I regulates growth hormone secretion and messenger RNA levels in human pituitary tumor cells. J Clin Endocrin Metab 63: 730-735
36. Yamashita S, Melmed S (1987) Insulin-like growth factor I regulation of growth hormone gene transcription in primary rat pituitary cells. J Clin Invest 79: 449-452
37. Yamashita S, Melmed S (1986) Insulin regulation of rat GH gene transcription. J Clin Invest 78: 1008-1014
38. Seeburg PH (1982) The human GH gene family. DNA 1: 2329-2349
39. Miller WL, Eberhardt NL (1983) Structure and evolution of the growth hormone gene family. Endocr Rev 4: 97-130
40. Yamashita S, Ong J, Melmed S (1987) Regulation of human GH gene expression by IGF-I in transfected cells. J Biol Chem 262: 13254-13257
41. Binoux M, Hossenlop P, Lussarre C, Hardquin N (1981) Production of insulin-like growth factors and their carrier by rat pituitary gland and brain explants in culture. FEBS Letts 124: 178-184
42. Fagin J, Melmed S (1987) Pituitary insulin-like growth factor I gene expression: in vitro and in vivo regulation by triiodothyronine and growth hormone. (in review)
43. Shupnick MA, Chin WW, Habener JF, Ridgeway EC (1985) Transcriptional regulation of the thyrotropin subunit genes by thyroid hormone. J Biol Chem 260: 2900-2905
44. Eberwine JH, Roberts JL (1982) Glucocorticoid regulation of pro-opiomelanocortin gene transcription in the rat pituitary. J Biol Chem 259: 2166-2170
45. Clemmons DR, Van Wyk JJ (1984) Factors controlling blood concentration of somatomedin C. Clin Endocrinol Metab 13: 113-143

MECHANISMS OF ACTION OF GROWTH FACTORS

L. Vicentini,* A. Pandiella and J. Meldolesi

Department of Pharmacology* and Scientific Institute S. Raffaele, University of Milan, Milan, Italy

INTRODUCTION

Understanding the mechanisms that control cell growth is a central problem in cell biology today. During the last few years, it has become increasingly clear that one or more defects in the physiological mechanisms that regulate cell proliferation can be the cause of tumor cell growth. The study of normal cell division is therefore important not only per se, but especially in relation to its pathological implications.

CELL CYCLE

The cell cycle can be divided into four periods: G_1, the gap between the previous nuclear division and the beginning of DNA synthesis; S, the period of DNA synthesis; G_2, the gap between DNA synthesis and nuclear division; and M, the period of mitosis, during which chromosomes split and segregate into daughter cells. Following this division cells can either re-enter G_1, and pass therefore through another cell cycle, or become quiescent and enter a G_0 phase. The growth factors push the quiescent cell out of G_0 into G_1 and cell division.

POLYPEPTIDE GROWTH FACTORS

Growth factors are polypeptides that bind to specific receptors located on the membranes of target cells (Fig. 1). Receptor binding is

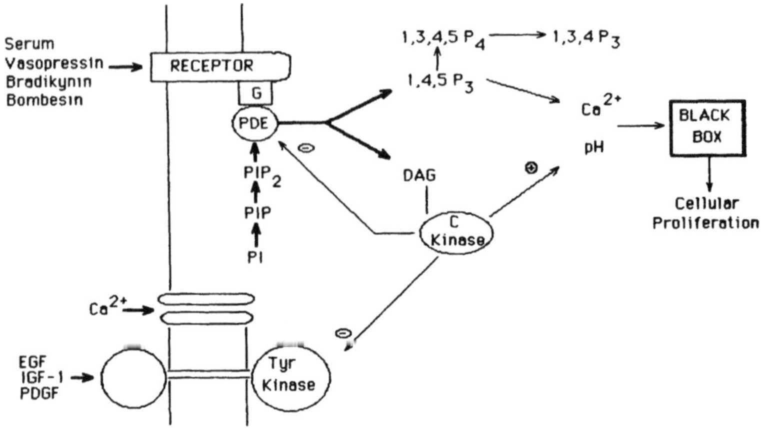

Figure 1. Early intracellular signals stimulated by the growth factors: PI turnover, C kinase activation, Ca^{2+} and pH rise, tyrosine kinase activation.
Abbreviations used: PI, phosphatidylinositol; PIP, phosphatidyl-inositol-4-phosphate; PIP_2, phosphatidylinositol-4,5-bisphosphate; DAG, diacylglycerol; G, GTP binding protein; PDE, phosphodiesterase.

believed to trigger the generation of intracellular signals that lead eventually (12-24 hr later) to DNA synthesis. Until recently, the nature of the biochemical signals generated at growth factor receptors and especially, the relationship between these signals and later nuclear events remained relatively unexplored. Developments in this field have been made possible by the use of cell cultures that can be arrested in the G_0 phase by serum deprivation. Re-addition of either serum or single growth factors to these quiescent cells generates signals resulting in mitogenic responses.

Na^+/H^+ EXCHANGE

One of the earliest biochemical events stimulated by growth factors is the stimulation of Na^+/H^+exchange (for a review see 1). Such a process is typically inhibited by the diuretic drug amiloride. The Na^+/H^+ exchange is an ubiquitous membrane-bound antiport system that is responsible for pH regulation. Its activation by growth factors results in a persistent alkalinization of the cytoplasm, which has been shown to be a permissive event allowing DNA synthesis to occur (2). In other words, alkalinization per se is not sufficient to cause growth stimulation, which requires other events to trigger it. At levels below a critical pH (typically around 7.2), however, such events become ineffective.

The Na^+ flux, which is stimulated within 30 sec, is not the only ionic movement stimulated by growth factors. An even faster response is the increase of cytosolic free Ca^{2+} concentration $[Ca^{2+}]_i$, which with many growth factors has been shown to occur immediately (<1 sec) after receptor activation (3).

Studies of this kind have been made possible by the introduction of the fluorescent probes Quin 2 and Fura 2, which allow the measurement of $[Ca^{2+}]_i$ in cells of any size (4,5). Most agonists that increase $[Ca^{2+}]_i$ have been shown to be coupled to the hydrolysis of membrane phosphoinositides. This metabolic reaction gives rise to two second messengers: inositol-1,4,5-trisphosphate (IP_3), which mobilizes Ca^{2+} from intracellular stores, and diacylglycerol (DAG), which activates protein kinase C (PKC) (for a recent review see 6). During the last few years, we have been among the first to show that different types of mitogens - serum, bradykinin, vasopressin, bombesin, PDGF - stimulate the generation of IP_3 within a few seconds after application to human and murine fibroblasts in culture (7,8).

PROTEIN KINASE C ACTIVATION

As we pointed out earlier, one of the consequences of polyphosphoinositide hydrolysis is the activation of PKC by diacylglycerol. Since growth factors induce polyphosphoinositide hydrolysis and thus activate PKC, a possible role for this enzyme in regulating early membrane events stimulated by mitogens was examined. The existence of a potent pharmacological tool, the tumor-promoter phorbol ester, phorbol-12 myristate, 13 acetate (PMA), which stimulates the activity of PKC, allowed us and others to study the role of PKC in the regulation of the Na^+/H^+exchange system. PMA was found to stimulate an amiloride-sensitive Na^+ flux in Swiss 3T3 fibroblasts, a murine immortalized cell line, and in WI-38, a human embryonal lung fibroblast cell line (9). The extent of Na^+ flux stimulation by PMA was smaller than that seen with serum (9). These data suggest that PKC is involved in the activation of the Na^+/H^+ exchange, at least in 3T3 and WI-38 cells. To our surprise, however, PMA alone failed to stimulate a Na^+ flux in adult human foreskin fibroblasts, but it induced synergistic responses when applied together with submaximal doses of the Ca^{2+} ionophore A23187 (9). It appears, therefore, that in human fibroblasts the simultaneous activation of PKC and Ca^{2+} mechanisms is needed to generate a full Na^+/H^+ exchange activation. Taken together, the results obtained with human fibroblasts and the murine 3T3 cell lines suggest that the regulation and sensitivity to PKC of the Na^+/H^+ antiport can vary considerably, possibly depending on the stage of differentiation and the proliferating activity of the various cell types investigated.

So far, the activation of PKC has been considered to be one of the signals that may participate in the triggering of the mitogenic response. However, the enzyme appears to be responsible also for a feedback inhibitory loop of the response. In fact, pretreatment of human fibroblasts with PMA for two min resulted in a significant inhibition of the Na^+ flux stimulated by serum, bradykinin and vasopressin (9). From the above results, we conclude that in human fibroblasts, serum growth factors as well as vasopressin and bradykinin, by interacting with their specific membrane receptors, stimulate the hydrolysis of phosphoinositides to generate IP_3 and DAG. The synergistic interaction of Ca^{2+}, mobilized by IP_3, and PKC, activated by DAG, results in the stimulation of Na^+/H^+ exchange, which induces alkalinization of the cytoplasm. Shortly thereafter, the negative feedback control mediated by PKC turns off the growth factor-stimulated response, thereby providing cells with a mechanism of autoregulation.

EPIDERMAL GROWTH FACTOR

Epidermal growth factor (EGF) is a mitogen for various cell types. Binding of EGF to its receptor leads to the activation of an enzyme, a protein kinase specific for tyrosine residues, which is intrinsic to the receptor molecule. The enzyme, activated by EGF binding, phosphorylates various cellular proteins, but the most prominent substrate is the EGF receptor itself (10). In the human epidermoid carcinoma cell line A431, which overexpresses EGF receptors, we showed that EGF stimulates intracellular Ca^{2+} rise and accumulation of inositol phosphates (11). Therefore EGF, together with PDGF and bombesin, belongs to the category of growth factors that seem to be coupled to more than one effector system (tyrosine kinase activation and phosphoinositide hydrolysis).

The EGF receptor is particularly sensitive to the inhibition brought about by pretreatment with the phorbol esters. PMA and other phorbol esters induce a significant decrease in the EGF-stimulated tyrosine kinase activity and a complete loss of the increased $[Ca^{2+}]_i$ and inositol phosphate accumulation (our unpublished results). Recently, it has been shown that a long pretreatment with phorbol esters results in a down regulation of PKC, accompanied by a consequent reduction of enzyme activity (12). This experimental protocol has been employed to study the role of PKC in the regulation of EGF-stimulated intracellular signals. We found that in long-term pretreated cells, the $[Ca^{2+}]_i$ rise induced by EGF is greater and much more persistent than in control cells. In agreement with the above results, the inositol phosphate accumulation stimulated by EGF is larger in cells pretreated over the long term with phorbol esters. These results demonstrate that in A431 cells, the activation of PKC induced by EGF is sufficient to trigger a feedback autolimited regulation of the EGF receptor function, similar to that discussed above for the other mitogens.

CONCLUSION

In conclusion, the early biochemical events stimulated by mitogens include: increase in phosphoinositide turnover, increase in cytosolic Ca^{2+} concentrations, activation of PKC, stimulation of a Na^+/H^+ antiport leading to cytosolic alkalinization, and increase in the tyrosin-specific protein kinase activity intrinsic to some growth factor receptors (EGF, PDGF, IGFI). These intracellular signals generated by growth factors, however, should not be considered independently from each other and from other intracellular events. In contrast, they are part of a complicated network of positive and negative controls of proliferation, which appears to operate continuously within the cells. The precise correlation between the early membrane and late nuclear events stimulated by growth factors is still obscure and represents the most exciting challenge for future research in this field.

ACKNOWLEDGMENTS

This work was supported by CNR special project Oncology and AIRC grant.

REFERENCES

1. Vicentini LM, Villereal ML (1986) Inositol phosphates turnover, cytosolic Ca^{++} and pH: putative signals for the control of cell growth. Life Sci 38: 2269-2276.
2. L'Allemain G, Paris S, Pouyssegur J (1984) Growth factor action and intracellular pH regulation in fibroblasts. J Biol Chem 259: 5809-5815.
3. Moolenaar WH, Tertoolen LGJ, De Laat SW (1984) Growth factors immediately raise cytoplasmic free Ca^{++} in human fibroblasts. J Biol Chem 259: 8066-8069.

4. Tsien RY, Pozzan T, Rink TJ (1982) Calcium homeostasis in intact lymphocytes: cytoplasmic free calcium monitored with a new, intracellularly trapped fluorescent indicator. J Cell Biol 94: 325-334.
5. Grynkiewicz G, Poenie M, Tsien RY (1985) A new generation of Ca^{++} indicators with greatly improved fluorescence properties. J Biol Chem 260: 3440-3450.
6. Berridge MJ (1987) Inositol trisphosphate and diacylglycerol: two interacting second messengers. Ann Rev Biochem 56: 159-193.
7. Vicentini LM, Villereal ML (1984) Serum, bradykinin and vasopressin stimulate release of inositol phosphates from human fibroblasts. Biochem Biophys Res Comm 123: 663-670.
8. Sturani E, Vicentini LM, Zippel R, Toschi L, Pandiella A, Comoglio PM, Meldolesi J (1986) PDGF-induced receptor phosphorylation and phosphoinositide hydrolysis are unaffected by proteinkinase C activation in mouse Swiss 3T3 and human skin fibroblasts. Biochem Biophys Res Commun 137: 343-350.
9. Vicentini LM, Villereal ML (1985) Activation of Na^+/H^+ exchange in cultured fibroblasts: synergism and antagonism between phorbol ester, Ca^{++} ionophore and growth factors. Proc Natl Acad Sci 82: 8053-8056.
10. Hunter T, Cooper JA (1985) Protein-tyrosine kinases. Ann Rev Biochem 54: 897-930.
11. Pandiella A, Malgaroli A, Meldolesi J, Vicentini LM (1987) EGF raises cytosolic Ca^{++} in A_{431} and Swiss 3T3 cells by a dual mechanism. Exp Cell Res 170: 175-185.
12. Rodriguez-Pena A, Rozengurt E (1984) Disappearance of Ca^{++}-sensitive, phospholipid-dependent protein kinase activity in phorbol ester-treated 3T3 cells. Biochem Biophys Res Comm 120: 1053-1059.

NEUROPEPTIDES AND NEUROTRANSMITTERS: INOSITOL PHOSPHOLIPID AND cAMP RESPONSES IN CULTURED BRAIN CELLS

J. Ham, J. R. Peters and M. F. Scanlon

Neuroendocrine Unit, Department of Medicine,
University of Wales College of Medicine,
Heath Park, Cardiff, Wales, U.K.

Neuropeptide and neurotransmitter receptors have been studied by measurement of direct binding of labeled ligands to either tissue slices or membrane homogenates. This procedure, however, does not allow differentiation between agonists and antagonists nor does it allow the underlying mechanisms involved in the ligand-receptor response to be investigated.

In recent years, improvements in tissue culture technology have allowed primary cerebral cultures to be used as a model to study both peptide secretion (1,2) and receptor binding (3). Furthermore, the defined conditions employed in tissue culture may allow the selection of specific cell types, thus enabling the localization of a particular receptor site to be more accurately identified than previously. An added advantage with cell cultures is that problems with ligand penetration and cell viability are minimized.

Using cultures of rat cerebral cortex, hypothalamus and brain stem as targets, we have investigated catecholamine and peptide receptor activation. Vasoactive intestinal peptide (VIP) has been used as a model to examine possible interactions between peptides and amines. The parameters we have chosen to measure are alterations in the secondary messengers cyclic AMP (cAMP) and inositol polyphosphates (IPs) because they give a more appropriate indication of biological activity than that provided by simple ligand-binding experiments.

CELL CULTURE

Cell cultures were prepared by the method of Lewis (4). Briefly, brains from 17 day old rat embryos were removed, and the hypothalamus, cortex and brain stem carefully dissected. Cells were dispersed using dispase and DNase I and plated into replicate wells at a seeding density of 10^6 cells/ml. Additional L-glutamine and KCl were added to the culture media to promote neuronal growth (5). The culture medium was replenished after 3 days and experiments performed on days 4 and 5.

CYCLIC AMP AND IPs DETERMINATION

Similar protocols were used for the cAMP and IPs experiments, although for the latter, cells were labeled with 2 μCi/ml [^{3}H]myo-inositol for 24 hr prior to the experiment. Cultures were prewashed with Earle's Balanced Salt Solution (EBSS), followed by a 20 min incubation in the presence of agonist (1-100 nM for peptides and 0.1 μM-0.1 mM for amines) and ascorbic acid. For IPs experiments 10 mM LiCl was included, and for the cAMP determinations 1 mM isobutylmethylxanthine (IBMX) was added. Total labeled IPs were eluted from anion-exchange (AG1-X8 formate) columns using the method of Berridge (6). Samples for cAMP estimation were acetylated and measured by radioimmunoassay (for review see ref. 7).

EFFECTS OF AMINES

The dose-related stimulations of cAMP and IPs formation by nor-adrenalin, dopamine and carbachol in hypothalamic cultures are shown in Figure 1. The action of noradrenalin can be mediated either by α-adrenergic receptors (IPs) or β-adrenergic receptors (cAMP). Dopamine stimulation of cAMP was only observed at a concentration of 0.1 mM; this, however, may represent a net effect of binding to both D_1 receptors (+ve for cAMP) (8) and D_2 receptors (-ve for cAMP) (9), both of which have been reported to exist extensively throughout the CNS. The involvement of IPs in dopamine receptor activation remains highly contentious, although in our hands a dose-responsive elevation was consistently observed. In rat striatal slices (10), dopamine stimulation of D_2 receptors resulted in a <u>decrease</u> in IPs and there have been conflicting reports on the effects of dopamine on IPs in the rat anterior pituitary (11,12).

Two other neurotransmitters, carbachol (a muscarinic agonist) and 5HT (results not shown), stimulate IPs turnover, the latter probably via a $5HT_2$ receptor. The amino acid neurotransmitters glutamate, glycine and GABA are inactive in terms of basal cAMP and IPs stimulation.

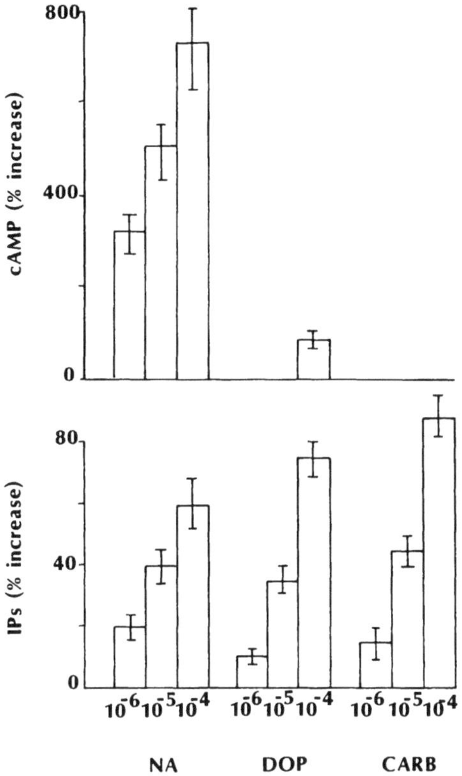

Figure 1. Dose-responsive stimulation of IPs and cAMP by noradrenalin, dopamine and carbachol in rat fetal hypothalamic cultures.

EFFECTS OF PEPTIDES

Two peptides widely distributed throughout the CNS, VIP and substance P, stimulate cAMP and IPs, respectively. Both corticotrophin releasing factor (CRF) and growth hormone releasing factor (GRF) stimulate cAMP turnover in brain cell cultures. CRF activation of adenylate cyclase in rat cerebral cultures has been linked to the release of pro-opiomelanocortin peptides (13) and GRF together with somatostatin (SRIF) is involved in the regulation of growth hormone secretion from the pituitary. The lack of effect of thyrotropin releasing hormone (TRH) on IPs formation was surprising since binding experiments have demonstrated the presence of TRH receptors throughout the CNS (14). SRIF and met-enkephalin (M-enk) decreased basal cAMP levels in cortical cells, but not in hypothalamic or brain stem cells; this may reflect differences in receptor number since SRIF receptors, for example, are most prominent in the cortex (15), with relatively few in the hypothalamus.

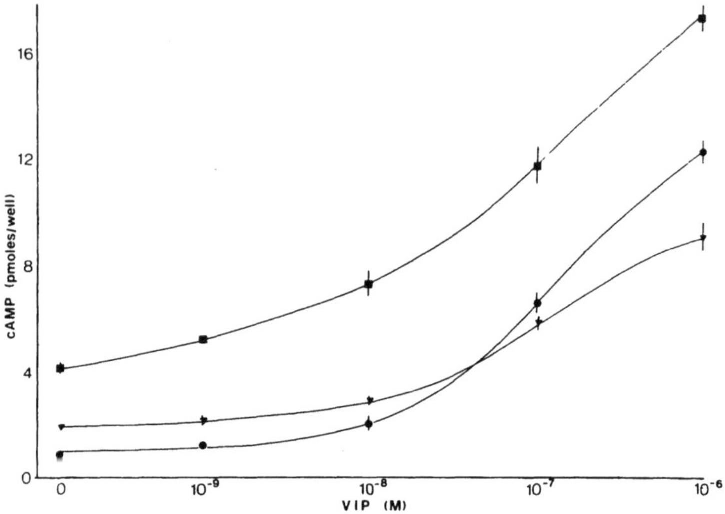

Figure 2. Dose-responsive stimulation of cAMP by VIP (●), VIP + 0.1 mM Dop (▼) and VIP + 10 μM Nor (■) in rat fetal hypothalamic cultures.

TABLE 1

IPs AND cAMP RESPONSES TO NEUROPEPTIDES AND NEUROTRANSMITTERS IN RAT FETAL HYPOTHALAMIC, CORTICAL AND BRAIN STEM CULTURES*

	Hypothalamus		Cortex		Brain Stem	
	IPs	cAMP	IPs	cAMP	IPs	cAMP
Dop	↑	↑↓	↑	↑	↑	↑
NA	↑	↑	↑	↑	↑	↑
5HT	↑	-	↑	-	↑	-
Ach	↑	-	↑	-	↑	-
CRF	-	↑	-	↑	-	↑
GRF	-	↑	-	↑	-	-
TRH	-	-	-	-	-	-
VIP	-	↑	-	↑	-	↑
Sub.P	↑	-	↑	-	↑	-
SRIF	-	-	-	↓	-	-
M.Enk	-	-	-	↓	-	-

*↑ stimulation, ↓ inhibition, - no effect.

AMINE AND PEPTIDE INTERACTIONS

The co-existence of amines and peptides (16,17) suggests their interactions may be important in regulating neuronal function. When dopamine was incubated together with VIP, an additive effect on cAMP levels was seen at low doses of VIP. In contrast, at high concentrations of VIP, dopamine caused some attenuation of the VIP response (Fig. 2). The mechanism of interaction between VIP and dopamine is unclear, but it is likely that dopamine D_1 receptor binding is predominant at low concentrations of VIP, whereas at maximally stimulated cAMP levels only the D_2 receptor binding is observed. This interaction between VIP and dopamine was also seen in cortical cultures.

Noradrenalin and VIP produced an additive effect on cAMP stimulation (Fig. 2). This was in contrast to the synergistic action reported in cortical (18) and hypothalamic slices (19). VIP-stimulated cAMP was unaffected by carbachol, 5HT or GABA, but was attenuated by adenosine, although adenosine itself did not decrease basal cAMP levels.

CONCLUSION

The neuropeptide and neurotransmitter responses outlined above are summarized in Table 1. In many cases, ligands can activate different receptor subtypes, each of which may only be distinguished by the use of specific agonists and antagonists. There have been a number of reports suggesting that cAMP may interact with the inositol phospholipid pathway; in platelets (19,20) and lymphocytes (21), for example, cAMP has a marked inhibitory effect on the hydrolysis of inositol phospholipids; yet in contrast to this, polyphosphatidylinositide metabolism is not altered by dibutryl cAMP in hepatocytes (21). These interactions between the adenylate cyclase and inositol phospholipid pathways suggest that different responses occur in different cell types and may furthermore play an important role in ligand receptor activation.

Finally, further studies are required to correlate these secondary message responses with physiological functions such as peptide release. This is particularly important to the understanding of the cosecretion of amine and peptide, whose combined actions may lead to a selective response at a specific cell population.

REFERENCES

1. Robbins RJ, Sutton RE, Reichlin S (1982) Sodium + calcium dependent somatostatin release from dissociated cerebral cortical cells in culture. Endocrinology 110: 496-499

2. Peterfreund RA, Vale W (1982) High molecular weight somatostatin secretion by cultured rat brain cells. Brain Res 239: 463-477
3. Ciaraldi T, Robbins R, Leidy JW, Thamin P, Berhanu P (1985) Insulin receptors on cultured hypothalamic cells: functional and structural differences from receptors on peripheral target cells. Endocrinology 116: 2179-2185
4. Lewis MD, Foord SM, Lewis BM, Hall R, Scanlon MF (1986) Differential production of SRIF 14 and 28 by foetal rat hypothalamic cells separated by velocity sedimentation. Neuroendocrinol 44: 125-131
5. Barochovsky O, Bradford HF (1987) Development of transmitter-releasing capacity in neuron-enriched tissue cultures. J Neurochem 48: 787-797
6. Berridge MJ, Dawson RMC, Downes CP, Heslop JP, Irvine RF (1983) Changes in the levels of inositol phosphates after agonist-dependent hydrolysis of membrane phosphoinositides. Biochem J 212: 473-482
7. Brooker G, Harper JF, Terasaki WL, Moylan RD (1979) In: Brooker G, Greengard P, Robinson GA (eds) Advances in Cyclic Nucleotide Res V.10. New York: Raven Press: pp 1-33
8. Bockaert J (1981) General characteristics, localization and adaptive responsiveness of neurotransmitter-sensitive adenylate cyclases in the central nervous system. In: Greengard P, Robinson G (eds) Advances in Cyclic Nucleotide Res. V 14. New York, Raven Press: pp 611-628
9. Enjalbert A, Bockaert J (1983) Pharmacological characterisation of the D_2 dopamine receptor negatively coupled with adenylate cyclase in rat anterior pituitary. Mol Pharmacol 23: 576-584
10. Pizzi M, D'Agostini F, Da Prada M, Spano PF, Haefely WE (1987) Dopamine D_2 receptor stimulation decreases the inositol triphosphate level of rat striatal slices. Eur J Pharmacol 136: 263-264
11. Enjalbert A, Sladeczek F, Guillon G, Bertraud P, Shu C, Epelbaum J, Garcia-Sairoz A, Jard S, Lombard C, Kordon C, Bockaert J (1986) Angiotensin II and dopamine modulate both cAMP and inositol phosphate productions in anterior pituitary cells. J Biol Chem 261: 4071-4075
12. Canonico PL, Jarvis WD, Judd AM, MacLeod RM (1986) Dopamine does not attenuate phosphoinositide hydrolysis in rat anterior pituitary cells. J Endocr 110: 389-393
13. Nikorlarakis KE, Almeida OFX, Herz A (1986) Stimulation of hypothalamic β-endorphin and dynorphin release by corticotropin-releasing factor (in vitro). Brain Res 399: 152-155
14. Patel YC, Baquiran G, Srikant CB, Posner BI (1986) Quantitative in vivo autoradiographic localization of [^{125}I-Tyr^{11}] somatostatin-14 and [Leu^{8}, D-Trp^{22}-^{125}I-Try^{25}] somatostatin-28 binding sites in rat brain. Endocrinology 119: 2262-2269
15. Leroux P, Pelletier G (1984) Radioautographic localisation of somatostatin-14 and somatostatin-28 binding sites in the rat brain. Peptides 5: 503-506
16. Eckenstein F, Baughman RW (1984) Two types of cholinergic innervation in cortex, one co-localized with vasoactive intestinal peptide. Nature 309: 153-155
17. Hendry SHC, Jones EG, de Felipe J, Schmechel D, Brandon C, Emson PC (1984) Neuropeptide-containing neurons of the cerebral cortex are also GABAergic. Proc Nat Acad Sci USA 81: 6526-6530
18. Magistretti PJ, Schorderet M (1984) VIP and noradrenalin act synergistically to increase cAMP in cerebral cortex. Nature 308: 280-282
19. Redgate ES, Deupree JD, Axelrod J (1986) Interaction of neuropeptides and biogenic amines on cAMP accumulation in hypothalamic nuclei. Brain Res 365: 61-69
20. Nishizuka Y (1984) The role of protein kinase C in cell surface signal transduction and tumour promotion. Nature 308: 693-698
21. Kaibuchi K, Takai Y, Ogawa Y, Kimura S, Nishizuka Y, Nakamura T, Tomomura A, Ichihara A (1982) Inhibitory action of adenosine 3', 5'-monophosphate on phosphatidyl-inositol turnover: difference in tissue response. Biochem Biophys Res Commun 104: 105-112

BRAIN HORMONES AND THE IMMUNE SYSTEM

C. J. Heijnen, A. Kavelaars and R. E. Ballieux*

Department of Immunology, University Hospital for Children and Youth "Het Wilhelmina Kinderziekenhuis," and *Department of Clinical Immunology University Hospital, Utrecht, The Netherlands

INTRODUCTION

For many years, immunologists have studied the structure and function of the immune system. In these studies, the fact has generally been ignored that the workhorses of this defense system, the lymphocytes and accessory cells like macrophages, are not operating in splendid isolation. Lately the concept of the immune system being interlinked with and influenced by other physiological systems in the body, has gained considerable interest. In relation to this, an interesting field of interdisciplinary research has developed which focuses on the interaction among the nervous, neuroendocrine, and immune systems. This emerging new discipline, often referred to as psychoneuroimmunology (1,2), behavioral immunology (3) or neuroimmunomodulation (4), has provided exciting data on the influence of the brain on immune functions. A number of studies have shown that psychosocial factors, such as stress, can alter immune reactivity. One frequently cited example is the effect of bereavement on immune functions in man, mentioned as early as 1884 in the British Medical Journal (5), and well documented in later years by Bartrop et al. (6) and Stein and colleagues (7). In addition, a variety of stressors have been reported to alter immune responses, both in man and in animals (8-10). In some studies, the involvement of hormones and/or neuropeptides has been shown (11). This will be elaborated upon in one of the following sections.

THE IMMUNE SYSTEM: SOME GUIDELINES FOR NON-IMMUNOLOGISTS

The activity of the immune system is characterized by two fundamental features: specificity and memory. These properties belong to the cells of the lymphoid system, including many lymphoid organs such as the thymus, spleen, bone marrow, lymphnodes and gut-associated lymphoid tissues. The lymphocyte is characterized morphologically by a relatively large, round nucleus and little cytoplasm. Individual lymphocytes possess surface receptors that can specifically recognize certain molecular structures - epitopes - present on antigens. The interaction of an epitope with the appropriate receptor leads to the activation of the lymphocyte. This in principle results in the development of the effector phase of the immune response. Two different types of immune responses can be distinguished: cell-mediated and humoral. The latter refers to the interaction of specific antibodies with the corresponding antigen, which usually leads to elimination of the foreign substance. Cellular immune responses are based on the cellular interaction with the antigen. These interactions may result in the killing of cellular antigens or in the production of lymphocyte-derived effector molecules, which may attract accessory cells (like phagocytic cells) to exert the effector function. On the basis of characteristic membrane-associated molecular structures, which can be recognized by certain analytic bioreagents ("monoclonal antibodies"), three main classes of lymphocytes can be distinguished: T lymphocytes, B lymphocytes, and a population lacking T or B cell specific structures, i.e. non-T/non-B cells. This last subset is heterogeneous and harbors among others the so-called natural killer (NK) cells, which are believed to be involved in the defense against virus-infected and cancer cells. Cell-mediated immune responses are the result of T cell activation by antigen, whereas humoral, antibody-mediated immune responses develop after antigenic stimulation of B lymphocytes (12).

Many studies concerning the influence of hormones on lymphocyte function use in vitro models in which lymphocytes are stimulated by antigen in cell cultures. The response is measured by analyses of such factors as antibody production, cell-mediated cytotoxicity or lymphoid cell proliferation. Since each lymphoid cell carries antigen receptors for only one specific epitope, the magnitude of the response after stimulation with an antigen is relatively small. Therefore, in the majority of investigations, the antigen is replaced by a mitogen, which activates T and/or B lymphocytes in a more general way, regardless of the specificity of the antigen-receptors of the cells. In the effector phase, the production of lymphokines (T cells) or immunoglobulins (B cells) is measured or the cellular proliferation is quantified. Mitogens commonly used are phytohaemagglutinin (PHA) and Concanavalin A (ConA) for T cells, and Pokeweed mitogen (PWM) for B cells (12).

Finally, it should be mentioned that the T cell compartment not only comprises T effector cells, but also two types of regulatory T cells: T helper (Th) and T suppressor (Ts) cells. These lymphocytes tune the level of immune reactivity. Th cells are essential for the initiation of the immune responses to most antigens. Furthermore, they provide growth factors that help to enlarge the magnitude of the response. Ts cells, as the name suggests, have the capacity to suppress immune reactions (12). It is easy to understand that a defective function at the Th cell level (e.g., patients suffering from AIDS) can result in severe immune deficiency. The absence or malfunctioning of Ts cells may lead to uncontrolled immune reactivity at the nonphysiological level.

THE INFLUENCE OF POMC-DERIVED PEPTIDES ON IMMUNE FUNCTIONS

It is obvious that within the framework of this paper, we can only deal with a few aspects of immunoregulation by brain hormones. Several recent publications should be consulted for a more extensive overview (13-15). The present contribution is limited to the effect of the POMC-derived peptides ACTH and endorphins on certain immune parameters.

Immunomodulation by ACTH

Recently, it has been shown by Smith et al. that human peripheral blood leukocytes possess receptors for ACTH that are similar in affinity and number to adrenal cell ACTH receptors (16).

The functional presence of binding sites for this neuropeptide had been shown earlier by the fact that ACTH is capable of modulating the immune response in vitro. Johnson et al. demonstrated that ACTH could inhibit the primary antibody response of murine B cells (17), whereas Alvarez-Mon et al. reported that ACTH could enhance the immunoglobulin synthesis of human tonsillar B cells (18). These apparently contradictory results could also be observed in experiments in which we determined the modulatory effects of ACTH on the proliferative response of human peripheral blood lymphocytes (19). ACTH in physiological concentrations could either enhance or inhibit the proliferative response of human peripheral blood lymphocytes. The reason why there is no uniform response pattern is still unclear. The type of peptide-induced modulation is, however, donor-specific.

Since the outcome of the proliferative response is the net result of multiple cellular events on which ACTH could have differential effects, we tried to measure the effect of the peptide on a more "early event" in lymphocyte activation, the ligand-induced increase of cytosolic free calcium ($[Ca^{2+}]_i$). When a mitogen or an antigen triggers the antigen T cell receptor complex, it induces the formation of second messengers necessary to transduce the signal from the exterior to the interior of the cell for the generation of a response to the stimulus. It has been shown that T cell receptor triggering leads to the hydrolysis of phosphatidylinositol 4,5-bisphosphate (PIP_2), resulting in the formation of the second messengers inositol trisphosphate (IP_3) and diacylglycerol (DG). IP_3 is responsible for the release of Ca^{2+} from intracellular stores, as well as for the influx of Ca^{2+} from the exterior of the lymphocyte by one of its metabolites (20).

The increase in $[Ca^{2+}]_i$ has been shown to be a prerequisite for the proliferative response of the lymphocyte to the stimulus. We were able to demonstrate that ACTH could interfere with the increase in $[Ca^{2+}]_i$ by the fact that it enhances the rise in $[Ca^{2+}]_i$ after activation of the lymphocyte with the T cell mitogen ConA (see Fig. 1).

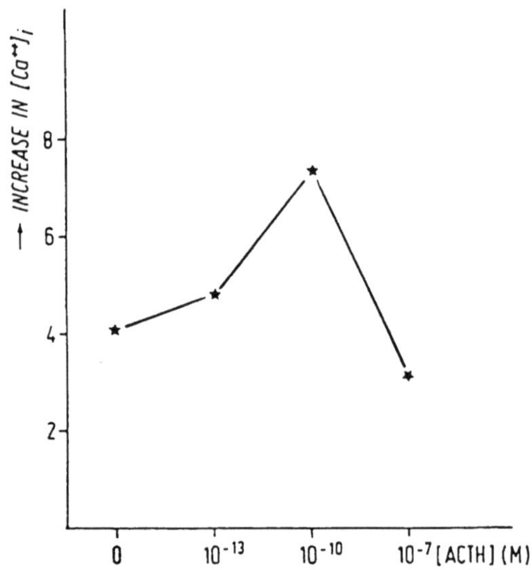

Figure 1. Modulation of the Concanavalin-A-induced increase in cytosolic free Ca^{2+} by ACTH. Human T cells were loaded with the Ca^{2+}-binding fluorescent dye Indo-1-AM and stimulated with 250 μg Concanavalin-A/ml. The fluorescence intensity was measured by means of a spectrophoto-fluorimeter at wavelengths of 355 nm for excitation and 410 nm for emission.

On the basis of these data, we would like to suggest that ACTH is an important modulator for the internal regulation of the immune system.

Immunomodulation by β-Endorphin

Recently it has been demonstrated that the endogenous opioid peptide β-endorphin (β-End) can, like ACTH, modulate several types of immune responses, which are listed in Table 1.

TABLE 1

MODULATORY EFFECTS OF β-ENDORPHIN ON THE IMMUNE RESPONSE IN VITRO

Immune Function	Effect	References
Natural killer cell activity	Enhancement	43-47
Antibody-dependent cellular cytotoxicity	No effect	47
Interferon-γ production	Enhancement	48
	Inhibition	49
T cell suppressor activity	Inhibition	50
Lymphocyte proliferation	Enhancement and inhibition	19, 51
	Enhancement	52
	Inhibition	53
Antibody synthesis	No effect	
	Enhancement	54
Mononuclear cell chemotoxic activity	Enhancement	55
Neutrophil migration	Enhancement	56

From this table we can deduce that the peptide can influence the functions of T cells, B cells, NK cells, granulocytes, etc. Although β-End appears to be a potent modulator of immune reactivity, it hitherto has been difficult to demonstrate the receptor(s) for this peptide on the surface of the cells. Mehrishi et al. have shown the presence of opiate receptors on lymphocytes (21). However, other authors failed to demonstrate opiate binding sites, and could only detect nonopiate binding receptors for β-End on lymphocytes (22,23). In addition, Schweigerer reported nonopiate binding of β-End to both terminal complexes of human complement, the membrane-derived terminal complex C5b-9(m) as well as the serum-derived SC5mb-9 complex (24). These data make it likely that β-End can bind to several receptor types on the membrane of the lymphocyte, leading to a complicated reaction pattern, which involves various responses on different levels of cellular activation. The molecule β-End, which consists of 31 amino acids, can easily be cleaved by proteolytic enzymes to γ-End (= β-End$_{1\text{-}17}$) and α-End (= β-End$_{1\text{-}16}$). We investigated the modulatory actions of β-End and α-End on the primary antibody response in vitro (see Fig. 2 and Ref. 25). When human peripheral blood cells are cultured for 6 days in the presence of Ovalbumin (OA), one can observe the generation of IgM anti-OA antibodies that are secreted by B cells. In this assay we could demonstrate that β-End$_{1-31}$ has an enhancing effect on the antibody response, whereas α-End (=β-End$_{1-16}$) has an inhibitory effect. When we removed the N-terminal amino acid tyrosin from α-End (=dtαEnd) and β-End (=dtβ-End), which means that the molecules no longer bind "classical" opiate receptors, it appeared that dtαEnd can no longer inhibit the response, whereas dtβ-End has the same (enhancing) effect as β-End$_{1-31}$.

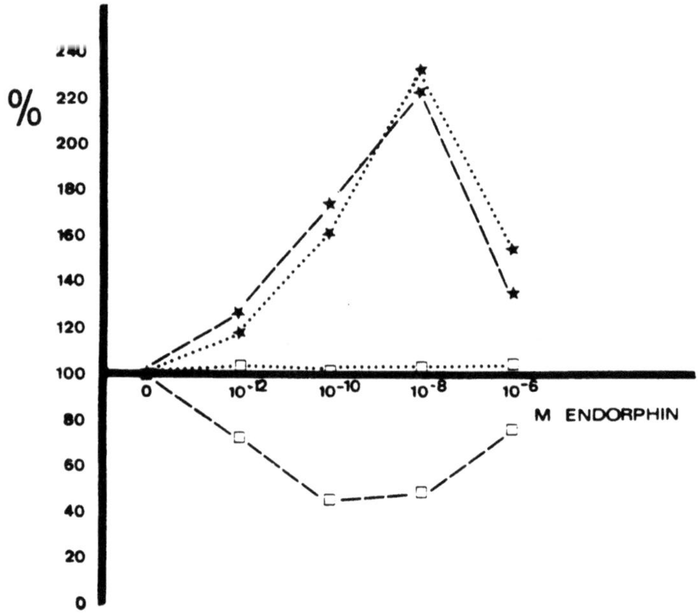

Figure 2. Peripheral blood lymphocytes were cultured with the antigen Ovalbumin in the presence or absence of endorphins: β-End ★ --- ★; dtβ-End ★ ··· ★; α-End □ --- □ or dtα-End □ ··· □. After 6 days, the antibody response was measured by means of a plaque-forming cell assay. At the Y-axis the PFC responses are given as percentages of the control response (without endorphins). At the X-axis the doses of endorphin are represented from 0-10^{-6}M. Reprinted from ref. 25, with permission.

When we tested the effect of α- and γ-End on T cell proliferation, it could be observed that a difference of one amino acid could already lead to a difference in modulatory activity (19).

In conclusion, we have to admit that this matter is far from being resolved. There is no clear structure-activity relationship of the peptides demonstrable. However, these results show clearly that endorphin can modulate the immune response in physiological concentrations. The type of the modulatory action depends on the structure of the peptide, the donor of the lymphocytes, and the assay used for the determination of immune reactivity.

PRODUCTION OF "NEUROPEPTIDES" BY CELLS OF THE IMMUNE SYSTEM

It is now generally accepted that the production of POMC-derived peptides, e.g., β-End and ACTH, is not only restricted to the CNS. These peptides have also been detected in the human placenta, pancreas, etc. and, which is important in the context of this paper, in cells of the immune system (26). In addition, it appears that lymphocytes are the only cell types, besides the pituitary, that actively secrete these peptides. The observation that receptors for these neuropeptides are not only present on mammalian brain cells, but also on cells of the immune system, has initiated investigations on the possible role of the "neuropeptides" as lymphokines for the immune system. Lymphokines are soluble mediators of induction signals between lymphocytes. Since the discovery of T cell growth factor or interleukin 2 (I1-2) (27), it has generally been assumed that different lymphokines are selective for distinct subpopulations of lymphocytes. However, there is now ample evidence that this may not be entirely correct. In a number of studies, it has been shown that lymphokines express multifunctional activity. I1-2, for example, has not only the capacity to promote growth of cells of the T helper lineage, but can also induce growth and differentiation of cytotoxic T cells (28) and of B cells (29). Furthermore, I1-2 has also been shown to induce the proliferation and differentiation of oligodendrogliacells (30), which play a role in the myelinization of peripheral nerves. I1-1, which is a factor produced by monocytes, is a prominent example of a soluble mediator acting on different target cells. It induces differentiation of fibroblasts, synovial cells, endothelium, epithelial cells, bone marrow cells, B lymphocytes and T lymphocytes, and it induces pituitary cells to secrete POMC-derived peptides (31). Recently, Uchara et al. showed that I1-1 can cause an increase in plasma ACTH, probably by inducing the release of endogenous CRF (32). In the light of these findings, it seems realistic to state that POMC-derived peptides, which structurally have been highly conserved in evolution, can be viewed as lymphokines that serve several targets and are multifunctional.

PRODUCTION OF POMC-DERIVED PEPTIDES BY MACROPHAGES

Lolait et al. have clearly shown by radioimmunoassay and immunofluorescence using antibodies to β-End and ACTH, that a subpopulation of murine splenic macrophages, expressing the MAC-1 epitope, produce immunoreactive (ir)-β-End and ir-ACTH as well as products of higher molecular weight (33). When homogenates of mouse splenic macrophages were applied to gel filtration, substances could be detected with molecular weights of 3500 (probably β-End) and 11,500 (putatively β-lipotropin), and with an even higher molecular weight (putatively the β-End-precursor proopiomelanocortin (POMC)). Using high performance liquid chromatography, the authors could show that β-End_{1-31} is produced predominantly, whereas they could also detect smaller amounts of β-End_{1-17} (γ-End), β-End_{1-16} (αEnd), β-End_{1-27} and acetylated forms of the molecules. The capacity of the murine macrophages to produce these molecules was not only determined on the basis of immunoreactivity with specific antisera, but also by demonstrating the presence of mRNA for POMC in the cells, using a complementary cDNA probe and northern blot analysis of cell homogenates (34).

The POMC-derived peptides were not secreted by the cells. However, no attempt was made to stimulate the splenic macrophage for the secretion of these peptides.

PRODUCTION OF POMC-DERIVED PEPTIDES BY LYMPHOCYTES

Smith et al. were the first investigators to show that virus-infected lymphocytes are capable of synthesizing ir-ACTH and ir-β-End (35). Structurally the ir-hormones are very similar to pituitary-derived neuropeptides on the basis of antigenicity, molecular weight and stability. Functionally, ir-ACTH that is isolated from the supernatants of cultures of the virus-infected lymphocytes, is capable of inducing corticosterone production when co-cultured with primary rat adrenal cells or mouse adrenal tumor cells. The ir-β-End binds to opiate receptors in vitro, and is capable of inducing analgesia and catatonia after intracerebral injection into mice. Subsequent to these studies, Westly et al. demonstrated that infection of lymphocytes with New Castle Disease virus (NDV) resulted in the expression of the pro-opiomelanocortin gene as determined by dot and Northern blot analysis of poly (A+) cytoplasmic RNA (36).

Apart from infection of the cells with lymphotropic viruses, POMC-derived peptides can also be produced by lymphocytes when they are stimulated by typhoid vaccine (37) or lipopolysaccharide from E. coli (38). Moreover, Smith et al. reported that not only the above-mentioned classical immunostimulants were capable of inducing POMC-derived peptide synthesis by lymphocytes, but also corticotropin releasing factor (CRF), which is known to regulate the pituitary-derived ACTH and endorphin production (39). These various stimuli all cause

TABLE 2

SECRETION OF ir-ACTH BY TRANSFORMED LYMPHOCYTES

Cell Source	ir-ACTH-secreting cells/10^6cells
EL-4 (murine thymoma)	11500
MCA-144 (simian thymoma)	2400
EBV clone 43 (human peripheral B cells)	13000
EBV clone 44 (human peripheral B cells)	97000

the production of these POMC-derived peptides. It appeared, however, that the processing of the POMC molecule depends on the nature of the stimulus. Smith et al. demonstrated that NDV and CRF caused the release from the lymphocytes of ir-$ACTH_{1\text{-}39}$ and irβ-End-$_{1\text{-}31}$, whereas stimulation of lymphocytes with LPS resulted in the production of ir-$ACTH_{1\text{-}22\ to\ 26}$, α-End (β-$End_{1\text{-}16}$) and γ-End (β-$End_{1\text{-}17}$) (40). In an elegant study, Harbour et al. proposed that the smaller fragments of ir-$ACTH_{1\text{-}22\ to\ 26}$, α-End and γ-End are proteolytic cleavage products of $ACTH_{1\text{-}39}$ and β-End, respectively (41). They could demonstrate the presence of an enzymatic activity in lysates of immune splenocytes after stimulation with LPS. From these data, we can conclude that the nature of the stimulus may determine the way the POMC molecules are processed. This may determine the ultimate modulatory capacity of the peptides produced. At this moment it is still hard to appreciate the physiological relevance of these phenomena. An important thing to be mentioned is that the quantity of the peptides produced apparently is low. Kavelaars et al. recently reported that only a small fraction of LPS-stimulated lymphocytes actively secretes the peptide (100 ACTH-secreting cells/10^6 lymphocytes) (42). However, EBV-transformed lymphocytes or tumor cells of lympoid origin have been shown to secrete these peptides far more efficiently (Table 2). The latter phenomenon indicates that lymphocyte-derived "brain"-peptides may play an important role in various pathological situations.

ACKNOWLEDGMENT

Part of the research in the authors' laboratories have been supported by a grant (#84.21) from the Dutch Asthma Foundation.

REFERENCES

1. Solomon GE (1984) Emotions, immunity and disease: a historical and philosophical perspective. In: Ballieux RE (ed) Breakdown in Human Adaption to Stress, vol. II. Boston: Martinus Nijhoff: 671-680
2. Ader R (ed) (1981) Psychoneuroimmunology. Academic Press, New York
3. Locke SE, Hornig-Rohan M (1983) Mind and immunity: behavioral immunology. Inst Adv Health, New York, NY
4. Spector NV (1980) The central state of the hypothalamus in health and disease: old and new concepts. In: Morgane PJ, Panksepp J (eds) Handbook of the Hypothalamus, vol 2: 453-517
5. Leader (1884) Brit Med J 1: 1163
6. Bartrop RW, Lazarus L, Luckhurst E, Kiloh LB, Penny R (1977) Depressed lymphocyte function after bereavement. The Lancet i: 834-836
7. Schleifer SJ, Keller SE, Camerino M, Thorton JC, Stein M (1983) Suppression of lymphocyte stimulation following bereavement. J Amer Med Assoc 250: 374-377
8. Ballieux RE, Heijnen CJ (1988) Stress and the immune response. In: Hellhammer DH, Florin I, Weiner H (eds) Neurobiological Approaches to Human Disease. Toronto: Hans Huber Publ.: pp 301-306
9. Dorian B, Garfinkel P, Brown B, Shore A, Gladman D, Keystone E (1982) Aberrations in lymphocyte subpopulations and function during psychological stress. Clin Exp Immunol 50: 132-138
10. Croiset G, Heijnen CJ, Veldhuis HJ, de Wied D, Ballieux RE (1987) Modulation of the immune response by emotional stress. Life Sci 40: 775-782
11. Shavit Y, Lewis JW, Terman GW, Gale RP, Liebeskind JC (1984) Opioid peptides mediate the suppressive effect of stress on natural killer cell cytotoxicity. Science 223: 188-190
12. Roitt I, Brostoff J, Male D (1985) Immunology. London: Gower Med Publ.
13. Berczi I (ed) (1986) Pituitary function and immunity. CRC Press, Boca Raton (Fla)
14. Berczi I, Kovacs K (1987) Hormones and immunity. MTP Press Ltd, Lancaster (UK)
15. Jankovic BD, Markovic BM, Spector NH (eds) (1987) Neuroimmune interactions. Ann NY Acad Sci, vol 496
16. Smith EM, Brosman P, Meyer WJ, Blalock JE (1987) An ACTH receptor on human mononuclear leukocytes. N Engl J Med 317: 1266-1269
17. Johnson HM, Smith EM, Torres BA, Blalock JE (1982) Regulation of the in vitro antibody response by neuroendocrine hormones. Proc Natl Acad Sci USA 79: 4171-4174
18. Alavarez-Mon M, Kehrl JH, Fauci AS (1985) A potential role for adrenocorticotropin in regulating human B lymphocyte functions. J Immunol 135: 3823-3826
19. Heijnen CJ, Zijlstra J, Kavelaars A, Croiset G, Ballieux RE (1987) Modulation of the immune response by POMC derived peptides: I Influence on proliferation of human lymphocytes. Brain Behav Immunol (in press)
20. Berridge MJ, Irvine RF (1984) Inositoltriphosphate a novel second messenger in cellular signal transduction. Nature 312: 315-321
21. Mehrishi JN, Mills IH (1983) Opiate receptors on lymphocytes and platelets in man. Clin Immunol Immunopathol 27: 240-249
22. Mendelsohn LG, Kerchner GA, Culwell M, Ades W (1985) Immunoregulation by opioid peptides. J Clin Lab Immunol 16: 125-129
23. Hazum E, Chang KJ, Cantrecasas P (1979) Specific non-opiate receptors for β-endorphins. Science 205: 1033-1035

24. Schweigerer L, Teschemacher H, Bhakdi S (1982) Interaction of human β-endorphin with the terminal SC5b-9 and "preterminal" SC5b-7 and SC5b-8 complexes of human complement. Life Sci 31: 2275-2278
25. Heijnen CJ, Bevers C, Kavelaars A, Ballieux RE (1986) Effect of α-endorphin on the antigen-induced primary antibody response of human blood B cells in vitro. J Immunol 136: 213-216
26. Blalock JE (1984) The immune system as a sensory organ. J Immunol 132: 1067-1070
27. Smith KA (1980) T-cell growth factor. Immunol Rev 51: 337-342
28. Gillis S, Ferm MM, Ou W, Smith KA (1978) T-cell growth factor: parameters of production and a quantitative microassay for activity. J Immunol 120: 2027-2031
29. Miyawaki T, Suzuki T, Butler JL, Cooper M (1987) Interleukin-2 effects on human B cells activated in vivo. J Clin Immunol 7: 277-287
30. Saneto RP, Altman A, Knobler RL, Johnson HM, de Vellis J (1986) Interleukin 2 mediates the induction of oligodendrocyte progenitor cell proliferation in vitro. Proc Natl Acad Sci 83: 9221-9225
31. Brown SL, Smith LR, Blalock JE (1987) Interleukin 1 and Interleukin 2 enhance proopiomelanocortin gene expression in pituitary cells. J Immunol 139: 3181-3183
32. Uchara A, Gottschall PE, Dahl PR, Arimura A (1987) Interleukin-1 stimulates ACTH release by an indirect action which requires endogenous corticotropin releasing factor. Endocrinol 121: 1580-1582
33. Lolait SJ, Lim ATW, Toh BH, Funder JW (1984) Immunoreactive β-endorphin in a subpopulation of mouse spleen macrophages. J Clin Invest 73: 277-280
34. Lolait SJ, Clements JA, Markwick AJ, Cheng C, McNally M, Smith AI, Funder JW (1986) Pro-opiomelanocortin messenger ribonucleic acid and posttranslational processing of β-endorphin in spleen macrophages. J Clin Invest 77: 1776-1779
35. Smith EM, Blalock JE (1981) Human lymphocyte production of corticotropin and endorphin-like substances: association with leukocyte interferon. Proc Natl Acad Sci USA 78: 7530-7534
36. Westly HJ, Kleiss AJ, Kelley KW, Wong PKY, Yuen PH (1986) Newcastle disease virus-infected splenocytes express the proopiomelanocortin gene. J Exp Med 163: 1589-1594
37. Blalock JE, Smith EM (1985) A complete regulatory loop between the immune and neuroendocrine systems. Fed Proc 44: 108-111
38. Harbour-McMenamin D, Smith EM, Blalock JE (1985) Bacterial lipopolysaccharide induction of leukocyte-derived corticotropin and endorphins. Infect and Immun 48: 813-817
39. Smith EM, Morrill AC, Meyer WJ, Blalock JE (1986) Corticotropin releasing factor induction of leukocyte-derived immunoreactive ACTH and endorphins. Nature 321: 881-882
40. Smith EM, Harbour-McMenamin D, Blalock JE (1985) Lymphocyte production of endorphins and endorphin-mediated immunoregulatory activity. J Immunol 135: 779S-782S
41. Harbour DV, Smith EM, Blalock JE (1987) A novel processing pathway for proopiomelanocortin in lymphocytes: Endotoxin induction of a new prohormone-cleaving enzyme. J Neurosci Res 18: 95-101
42. Kavelalars A, Heijnen CJ, Ballieux RE (1987) Secretion of ir-ACTH by cells of lymphoid origin: detection at the single cell level (1987) In: Proceedings of the thirty-fifth colloquium Protides of the Biological Fluids. Ed: Peeters H, pp 201-204
43. Kay N, Morley JE, van Ree JM (1987) Enhancement of human natural killing function by non-opioid fragments of β-endorphin. Life Sci 40: 1083-1087
44. Morley JE, Kay NE, Solomon GF, Plotnikoff NP (1987) Neuropeptides: conductors of the immune orchestra. Life Sci 41: 527-544
45. Kay N, Allen J, Morley JE (1984) Endorphins stimulate normal human peripheral blood lymphocyte natural killer activity. Life Sci 35: 53-59
46. Mathews PM, Froelich CJ, Sibbitt WL, Bankhurst AD (1983) Enhancement of natural cytotoxicity by β-endorphin. J Immunol 130: 1658-1662
47. Froelich CJ, Bankhurst AD (1984) The effect of β-endorphin on natural cytotoxicity and antibody dependent cellular cytotoxicity. Life Sci 35: 261-265

48. Brown SL, van Epps DE (1986) Opioid peptides modulate production of Interferon γ by human mononuclear cells. Cell Immunol 103: 19-26
49. Peterson PK, Sharp B, Gekker G, Brummitt C, Keane WF (1987) Opioid-mediated suppression of interferon-γ production by cultured peripheral blood mononuclear cells. J Clin Invest 80: 824-831
50. McCain HW, Lamster IB, Bilotta J (1986) Modulation of human T cell suppressor activity by beta endorphin and glycyl-L-glutamine. Int J Immunopharmacol 8: 443-446
51. Fontana L, Fattorossi A, D'Amelio R, Migliorati A, Perricone R (1987) Modulation of human concanavalin A-induced lymphocyte proliferative response by physiological concentrations of β-endorphin. Immunopharmacol 13: 111-115
52. Gilman SC, Schwarz JM, Milner RJ, Bloom FE, Feldman JD (1982) Beta-endorphin enhances lymphocyte proliferation. Proc Natl Acad Sci USA 79: 4226-4230
53. McCain HW, Lamster IB, Bozzone JM, Grbic JT (1982) β-endorphin modulates human immune activity via non-opiate receptor mechanisms. Life Sci 31: 1619-1632
54. Heijnen CJ, Ballieux RE (1986) Influence of opioid peptides on the immune system. Advances, Institute for the Advancement of Health 3: 114-121
55. Van Epps DE, Saland L (1984) β-endorphin and met-enkephalin stimulate human peripheral blood mononuclear cell chemotaxis. J Immunol 132: 3046-3053
56. Fischer EG, Falke NE (1984) β-endorphin modulates immune functions. Psychother Psychosom 42: 195-204

NEUROENDOCRINE ASPECTS OF CARDIOVASCULAR CONTROL: THE BRAIN RENIN-ANGIOTENSIN SYSTEM

D. Ganten, K. Lindpaintner, R. E. Lang and Th. Unger

German Institute for High Blood Pressure Research and
Department of Pharmacology, University of Heidelberg,
Heidelberg, F.R.G.

We will briefly discuss some general principles of cardiovascular regulation and question why neuropeptides merit special attention as regulators of the complex homeostatic system. We plan to touch on aspects of the evolution of peptides, their genetics, synthesis and processing. The renin-angiotensin system will serve as an example of a well characterized enzyme-peptide system, which occurs both as a hormonal and as a neuropeptidergic system clearly involved in cardiovascular control and hypertension.

GENERAL PRINCIPLES OF CARDIOVASCULAR REGULATION

The left heart pumps the blood through the aorta into the periphery to supply the tissue with oxygen and nutrients as well as with hormonal messengers. Blood pressure (BP) is simply and precisely described as the product of pump performance - that is, cardiac output (CO) - and the total peripheral resistance (TPR) produced by the blood vessels. Disorders of blood pressure regulation, specifically abnormally high blood pressure or hypertension, can thus be explained by an increased cardiac output or by increased resistance from the constricted blood vessels.

The kidney is an important organ for blood pressure control because of its key role in volume regulation, which in turn has a major impact on cardiac output. The brain exerts a major integrative function in cardiovascular homeostasis operating both directly through neuronal circuits, and more diffusely through the hormonal endocrine system. The heart, blood vessels and kidney have long been the focus of hypertension research and therapy, probably because of the vicious circles triggered by end organ damage, which is the result of untreated hypertension. The cardiovascular consequences of hypertension - namely cardiac hypertrophy and heart failure, arteriosclerosis, kidney damage and stroke - cause about 50% of all deaths in western societies.

Research in cardiovascular control mechanisms is therefore not trivial and not simply an academic exercise, but rather is of the utmost clinical importance. Better understanding of the causes of hypertension will provide us with improved and more specific pharmacological treatment, and in the long run, hopefully will help us to prevent the disease.

PEPTIDES AS CHEMICAL MESSENGERS OF THE CARDIOVASCULAR SYSTEM

Why are peptides of particular interest as chemical messengers of the cardiovascular system? Much can be learned from the evolution of hormones and neurotransmitters. Cyclic AMP was one of the first molecules to serve a hormone-like function as illustrated by the amoeboid form of the slime mold, which releases cAMP as a signal to coalesce under conditions of carbohydrate deprivation. It has been pointed out that cAMP is probably too unstable a messenger to operate over the greater distances (cm-meters) required in higher organisms. The intercellular communication systems represented by contemporary peptide hormones have taken about 3×10^9 years to evolve. The linear polymer of approximately 20 dissimilar units, the amino acids, offers an almost infinite number of permutations and combinations. Hydrophilic peptides have evolved hydrophobic leader sequences or signal peptides to overcome initial transport difficulties, also facilitating regulated sequestration in vesicles, exocytosis and secretion (see ref. 1).

The close homology in amino acid sequence seen within and between families of peptides suggests strongly that they have evolved through multiple gene duplications. As Niall has pointed out in an excellent review, gene duplication is the name of the game. The usefulness of gene duplication in the evolution of peptide hormones is apparent when one considers that the original gene can continue to be expressed in the form of the ancestral original hormone, whereas the redundant copy is free to mutate to produce a new chemical messenger with its own target tissue and new function. The opioid peptides, the pancreatic polypeptides including NPY, and the pituitary peptides oxytocin and vasopressin as well as renin demonstrate this point well. In this context, the high degree of sequence homology encountered between peptide hormones is hardly surprising. Another evolutionary principle appears to be that nature conserves structure not function. In fact, the limited number of mutations provides us with an exact evolutionary clock indicating the time at which new peptides have evolved from their primordial protein. Not a valid argument for their importance, perhaps, but an impressive fact is the quantity of peptides that have been discovered in recent years and the new ones that continue to be described at rapid speed. This progress has been made possible by molecular biological techniques that allow prediction of new peptides from a given genetic code (or fragments of it) before they are actually sequenced.

If we look at the chemical messengers in the central nervous system, we have about ten classical transmitters - the catecholamines, biogenic amines and amino acids - as opposed to over 50 different neuropeptides, including previously well established circulating hormones such as angiotensin, atrial natriuretic peptide, vasopressin and oxytocin. Calcitonin and the related peptide CGRP, the opioid family of peptides, gastrointestinal peptides, pituitary peptides and their releasing factors, and miscellaneous peptides such as the kinins, bombesin, gallanin and others all occur as neuropeptides in the brain. There is evidence supporting a role in central cardiovascular control for angiotensin, opioid peptides, substance P, neuropeptide Y, vasopressin, atrial natriuretic peptide, kinins, corticotropin releasing factor, bombesin, somatostatin and other peptides.

These peptides have been localized in brain areas known to be important to blood pressure regulation. Specific high-affinity peptide receptors have also been discovered. Upon central administration, these peptides produce cardiovascular effects, partly by interacting with other blood pressure-controlling neuroregulators, e.g. catecholamines or GABA. Central inhibition of brain peptide synthesis or interaction with competitive antagonists at the receptor site results in marked cardiovascular effects. Altered peptide levels and activity of synthesizing enzymes, as well as supersensitivity to the pressor action of some brain peptides, have been described in experimental models of hypertension.

Neuropeptides have been found to colocate with each other or with monoamine transmitters, thereby increasing the variety of chemical signals that a neuron can utilize when communicating with other cells. If these transmitters are packaged together within a single vesicle in the nerve terminal, it is possible that a "synaptic cocktail," consisting of peptides, monoamines, and amino acids, can be simultaneously released. In general, peptide effects occur over a longer time course and over greater distances ("neuro-hormonal" action) than is the case for other neurotransmitters. Fundamental differences exist among biogenic amines, amino acids, and peptides in their mode of synthesis and metabolism, and peptides are not recycled. Differences in rate of synthesis, transport and replenishment of transmitters make it possible that similar repetitive stimuli may elicit quite different postsynaptic patterns in peptidergic as opposed to nonpeptidergic neurons.

All peptides are synthesized as high molecular weight precursors. By enzymatic cleavage, several active peptides can be released from one precursor and act at different receptors. The regulation of each of these "concerted action" peptide systems by a single gene provides for a complex regulatory output to a number of biological systems (heart, kidney, autonomic nervous system, hormone release) as well as for modulation of higher integrative functions, such as pain perception, salt appetite, thirst and hunger in response to perturbations of homeostatic balance. The recognized association of cardiovascular disease and health-damaging behavioral patterns invites speculation that these peptide systems, in addition to their direct effects on blood pressure, may play a role at a molecular level in such complex behavioral aspects of cardiovascular disease. We will limit our discussion to specific consideration of one peptide that is of particular interest for cardiovascular control, namely angiotensin, a plasma hormone that has been shown to be a neuropeptide in the brain, but which offers new insights into the pathophysiology of hypertension and exemplary possibilities for its treatment.

THE BRAIN RENIN-ANGIOTENSIN SYSTEM

The renin-angiotensin system is well defined with respect to all the features of specific interest: evolution, genetics, molecular biology, synthesis, processing, localization and finally specific pharmacological intervention and therapy.

From the high molecular weight precursor angiotensinogen, renin cleaves the inactive decapeptide angiotensin I (AI), which is converted into the effector peptide angiotensin II (AII), which acts on specific angiotensin receptors on the arteries to produce vasoconstriction, and on the adrenal gland to release aldosterone. This leads to sodium retention and increased blood volume (2). Increases in volume, cardiac output and total peripheral resistance then lead to higher blood pressure and eventually hypertension.

Converting enzyme inhibitors have been developed as therapeutic agents for the treatment for hypertension and have proved extremely effective. Surprisingly, however, their blood pressure lowering effect is not correlated to the plasma renin levels.

It was therefore of interest to find whether angiotensin, in addition to being a circulating plasma hormone, might occur in tissue and in the brain as a neuropeptide and be involved in cardiovascular control (3,4). Interference with the neuropeptidergic AII among others might then be of importance for the blood pressure lowering action of converting enzyme inhibitors.

In the following, we will demonstrate that renin-angiotensin is not only a classic endocrine, but truly a neuropeptidergic system as well by briefly presenting evidence that angiotensin is locally synthesized in brain tissue (see also ref. 5).

The renin gene has been cloned and characterized. The genomic gene structure has nine exons. The molecule has probably arisen by incomplete crossing over and gene duplication, exons 2 through 5 having striking sequence homology with exons 6 through 9. The biosynthetic pathway from "pre-pro" to "pro" and active "1-chain" renin is well established.

A crucial question concerns whether the renin messenger RNA is present in the brain. Using radioactive synthetic cRNA to hybridize renin mRNA on a histological brain section, we found a hybridizing signal with the endogenous brain mRNA in the lateral hypothalamus of rats. Control experiments showed that this is probably a specific reaction product, because using the same methodology, renin cRNA also hybridized with its specific messenger RNA in the afferent arteriol of the glomerulum in the kidney where it is known to be located.

A further control was the molecular characterization of renin mRNA by electrophoresis on agarose gel and subsequent hybridization with renin cRNA under stringent conditions, a so-called Northern blot, confirming the presence of a 1.6 kilo base renin mRNA in the brain, which is identical to kidney renin mRNA. These experiments showed that the renin gene is expressed in the brain - in other words, renin is locally produced (6-10).

The high molecular weight precursor of AII, angiotensinogen, has also been cloned and it was therefore possible to probe for specific angiotensinogen mRNA by using radioactively labeled synthetic cRNA in the brain (11,12). In such Northern blot experiments, it was clearly demonstrated that large amounts of angiotensinogen are present in brain tissue and in the liver, its classical site of synthesis, as well as in the heart atrium and ventricles, the adrenal gland and the kidney. In situ hybridization experiments showed clearly that angiotensinogen cRNA hybridization in the paraventricular nucleus was also the reaction product of angiotensinogen with renin and that the converting enzyme, AII, is highly localized. Thus, not only are renin and angiotensinogen present in brain tissue, but immunohistochemistry has clearly identified AII in neurons of the brain (7,10).

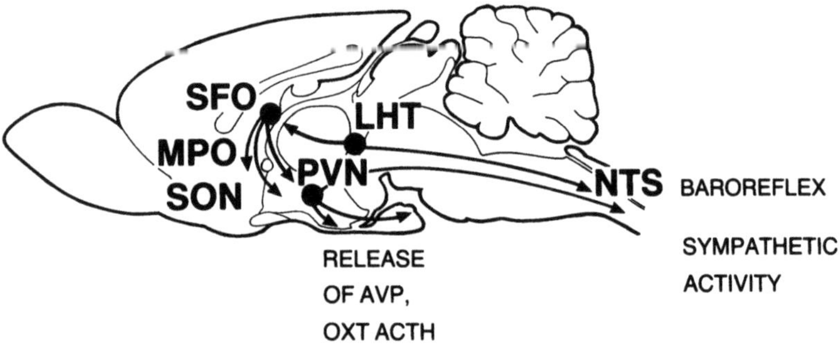

Figure 1. Schematic outline of angiotensin synthesis sites (black circles) and projections (arrows) in the rat brain. SFO: subfornical organ; MPO: median preoptic area; SNO: supraoptic nucleus; PVN: paraventricular nucleus; LHT: lateral hyperthalamus; NTS: nucleus tractus solitarii. All these areas are important cardiovascular control centers. The mechanisms which may be influenced by central angiotensin are equally indicated (see 13-15).

In collaboration with Fuxe, Lind and others, we have carefully mapped brain AII pathways by immunocytochemistry (see 8, 13-15). A summary of these studies in Figure 1 shows the sites of synthesis (black circles) and the neuronal projections of angiotensin (arrows). There is a dense network in the lateral hypothalamus (LHT), the subfornical organ (SFO), the medial preoptic area (MPO), the paraventricular nucleus (PVN) and projections into the pituitary and the spinal cord. These areas are important cardiovascular centers. To be of functional significance, they have to contain specific angiotensin receptors. Computerized receptor autoradiography studies by several authors indeed showed a high density of angiotensin receptors in the subfornical organ, the nucleus paraventricularis and periventricularis of the hypothalamus, nucleus suprachiasmaticus, lateral optical tract and also the amygdala.

A further test of the biological relevance was the stimulation of the subfornical organ by local application of AII. Angiotensin resulted in increased spiking activity of the neurons, which was specific for the angiotensin receptor and could be blocked by the angiotensin receptor antagonist saralasin. It is noteworthy that the angiotensin receptor antagonist alone reduced spontaneous firing of the neuron. The receptor stimulation results in an increase in blood pressure. Central (icv) AII has a much more prolonged duration than peripheral application of angiotensin. It has also been found that a major portion of the central angiotensin blood pressure response is due to the release of vasopressin since the blood pressure rise as well as the decreases in renal and mesenteric blood flow can be blocked efficiently by a V1 vasopressin receptor antagonist (see 5,8,16).

In summary, the essential components of the renin-angiotensin system have been identified in brain tissue: angiotensinogen, renin, AII and angiotensin receptors. Furthermore, central angiotensin receptor stimulation results in blood pressure increases. It can thus be concluded that there is an intrinsic functional renin-angiotensin system in the brain. Activity measurements of renin and angiotensin have shown that the system is stimulated in hypertensive animals (17,18).

The key experiment to test the system's contribution to the elevation of blood pressure is an examination of the inhibition of endogenous angiotensin synthesis in the brain. Unique for neuropeptides, this can be done at all levels of peptide synthesis and action by the inhibition of renin and the converting enzyme and by angiotensin receptor blockade. Inhibition of renin with pepstatin resulted in a retardation of blood pressure development. Central inhibition of angiotensin receptors in the same spontaneously hypertensive rats decreased blood pressure even more markedly after nephrectomy when circulating angiotensin is completely excluded (19-24).

We had mentioned before that low plasma renin patients respond well to inhibition of the converting enzyme. This was therefore also tested experimentally. Like the renin and angiotensin receptor blockade, the central converting enzyme inhibition by captopril clearly lowered blood pressure in spontaneously hypertensive rats. If this was indeed a central effect, as is probable because of the parallel decrease in converting enzyme activity centrally, then AII levels should also be lowered. This was indeed the case. AII decreased and AI accumulated following brain converting enzyme inhibition in these hypertensive rats (17).

In addition to its hormonal action, angiotensin can thus now be considered one of the well established neuropeptides. It is localized in cardiovascular centers of the brain and pharmacological interventions confirm its involvement in blood pressure control and maintenance of hypertension. Angiotensin provides us with a first practical example that it is possible to interfere specifically with a complex enzyme peptide system and improve treatment of a widespread disease: hypertension. It can be expected that substances will be developed to interfere with other peptide systems in a similar way.

REFERENCES

1. Niall HD (1980) The evolution of peptide hormones. In: Endocrinology 1980, Proceedings of the VIth International Congress of Endocrinology, Melbourne, Australia, Feb. 10-16, 1980, Australian Academy of Sciences, Amsterdam: Elsevier Press: pp. 13-18
2. Peach M (1977) Renin-angiotensin system: biochemistry and mechanism of action. Physiol Reviews 57: 313-370
3. Fischer-Ferraro C, Nahmod VE, Goldstein DJ, Finkielman S (1971) Angiotensin and renin in rat and dog brain. J Exp Med 133: 353-361
4. Ganten D, Minnich JL, Granger P, Hayduk K, Brecht HM, Barbeau A, Boucher R, Genest J (1971) Angiotensin-forming enzyme in brain tissue. Science 173: 64-65
5. Ganten D, Printz M, Phillips MI, Scholkens BA, eds 1982 The Renin Angiotensin System in the Brain. Experimental Brain Research (Suppl 4). New York: Springer
6. Dzau VJ (1984) Vascular renin-angiotensin: a possible autocrine or paracrine system in control of vascular function. J Cardiovasc Pharmacol 6: S377-S382
7. Dzau VJ, Ingelfinger J, Pratt RE, Ellison KE (1986) Identification of renin and angiotensinogen messenger RNA sequences in mouse and rat brains. Hypertension 8: 544-548
8. Ganten D, Lang RE, Lehmann E, Unger Th (1984) Brain angiotensin: on the way to becoming a well-studied neuropeptide system. Biochem Pharmacol 33: 3523-3528
9. Hirose S, Yokosawa H, Inagami T, Workman KJ (1980) Renin and prorenin in hog brain: ubiquitous distribution and high concentration in pituitary and pineal. Brain Res 191: 489-499
10. Suzuki F, Hellmann T, Murakami K, Ludwig G (1987) Expression of the genes for renin and angiotensinogen in tissues of rats. Naunyn-Schmiedebergs Arch Pathol 335: R66
11. Campbell DJ, Bouhnik J, Menard J, Corval P (1984) Identity of angiotensinogen precursors of rat brain and liver. Nature 308: 206-208
12. Nakanishi S, Ohkubo H, Nawa H, Kitamura N, Kageyama R, Ujihara M (1983) Angiotensinogen and kininogen: cloning and sequence analysis of the cD-NAS. Clin Exp Hyper A5(7/8): 997-1003
13. Lind RW, Swanson LW, Ganten D (1985) Organization of angiotensin II immunoreactive cells and fibers in the rat central nervous system. neuroendocrinology 40: 2-24

14. Lind RW, Swanson LW, Bruhn TO, Ganten D (1985) The distribution of angiotensin II-immunoreactive cells and fibers in the paraventriculo-hypophysial system of the rat. Brain Res 338: 81-89
15. Ganten D, Fluxe K, Phillips MI, Mann JFE, Ganten U (1978) The brain isorenin-angiotensin system: Biochemistry, localization and possible role in drinking and blood pressure regulation. In: Ganong WF, Martini L (eds) Frontiers in Neuroendocrinology, Vol. 5, New York: Raven Press: pp. 61-99
16. Rettig R, Lang RE, Rascher W, Unger Th, Ganten D (1982) Brain peptides and blood pressure regulation. Clin Sci 63: 269s-283s
17. Ganten D, Hermann K, Bayer C, Unger Th, Lang RE (1983) Angiotensin synthesis in the brain and increased turnover in hypertensive rats. Science 221: 869-871.
18. Schelling P, Meyer D, Loos HE, Speck G, Phillips MI, Johnson AK, Ganten D (1982) A micromethod for the measurement of renin in brain nuclei: its application in sponteously hypertensive rats. Neuropharmacology R1: 455-463
19. Berecek KH, Okuno T, Nagahama S, Oparil S (1983) Altered vascular reactivity and baroreflex sensitivity induced by chronic central administration of captopril in the spontaneously hypertensive rat. Hypertension 5: 689-700
20. Evered MD, Robinson MM, Richardson MA (1980) Captopril given intracerebroventricularly, subcutaneously or by gavage inhibits angiotensin converting enzyme activity in the rat brain. Eur J Pharmacol 68: 443-449
21. Mann JFE, Phillips MI, Dietz R, Haebara H, Ganten D (1978) Effects of central and peripheral angiotensin blockade in hypertensive rats. Am J Physiol 234: 629-637
22. McDonald W, Wickre C, Aumann S, Ban D, Moffit B (1980) The sustained antihypertensive effect of chronic cerebroventricular infusion of angiotesnin antagonist in spontaneously hypertensive rats. Endocrinology 107: 1305-1308
23. Unger Th, Kaufmann-Buhler I, Scholkens B, Ganten D (1981) Brain converting enzyme inhibition: a possible mechanism for the antihypertensive action of captopril in spontaneously hypertensive rats. Eur J Pharmacol 70: 467-478
24. Unger Th, Ganten D, Lang RE (1983) Pharmacology of converting enzyme inhibitors: new aspect. Clin Exp Hypertension A5(7/8): 1333-1354

CONTROL OF FLUID BALANCE

J. T. Fitzsimons

The Physiological Laboratory, Cambridge University
Cambridge, England, U.K.

CONSTANCY OF THE INTERNAL ENVIRONMENT

For more than a century, physiologists have been inspired by Claude Bernard's concept that constancy of the internal environment is necessary for the free and independent existence that characterizes the life of the higher vertebrates (1). W.B. Cannon, who introduced the term homeostasis to describe "the coordinated physiological processes which maintain most of the steady states in the organism. . .," was much influenced by this concept (2). Throughout his book *The Wisdom of the Body* (1932), he refers to the importance of homeostatic processes in maintaining the constancy of the internal environment, and in the first chapter he approves of J.S. Haldane's assessment of Bernard's celebrated statement: "No more pregnant sentence was ever framed by a physiologist" (2).

Constancy of the internal environment is normally achieved by the kidneys, but thirst and sodium appetite become important when water and salt are lacking. Osmolality is regulated to within 1 or 2% of its normal value. In the long-term, fluid volume depends on the quantity of solute in the compartment. In the short-term, where solute content remains fairly constant, keeping the osmolality within normal limits by controlling water intake and excretion automatically controls volume. Cellular volume depends mainly on the potassium content of the cells, which is determined by the transport activity and selective permeability of the cell membranes. Extracellular fluid volume is set by the amount of sodium, which, together with accompanying anions, accounts for over 95% of the extracellular solute. The tolerance for changes in the amount of sodium in the body is much greater than it is for changes in its concentration. An increase in sodium intake leads to sodium retention until the balance between sodium intake and excretion is restored. There may be a considerable gain in body weight as water is retained to maintain normal plasma osmolality. A low sodium diet has the opposite effect, causing an initial loss of body weight until balance between sodium intake and excretion is restored.

Plasma and interstitial fluid are closely coupled through the Starling filtration-reabsorption system, so that it is usually enough to control plasma volume in order to control extracellular volume. Plasma volume is measured by stretch receptors, located at certain strategic places in the circulation, notably where the veins enter the atria, but elsewhere as well. Although not impinging directly on the control of fluid balance, the control of the distribution of cardiac output by sympathetic vasoconstriction, particularly in the splanchnic circulation, and the ability to mobilize tissue fluid through the Starling filtration-reabsorption system, are important buffers in hypovolaemia for ensuring adequate blood flow to vital regions of the body. Since there is roughly twice as much interstitial fluid as there is plasma, the value of this immediately available fluid is obvious.

THE CONTROL OF FLUID INTAKE

Mammals show a basic urge to drink water even if at the time of drinking they do not need it. Rats infused with water in excess of what they need by routes that bypass the mouth continue to drink about a third or more of the amounts they drank previously. This drinking cannot be in response to an existing deficit of body water. It is the way of providing water in advance of a future need and it reflects the importance of drinking in the life of the animal. The amount of water drunk depends on habit, diet, climate, activity, etc. The basic urge to drink in the absence of need means that normally the kidneys determine fluid balance. Little is known of the mechanisms that control such drinking other than they appear to be governed by an autonomous nychthemeral rhythm, entrained with, but independent of, the activity and feeding rhythms.

Important though this type of drinking may be, it is also vital that there should be thirst mechanisms to respond to deficits of body water when these occur. No matter how efficient the kidneys and other water-conserving mechanisms of the body are, they cannot provide the water needed to replace inevitable losses. It is essential that the dehydrated person looks for and drinks enough water. The sensation of thirst ensures that this happens. The two main deficit signals to increased thirst are cellular dehydration and hypovolaemia. Hypovolaemia also causes increased sodium appetite, but this drive is less easy to assess, particularly in man where salt intake is influenced by culture and dietary habits.

Thirst Caused by Cellular Dehydration

When water is lacking or cell-dehydrating solutes such as NaCl are taken, water is lost from the cells and the subject becomes thirsty. Animals started to drink after slow intravenous infusions of hyperosmotic solutions of substances that dehydrate cells when the plasma osmolality had increased by 1-2%. Similar degrees of cellular dehydration in man also arouse thirst. When larger amounts of dehydrating solutes were given to nephrectomized rats, the amounts drunk were fairly precisely those required to restore the cellular water content to normal, but hyperosmotic solutions of solutes that penetrate cells were much less effective (3). Animals with functioning kidneys drank less water than nephrectomized rats because renal osmoregulation as well as drinking helped restore the plasma osmolality and cellular water to normal. In rats and many other animals, the predictions of an osmometric theory of thirst are therefore fully met and the mechanism is remarkably sensitive.

Since increased water intake caused by hyperosmotic solutions of dehydrating solutes is related to the degree of cellular dehydration produced, it is generally believed that the receptors responsible are osmoreceptors, similar to those which control release of vasopressin. In many animals, e.g. dog, this appears to be true because hyperosmotic solutions of NaCl or sucrose were equally effective at causing drinking when injected into the basal forebrain or cerebral ventricles (4). However, in goat (5) and sheep (6) evidence from intracranial injection experiments suggests the possibility of sodium-sensitive receptors; larger drinking and antidiuretic responses were obtained with hypertonic solutions containing sodium than with those which did not. A difficulty with this theory is that increases in sodium concentration of the cerebrospinal fluid produced by systemic infusion of some hyperosmotic solutions did not cause increased drinking.

Thirst osmoreceptors are located in the basal forebrain. Lesions of the anterior wall of the third ventricle caused permanent adipsia and failure to release vasopressin (7). These lesions involve one of the circumventricular organs, the organum vasulosum of the lamina terminalis (OVLT). In dog, drinking and vasopressin release in response to hypertonic NaCl were disrupted but not eliminated by discrete lesions of the OVLT, and there was a greater rise in plasma osmolality than normal in water deprivation (8).

Hypovolaemic Thirst

Extracellular dehydration causes thirst followed later by an increase in sodium appetite. These responses are caused mainly by the diminution in blood volume produced. The angiotensin that is formed contributes to the responses, but in neither behavior is angiotensin the sole stimulus. Decreases in blood volume of 10% or more produce significant increases in water intake on their own, though smaller decreases augment the responsiveness to cellular dehydration. Hypovolaemia is the main cause of thirst after blood loss, severe vomiting or diarrhea, and after procedures enforced on the circulation that mimic the effects of blood loss (3). Hypovolaemia makes a variable contribution to the thirst of water deprivation depending on species (9).

The receptors for hypovolaemic thirst are stretch receptors in the walls of the blood vessels and heart. Nervous discharge from the receptors inhibits drinking so that when the circulation is underfilled, the decrease in receptor discharge results in increased water intake (10). A commonly used procedure to attenuate receptor discharge is to reduce venous return to the heart by obstructing the inferior vena cava. This has been done repeatedly and in a graded fashion in the dog by inflating a balloon catheter previously inserted into the vena cava through a femoral vein and positioned with its inflatable tip above the renal veins (11). Within half an hour of obstructing the caval blood flow, drinking and antidiuresis occurred. Water was retained in direct proportion to the haemodynamic changes produced. There was also a small and delayed increase in sodium appetite after the initial increase in water intake. The reverse effect has been obtained by stimulating a proportion of the cardiovascular stretch receptors by inflating a balloon placed at a pulmonary vein-atrial junction in dog (12), or at the superior vena cava-atrial junction in rat (13); drinking to various thirst stimuli was inhibited and urine flow increased.

Receptors on the low-pressure side of the circulation, particularly those in the pulmonary vein-atrial junctions, are ideally sited to register the fullness of the circulation where it matters, i.e. where it influences cardiac output. But even when stretch receptors at the entrance to the heart are stimulated discretely, it is likely that other receptors in both the low-pressure and high-pressure sides of the circulation will be affected by the haemodynamic changes that inevitably occur as a result of the stimulation. Arterial baroreceptors are certainly involved since drinking in response to aortic obstruction just above the kidneys (14), or systemic infusion of angiotensin II (15), was partly suppressed by the rise in blood pressure also produced. The possible involvement of other types of receptors such as the carotid and aortic body chemoreceptors, which are stimulated by hypoxia in severe blood loss, has not been studied. It has also been suggested that there are receptors sensitive to changes in extracellular volume in the interstitial space.

Sodium Appetite in Hypovolaemia

After the initial thirst, hypovolaemia also causes a delayed increase in sodium appetite. This is important because sodium is needed to restore the extra-cellular volume and plasma volume to normal. Potassium deficiency may also develop in the hypovolaemia of severe diarrhea or vomiting, but need for potassium is rarely as pressing as it is for sodium; potassium is more abundant in the body and there is enough of it in most diets to meet the body's needs. A specific appetite for potassium has not been described. Herbivores, whose diet generally lacks sodium, show a well-marked and specific appetite for sodium, as do many animals when they become sodium-deficient (16). It is more difficult to demonstrate sodium appetite in man whose diet normally contains an excess of sodium, but many cases have been described.

The immediate stimulus to increased sodium appetite is uncertain. In the sodium-deficient sheep, lowering the sodium concentration in the cerebrospinal fluid by infusing hyperosmotic mannitol caused increased sodium intake (17), but in the omnivorous rat, changes in cerebrospinal sodium concentration did not affect sodium intake (18). Rice and Richter were the first to show that rats increase their intake of hypertonic NaCl solutions after treatment with larger than replacement doses of deoxycorticosterone, despite the sodium retention also produced (19). Since mineralocorticoid secretion is increased in sodium deficiency, this effect seems physiologically fitting. However, mineralocorticoids are not essential because increased sodium appetite also occurs in the rat made sodium deficient by removal of the adrenal glands.

Since angiotensin II rises in sodium deficiency and has a direct stimulating effect on sodium appetite, it is reasonable to postulate a role for it in sodium appetite. When the adrenals are present, angiotensin could do this by stimulating mineralocorticoid secretion, but in the absence of the adrenals it could act directly on the central nervous system since intracranial angiotensin is known to stimulate sodium appetite. The same cardiovascular receptors that operate in hypovolaemic thirst may also inhibit sodium appetite. In rat, distension of the superior vena cava-atrial junction attenuated the sodium appetite caused by hyperoncotic peritoneal dialysis or deoxycorticosterone treatment (20). This result complements the effect of caval obstruction, mentioned above, in which sodium appetite was stimulated. Yet another factor that could be involved in hypovolaemic thirst and sodium appetite is atrial natriuretic peptide which, when injected into the cerebral ventricles, inhibited intracranial angiotensin-induced drinking (21). Reduced levels in hypovolaemia may have a similar stimulating effect as reduced nervous discharge from cardiovascular stretch receptors.

Angiotensin and Drinking Behavior

Angiotensin II administered systemically or intracranially is a potent stimulus to thirst in many mammals, birds, reptiles and even some bony fish (3). After intracranial administration, drinking occurs within a minute or two of the injection, it is dose-dependent and the amounts drunk over the next 15-30 minutes may approach what the animal would normally drink in the next 24 hours. Increased drinking occurs when angiotensin is infused intravenously at rates that produce plasma levels similar to those that occur in dehydration or moderate blood loss. The site of action of circulating angiotensin is the subfornical organ (SFO), which, like the OVLT, lacks a blood-brain barrier. Angiotensin binding has been demonstrated here. Electrophysiological techniques have demonstrated angiotensin-sensitive neurons in the SFO. When the SFO was destroyed drinking to intravenous angiotensin was attenuated (22).

It is well established that renin released from the kidneys by isoprenaline or by interference with the blood supply to the kidney causes increased water intake. However, the contribution of endogenously formed angiotensin II to increased drinking is difficult to assess because there are also non-angiotensin controls which alone can maintain drinking. Two experimental approaches have been used to study this: (1) the source of renal renin is removed by nephrectomy, (2) the action of angiotensin II is blocked or its formation prevented using competitive antagonists of the renin-angiotensin cascade. Neither approach is entirely satisfactory because non-renin thirst mechanisms, or absent or altered renal function, affect drinking. However, these approaches have shown an undoubted renin contribution to the increased drinking of renal ischaemia and hypovolaemia.

Restricting the blood supply to the remaining kidney after unilateral nephrectomy in the dog caused increases in plasma renin activity and water intake (23). Urine flow was little altered so that the fluid balance became positive. Renal ischaemia produced by occluding the rat's abdominal aorta between the renal arteries caused polydipsia and increased NaCl intake which rapidly disappeared after removing the ischaemic kidney or giving captopril (24). Inflating a balloon implanted in the abdominal aorta above the renal arteries also caused drinking and a fall in urine flow (14). The increase in plasma renin was responsible for the increased water intake since drinking was abolished by saralasin.

In rat, removing the source of renal renin by nephrectomy, or blocking the action of angiotensin, reduced drinking caused by caval ligation, suggesting that renal renin contributes to the response (3). In dog, drinking in response to inferior vena caval obstruction was reduced but not abolished by simultaneous intravenous infusion of saralasin (11,25). Experiments in the rat with captopril support the hypothesis that angiotensin contributes to drinking following caval obstruction. After captopril, drinking was enhanced by small subcutaneous doses of captopril and reduced to below control values by large doses (26). The enhancement of drinking after small doses was caused by the increase in systemic angiotensin I brought about by eliminating the negative feedback of angiotensin II on renin secretion. The angiotensin I reaches the brain where it is converted to angiotensin II, causing increased drinking. After large doses, enough captopril penetrates the brain to prevent conversion of angiotensin I to angiotensin II in the brain as well as in the periphery so that drinking is blocked. This effect could be mimicked by giving a small subcutaneous dose combined with repeated intracerebroventricular doses; not only did the low dose enhancement disappear, but the basal caval obstruction drinking response was also reduced.

Intracranial administration of angiotensin or its precursors also causes increased sodium intake, which follows the increased water intake but is not secondary to it. Nor is it secondary to the increased sodium loss that also occurs. In view of the different time courses of action of angiotensin in the two behaviors, it seems likely that different types of mechanisms are involved. There is evidence of a synergistic action on sodium appetite between angiotensin and mineralocorticoids (27). Experiments with captopril suggest that endogenously formed angiotensin is involved in sodium appetite. When normal rats allowed continuous access to water and hypertonic NaCl were given captopril for several days they increased their intake of NaCl (28,29). When captopril treatment is prolonged, it is thought that the high concentration of circulating angiotensin I gains access to the brain where, owing to converting enzyme biosynthesis and captopril metabolism, a sufficient amount is converted to angiotensin II to stimulate sodium appetite.

Further evidence for the involvement of endogenously formed angiotensin in sodium appetite is the effect of captopril on the NaCl intake of rats with a pre-existing sodium appetite. This effect depends on dose, and on whether or not the appetite is associated with increased renin secretion. In rats showing an increase in sodium appetite following adrenalectomy, and which were therefore hyper-reninaemic, low doses of captopril caused immediate increases in intakes of water and hypertonic NaCl, whereas high doses inhibited intake of both fluids after an initial increase in water intake (30). Both doses of captopril caused increases in plasma renin and angiotensin I, but only after low doses was there sufficient cerebral conversion of angiotensin to stimulate thirst and sodium appetite. By contrast, neither dose of captopril affected the sodium appetite of deoxycorticosterone-treated rats in which renin remains suppressed even after captopril. Other examples suggesting that endogenously formed angiotensin stimulates sodium appetite are the inhibition by large doses of captopril of the increased water and NaCl intakes of the renal hypertensive rat, and the decrease in sodium appetite of the spontaneously hypertensive rat produced by intracranial infusion of captopril (31).

The Role of Angiotensin in Thirst and Sodium Appetite

Circulating angiotensin plays a role in hypovolaemic thirst and sodium appetite, but the extent and the circumstances of the contribution and whether cerebral renin is also involved remain uncertain. Angiotensin is not essential in either behavior. Hypovolaemic thirst is attenuated but not abolished by angiotensin blockade, and the nephrectomized rat will drink water in response to some types of hypovolaemic stimuli. Angiotensin cannot play a role in spontaneous day-to-day drinking nor in drinking caused by cellular dehydration because in neither case is circulating angiotensin increased. Sodium appetite can be aroused by mineralocorticoid treatment even though the renin-angiotensin system is completely inhibited by such treatment. Circulating angiotensin is therefore not the exclusive stimulus of hypovolaemic thirst and sodium appetite. There is a component, presumably neural, which does not depend on renal renin and which cannot be eliminated by peripheral angiotensin blockade.

THE CONTROL OF EXCRETION

Vasopressin and the Countercurrent Mechanism: A Terrestrial Adaptation

The life-saving response in dehydration is increased water intake, but it is important also to economize on the use of water. Vasopressin acting on the kidney through the countercurrent mechanism, allows excretion of obligatory waste products in minimal quantities of water. When water supplies fail, further economy of water is possible by increasing plasma vasopressin. However, the urine is normally quite concentrated so that the potential for reduction in free water clearance is small. Even with maximal urinary concentration, there are continued obligatory water losses through the kidneys, lungs and skin which must be replaced by water intake.

Although it is essential for a terrestrial animal to conserve water, it is equally important for it to be able to get rid of surplus water. Without the ability to inhibit vasopressin secretion and allow urine flow to increase, there is a risk of water intoxication. This may occur in the syndrome of inappropriate secretion of antidiuretic hormone, or in diabetes insipidus patients being treated with antidiuretic agents, because in neither case is antidiuretic activity under the control of osmoreceptors.

Vasopressin

Vasopressin is a nonapeptide consisting of a ring of six amino acids with a disulphide bridge between two cysteine residues, and a side chain of three amino acids with the strongly basic arginine in position 8. The ring is essential for biological activity, but not the disulphide bridge as was once thought. Vasopressin and its carrier protein are synthesized as a composite precursor in the magnocellular neurons of the supraoptic and paraventricular nuclei. The vasopressin gene is 1.85 kilobase pairs long and consists of three exons (32). Exon A encodes vasopressin itself, part of the N terminus of the neurophysin, an intervening sequence of three amino acids between them and a signal peptide. Exon B encodes an internal part of the neurophysin molecule which is constant between species. Exon C encodes the C terminus of neurophysin, a glycopeptide (not possessed by oxytocin), and one arginine residue between them.

Prohormone is transported along the hypothalamo-hypophysial tracts in neurosecretory granules where it is cleaved into vasopressin, neurophysin and glycopeptide, and the C terminal glycine of vasopressin undergoes amidation. Active hormone, neurophysin and glycopeptide reaching the pars nervosa are stored there until nervous impulses depolarize the terminations of the hypothalamo-hypophysial tracts, causing calcium-dependent exocytosis of hormone into the general circulation. The entire contents of the neurosecretory granules are released into the blood stream but only vasopressin has been found to have biological activity outside the neuron.

The antidiuretic effect of vasopressin is mediated by cyclic AMP, but in most other sensitive tissues vasopressin only causes an increase in intracellular calcium. Two types of vasopressin receptor have been described, V_2 in the kidney coupled to adenylate cyclase and V_1 in extrarenal tissues coupled to mechanisms that affect calcium. The variety of biological effects of vasopressin has encouraged the search for analogues of vasopressin, and specific agonists and antagonists are now available (33). An example is deamino-D-arginine vasopressin, which is a more potent and longer acting antidiuretic agent than vasopressin but has very little pressor activity. It can also be given by mouth.

Physiological Role of Vasopressin

The most important function of vasopressin is to regulate the effective extracellular osmolality by causing the kidney to conserve water. Vasopressin is usually regarded as being antidipsogenic through its water-retaining action on the kidney. The dog is unusual among laboratory animals in that vasopressin has a direct stimulating action on thirst (34). It may therefore contribute directly to the thirst of cellular dehydration. This is appropriate because increased vasopressin secretion generally means a need for water which is not met by simply reducing urine flow.

Changes in osmolality are detected by osmoreceptors which respond to changes in their water content by stimulating or inhibiting vasopressin release. The idea of receptors responsive to their own water content and which can therefore represent all the other cells of the body was clearly stated by Verney (35), who introduced the term osmoreceptor to describe them. Small amounts of hypertonic NaCl were shown to cause antidiuresis when injected into the carotid artery, but the same amounts given intravenously, or intracarotid injection of hyperosmotic solutions of solutes that did not dehydrate cells, were ineffective. The antidiuresis was shown to depend on the posterior pituitary and could be matched with the appropriate dose of pituitary extract.

The nature and site of osmoreceptors have been debated ever since Verney's original experiments. By selective ligation of intracranial blood vessels, Verney narrowed down the location to the supraoptic region. Much evidence gained since then supports the view that neurons here function as osmoreceptors. Electrophysiological analysis has shown that the supraoptic neurons have many neural imputs, some of which themselves respond to changes in osmolality or sodium concentration (36). During release of vasopressin the magnocellular neurons show a phasic pattern of activity but it is not known whether the periodicity is synaptically driven or whether it is a property of the vasopressinergic cells themselves.

Some evidence supports the view that there are osmoreceptors in the liver. For example, distilled water infused into the portal vein of a conscious rat caused a diuresis whereas hypertonic NaCl caused an antidiuresis. The same infusions in the inferior vena cava had no effect on urine flow. Some electrophysiological evidence supports the idea since afferent activity in hepatic branches of the vagi is increased by moderate changes in osmolality of the fluid perfusing the liver. However, much of the evidence for osmoreceptors outside the central nervous system is inconclusive (37). The diversity of cardiovascular effects elicited by osmolality changes, with the strong likelihood that these produce changes in sympathetic activity to the kidney, suggest the possibility that the osmosensitivity is non-specific. Final judgment must await a full stimulus-response analysis with hormone assays.

In man a loss of just 2% of total body water causes near maximal antidiuresis and a desire to drink (38). An increase in body water of this amount causes water diuresis. The threshold for vasopressin secretion is about 280 mosmol/kg, below which vasopressin is undetectable and urine flow and dilution maximal. The threshold for thirst is about 295 mosmol/kg at which level there is enough vasopressin (about 6 pg/ml) to produce maximal urinary concentration of 1200 mosmol/kg or more and minimal volume. In normally hydrated man, plasma osmolality is 287 mosmol/kg, plasma vasopressin 3 pg/ml and urinary concentration 600 mosmol/kg.

Secretion of vasopressin and thirst are also affected by changes in blood volume and/or pressure, mediated by vascular stretch receptors in the high and low pressure sides of the circulation. Vasopressin is little affected by small changes in blood volume or pressure, but changes larger than about 10% may produce hormone levels many times those required to produce maximal antidiuresis. There is a massive release of hormone in severe blood loss which is a significant factor in the restoration and maintenance of vascular tone. Changes in blood volume or pressure large enough to affect vasopressin secretion appear to act by shifting the set of the osmoregulatory system so as to increase the effect of a given osmotic stimulus on vasopressin secretion. A change in the slope or sensitivity of the osmotic response may also occur but this effect is less consistent.

Control of Sodium Excretion

The control of sodium excretion involves intrarenal as well as nervous and hormonal mechanisms. Physical factors affect glomerular filtration and tubular reabsorption. Because stabilization of the glomerular capillary pressure and flow by autoregulation is imperfect, changes in perfusion pressure can affect urine flow by intrarenal mechanisms. An increase in perfusion pressure to the isolated kidney causes a pressure diuresis and natriuresis, and a fall has the reverse effect. The plasma colloid osmotic pressure also affects sodium excretion since it is one of the factors that determine the balance of Starling forces in the glomerular and peritubular capillaries. Overtransfusion with isotonic saline causes dilution of the plasma proteins and a fall in the colloid osmotic pressure. The glomerular filtration rate and filtration fraction are increased, and the fractional tubular reabsorption is decreased, which rid the body of the saline load.

In the intact animal, reflex changes in sympathetic nerve activity make the pressure-natriuresis relation much steeper than for the isolated kidney (39).

These reflex responses operate over a longer time scale than the direct effect of pressure on filtration. They arise from altered stimulation of stretch receptors in the low and high pressure sides of the circulation. A fall in central venous pressure results in increased sympathetic tone to the kidney, which has complex effects on renal function. In general these effects are antinatriuretic through decreased glomerular filtration, stimulation of the renin-angiotensin-aldosterone system, and redistribution of blood flow within the kidney. The reflex effects arising from the carotid and aortic baroreceptors from an acute fall in arterial pressure are similar and also contribute to decreased urine flow and sodium excretion when the hypovolaemia is severe.

De Bold and his colleagues have demonstrated the presence of a factor in atrial musculature which causes a powerful natriuretic response when injected intravenously (40). The natriuresis is caused by a peptide or peptides which have been sequenced and for which assays have been developed. Atrial natriuretic peptides are found in secretory granules in the walls of the atria of some species including man. They are released by atrial distension, but the mechanism of their release is unknown. They inhibit renin release and cause reduced sodium reabsorption in the distal part of the nephron. They are also potent vasodilators and are therefore ideal substances to combat rises in blood pressure produced by increases in blood volume. It is uncertain whether there is tonic release at normal blood volumes which is inhibited in hypovolaemia. The possible physiological role of these peptides in sodium balance is being actively studied. The relation between atrial peptides and other less well characterized natriuretic factors is unknown. There are also many other substances, such as prolactin, the prostaglandins and the kinins, which may be involved in sodium balance control. One view of the roles that some of the mechanisms may play in the control of sodium balance is that the sympathetic nerves and the renin-angiotensin-aldosterone system are important in sodium deprivation, whereas increased atrial natriuretic factor may be important in sodium loading and volume expansion (41). However, this must be regarded as a tentative and incomplete story.

CONCLUSION

The importance of the central nervous system in achieving constancy of the internal environment, as stated long ago by Claude Bernard, is not in doubt. Vasopressin release, renal haemodynamics, thirst and sodium appetite depend to a very large extent on nervous mechanisms. Since Bernard's day, it has become clear that many aspects of body fluid homeostasis also depend on endocrine mechanisms. The discovery of atrial natriuretic peptides is the most recent exciting development, but the classical Verney osmoreceptor-vasopressin mechanism remains the best understood of all endocrine controls of the body fluids. Although, necessarily, the central nervous system retains its pre-eminent role in the control of fluid intake, even here there is good evidence that angiotensin, mineralocorticoids and some of the reproductive hormones are involved. Fluid balance control both in its behavioral and renal aspects is a complex neuroendocrine phenomenon.

REFERENCES

1. Bernard C (1878) Lecons sur les phenomenes de la vie communs aux animaux et aux végétaux, vol. 1. Paris: Baillière
2. Cannon WB (1932) The Wisdom of the Body. London: Kegan Paul, Trench, Trubner
3. Fitzsimons JT (1979) The Physiology of Thirst and Sodium Appetite. Monographs of the Physiological Society, No. 35. Cambridge: Cambridge University Press.
4. Thrasher TN, Jones RG, Keil LC, Brown CJ, Ramsay DJ (1980) Drinking and vasopressin release during ventricular infusions of hypertonic solutions. Am J Physiol 28: R340-R345
5. Andersson B (1978) Regulation of water intake. Physiol Revs 58: 582-603
6. McKinley MJ, Denton DA, Leksell L, Tarjan E, Weisinger RS (1980) Evidence for cerebral sodium sensors involved in water drinking in sheep. Physiol and Behav 25: 501-504
7. Johnson AK, Buggy J (1978) Periventricular preoptic-hypothalamus is vital for thirst and normal water economy. Am J Physiol 234: R122-R129
8. Thrasher TN, Keil LC, Ramsay DJ (1982) Lesions of the organum vasculosum of the lamina terminalis (OVLT) attenuate osmotically induced drinking and vasopressin secretion in the dog. Endocrinology 110: 1837-1842
9. Rolls BJ, Rolls ET (1982) Thirst. Cambridge: Cambridge University Press
10. Kozlowski S, Szczepańska-Sakowska E (1975) Mechanisms of hypovolaemic thirst and interactions between hypovolaemia, hyperosmolality and the antidiuretic system. In: Peters G, Fitzsimons JT, Peters-Haefeli L (eds) Control Mechanisms of Drinking. New York: Springer Verlag: pp 15-35
11. Fitzsimons JT, Moore-Gillon MJ (1980) Drinking and antidiuresis in response to reductions in venous return in the dog: neural and endocrine mechanisms. J Physiol 308: 403-416
12. Moore-Gillon MJ, Fitzsimons JT (1982) Pulmonary vein-atrial junction stretch receptors and the inhibition of drinking. Am J Physiol 242: R452-R457
13. Kaufman S (1984) Role of right atrial receptors in the control of drinking in the rat. J Physiol 349: 389-396
14. Fitzsimons JT, Moore-Gillon MJ (1981) Renin-dependence of drinking induced by partial aortic obstruction in the dog. J Physiol 320: 423-433
15. Robinson MM, Evered MD (1986) Angiotensin II and arterial pressure in the control of thirst. In: de Caro G, Epstein AN, Massi M (eds) The Physiology of Thirst and Sodium Appetite. New York: Plenum: pp 193-198
16. Denton DA (1982) The Hunger for Salt. New York: Springer Verlag
17. Weisinger RS, Considine P, Denton DA, Leksell L, McKinley MJ, Mouw DR, Muller AF, Tarjan E (1982) Role of sodium concentration of the cerebrospinal fluid in the salt appetite of sheep. Am J Physiol 242: R51-R53
18. Epstein AN, Zhang DM, Schultz J, Rosenberg M, Kupsha P, Stellar E (1984) The failure of ventricular sodium to control sodium appetite in the rat. Physiol and Behav 32: 683-686
19. Rice KK, Richter CP (1943) Increased sodium chloride and water intake of normal rats treated with desoxycorticosterone acetate. Endocrinology 33: 106-115

20. Kaufman S (1986) Control mechanisms of salt appetite. In: deCaro G, Epstein AN, Massi M (eds) The Physiology of Thirst and Sodium Appetite. New York: Plenum: pp 459-463
21. Nakamura M, Katsuura G, Nakao K, Imura H (1985) Antidipsogenic action of human atrial natriuretic polypeptide administered intracerebroventricularly in rats. Neuroscience Lett 58: 1-6
22. Simpson JB (1981) The circumventricular organs and the central actions of angiotensin. Neuroendocrinology 32: 248-256
23. Liard JF, Cowley AW, McCaa RE, McCaa CS, Guyton AC (1974) Renin, aldosterone, body fluid volumes, and the baroreceptor reflex in the development and reversal of Goldblatt hypertension in conscious dogs. Circulation Res 34: 549-560
24. Costales M, Fitzsimons JT, Vijande M (1984) Increased sodium appetite and polydipsia induced by partial aortic occlusion in the rat. J Physiol 352: 467-481
25. Thrasher TN, Keil LC, Ramsay DJ (1982) Hemodynamic, hormonal and drinking responses to reduced venous return in the dog. Am J Physiol 243: R354-R362
26. Fitzsimons JT, Elfont RM (1982) Angiotensin does contribute to drinking induced by caval ligation in the rat. Am J Physiol 243: R558-R562
27. Fluharty SJ, Epstein AN (1983) Sodium appetite elicited by intracerebroventricular infusion of angiotensin II in the rat: II synergistic interaction with systemic mineralocorticoids. Behav Neuroscience 97: 746-758
28. Fregly MJ (1980) Effect of the angiotensin converting enzyme inhibitor, captopril, on NaCl appetite of rats. J Pharmacol and Exp Therap 215: 407-412
29. Elfont RM, Epstein AN, Fitzsimons JT (1984) Involvement of the renin-angiotensin system in captopril-induced sodium appetite in the rat. J Physiol 354: 11-27
30. Elfont RM, Fitzsimons JT (1985) The effect of captopril on sodium appetite in adrenalectomized and deoxycorticosterone-treated rats. J Physiol 365: 1-12
31. Di Nicolantonio R, Hutchinson JS, Mendelsohn FAO (1982) Exaggerated salt appetite of spontaneously hypertensive rats is decreased by central angiotensin-converting enzyme blockade. Nature. 298: 846-848
32. Richter D (1983) Vasopressin and oxytocin are expressed as polyproteins. TIBS 8: 278-281
33. Jard S (1985) Vasopressin receptors. Front Horm Res 13: 89-104
34. Szczepanska-Sakowska E, Sobocinska J, Sadowski B (1982) Central dipsogenic effect of vasopressin. Am J Physiol 242: R372-R379
35. Verney EB (1947) The antidiuretic hormone and the factors which determine its release. Proc Royal Soc B135: 25-106
36. Leng G, Mason WT, Dyer RG (1982) The supraoptic nucleus as an osmoreceptor. Neuroendocrinology 34: 75-82
37. Bie P (1980) Osmoreceptors, vasopressin and the control of renal water excretion. Physiol Rev 60: 961-1048
38. Robertson GL (1985) Regulation of vasopressin secretion. In: Seldin DW, Giebisch G (eds) The Kidney: Physiology and Pathophysiology, Vol. 2. New York: Raven Press: pp 869-884
39. Guyton AC (1980) Arterial Pressure and Hypertension. Philadelphia: Saunders
40. De Bold AJ, Borenstein HB, Veress AT, Sonnenberg H (1981) A rapid and potent natriuretic response to intravenous injection of atrial myocardial extract in rats. Life Sci 28: 89-94
41. Navar, LG (1984) Renal regulation of body fluid balance. In: Staub NC, Taylor AE (eds) Edema. New York: Raven Press: pp 319-352

ENDOGENOUS Na^+, K^+-ATPase INHIBITORS

M. D. Lewis, S. E. Thomas, J. M. Kendall,
G. Spurlock, M. A. Mir and M. F. Scanlon

Department of Medicine, University of Wales College of Medicine,
Heath Park, Cardiff, Wales, U.K.

INTRODUCTION

Evidence began to mount during the 1960s for the existence of an endogenous regulator of sodium transport. De Wardener demonstrated that extracellular fluid volume (ECFV) expansion by saline infusion in dogs produced a marked increase in urinary sodium excretion (1). He showed that this was due to the presence of a circulating factor because blood from a ECFV-expanded dog could induce natriuresis in a recipient dog. The involvement of all known regulators of sodium and water balance was excluded, and de Wardener suggested that the circulating factor was the "natriuretic hormone." Such a factor could act by inhibiting renal Na^+, K^+-ATPase, thus inhibiting active reuptake of sodium with a resultant increase in urinary sodium.

The factor released in response to ECFV expansion or salt loading may be involved in certain forms of hypertension. Dahl et al. salt loaded a salt-sensitive strain of rat that was in parabiosis with a salt-resistant rat (2). This caused the salt-resistant rat to become hypertensive as well as the salt-sensitive rat. Dahl concluded from this that high sodium intake promotes the release of a natriuretic factor with hypertensinogenic capacity. In partial support of this, increased Na^+, K^+-ATPase inhibitory activity has been detected in the plasma of some hypertensive patients (3,4). Overbeck demonstrated that inhibition of Na^+, K^+-ATPase caused increased tension in vascular smooth muscle leading to an increase in blood pressure (5). Blaustein suggested that vascular smooth muscle contained a

Na^+-Ca^{2+} exchanger and that active calcium efflux was driven by the transmembrane sodium gradient (6). Inhibition of Na^+, K^+-ATPase would increase intracellular sodium with decreased calcium efflux, causing increased intracellular calcium with resultant vascular contraction. There is some experimental evidence to support this hypothesis (7), but it is still somewhat contentious.

The probable source of the circulating Na^+, K^+-ATPase inhibitor is the hypothalamus. Lesions of the hypothalamus anterior and ventral to the third ventricle (AV3V) prevent the responses to volume expansion, whereas nonlesioned rats show normal increases in blood pressure, glomerular filtration rate, natriuresis and circulating levels of a sodium transport inhibitor following ECFV (8,9). Several groups have partially purified an inhibitor from the hypothalamus, but there is considerable disagreement as to the nature of this factor (10,11). De Wardener has shown a Na^+, K^+-ATPase inhibitor in rat hypothalamus that is increased in rats on a high salt diet (12).

METHODS AND RESULTS

Purification of an Active Sodium Transport Inhibitor (ASTI) from Hypothalamic Cell Culture Medium

Since fetal rat hypothalamic cell cultures synthesize and release a variety of known hypothalamic peptides, we investigated whether such cultures also produced the proposed sodium transport inhibitor. We have shown that hypothalamic, but not cerebral cortical or control culture, media inhibits the active efflux rate of [^{22}Na] from human erythrocytes and that this inhibitory activity was increased in response to depolarization of the cells in a calcium-dependent manner (13). Hypothalamic culture medium also inhibits kidney Na^+, K^+-ATPase directly and displaces [^{3}H]ouabain binding from erythrocytes (14). We have subsequently attempted to purify and characterize the active sodium transport inhibitor (ASTI), using hypothalamic culture medium as the source material.

Culture medium from monolayer fetal rat hypothalamic cell cultures was pooled, heat treated and lyophilized for further purification. We have modified our purification procedure from that described previously (14). In order to increase yield by the prevention of adsorption of ASTI to the Sephadex and to allow concentration of large volumes of column fractions, culture medium is run on a Sephadex G-15 column in 20% methanol. ASTI runs at a position consistent with a molecular weight of 1000. The active fractions from several G-15 runs were pooled, lyophilized and then subjected to three different reversed phase HPLC separations. ASTI elutes off the third run of a shallow gradient of increasing acetonitrile in 0.1% trifluoroacetic acid from a C-18 column as a single peak at 35% acetonitrile. We have used this material for amino acid and sequence analysis.

Characterization of ASTI

We have previously shown that ASTI is heat stable (80°C, 10 min), but its activity is destroyed by a variety of proteases (14). We regard this as strong evidence that ASTI is peptidic. This is consistent with its chromatographic properties on Sephadex G-15 and reversed phase HPLC. We have carried out an amino acid analysis and have partially sequenced the material from the final HPLC run with manual derivatization and HPLC analysis of the DABTH derivatized amino acids using the method of Chang (15). We have found ASTI to be a peptide of nine amino acids, but we still have to confirm the sequence and show that the synthetic peptide has similar biological properties.

Vasoactive Properties

We have shown ASTI to produce a dose-dependent increase in tension of de-endothelialized rabbit aortic strips (16). The vasoactive property of ASTI can be destroyed by prior treatment with prolidase, showing that the vasoactive properties are due to a peptide. ASTI has also been shown to potentiate the vasoconstrictor effect of noradrenaline. Pretreatment of the aortic strip with ASTI causes a significant shift to the left of the subsequent noradrenaline dose-dependent increase in tension. ASTI increases the calcium uptake of aortic strips compared to control, which could explain the potentiation of the noradrenaline effect produced by ASTI. Pretreatment of the strip will increase intracellular calcium stores, therefore subsequent noradrenaline stimulation will release a greater amount of calcium, causing a larger increase in tension compared to an untreated aortic strip.

Presence of ASTI in Adult Rat Hypothalamus and Posterior Pituitary

Extracts of adult rat hypothalamus and posterior pituitary have been subjected to the same purification regime as that used for hypothalamic culture medium. ASTI activity was found in both tissues having the same chromatographic properties as that isolated from hypothalamic culture

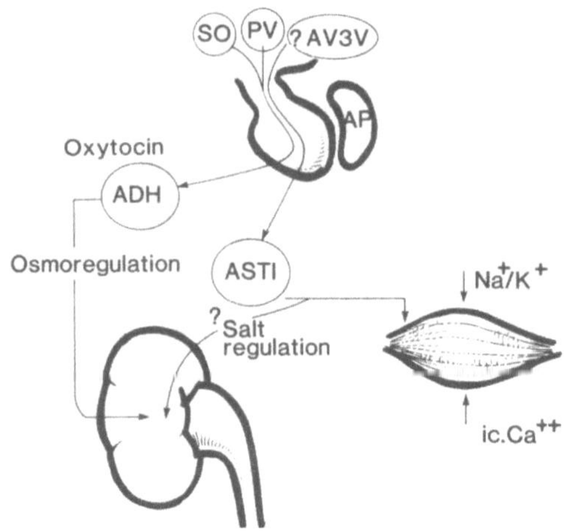

Figure 1. Hypothetical outline of ASTI as a posterior pituitary hormone with salt regulating and vasoconstrictive properties.

SO = Supraoptic nucleus
PA = Paraventricular nucleus
AP = Anterior Pituitary

medium. The yield from posterior pituitary is approximately four times that of the hypothalamus (per gram of lyophilized tissue extract).

DISCUSSION

Fetal rat hypothalamic cell cultures release a Na^+, K^+-ATPase inhibitor, which we have called ASTI. ASTI inhibits active sodium efflux from erythrocytes by inhibiting Na^+, K^+-ATPase directly and displaces ouabain bound to erythrocytes. Its activity can be completely destroyed by protease treatment. We have partially sequenced the molecule, which we believe to be a nine amino acid peptide. The presence of ASTI in the hypothalamus and posterior pituitary of adult rat leads us to propose that it is synthesized in the hypothalamus before being transported to the posterior pituitary for storage and subsequent release into the circulation (Fig. 1).

ASTI has vasoconstrictor properties that appear to act via increasing intracellular calcium levels. It may also play some part in the control of renal sodium excretion and may be implicated in certain forms of hypertension.

REFERENCES

1. de Wardener HE, Clarkson EM (1985) Concept of natriuretic hormone. Physiol Rev 65: 658-759
2. Dahl LK, Knudsen KD, Iwai J (1969) Humoral transmission of hypertension: evidence from parabiosis. Circ Res 24/25 (Suppl I): I21-I33
3. de Wardener HE, MacGregor GA, Clarkson EM, Alaghband-Zadeh J, Bitensky L, Chayen J (1981) Effect of sodium intake on ability of human plasma to inhibit renal Na^+-K^+ adenosine triphosphatase in vitro. Lancet i: 411-412
4. Hamlyn JM, Ringel R, Schaeffer J, Levinson PD, Hamilton BP, Kowarski AA, Blaustein MP (1982) A circulating inhibitor of (Na^+ + K^+)ATPase associated with essential hypertension. Nature 300: 650-652
5. Overbeck HW (1972) Vascular responses to cations, osmolarity, and angiotension in renal hypertensive dogs. Am J Physiol 223: 1358-1364
6. Blaustein MP (1977) Sodium ions, calcium ions, blood pressure regulation and hypertension: a reassessment and a hypothesis. Am J Physiol 232: C167-173
7. Blaustein MP (1984) Sodium transport and hypertension. Hypertension 6: 445-453
8. Bealer SL, Haywood JR, Gruber KA, Buckalew VM, Fink GD, Brody MJ, Johnson AK (1983) Preoptic-hypothalamic periventricular lesions reduce natriuresis to volume expansion. Am J Physiol 244: R51-57
9. Bealer SL (1983) Hemodynamic mechanisms in CNS-induced natriuresis in the conscious rat. Am J Physiol 244: F376-382
10. Haupert GT, Sancho JM (1979) Sodium transport inhibitor from bovine hypothalamus. Proc Natl Acad Sci USA 76: 4658-4660
11. Akagawa K, Hara N, Trukada Y (1984) Partial purification and properties of the inhibitors of Na,K-ATPase and ouabain-binding in bovine central nervous system. J Neurochem 42: 775-780
12. Alaghband-Zadeh J, Fenton S, Hancock K, Millett J, de Wardener HE (1983) Evidence that the hypothalamus may be a source of a circulating Na^+-K^+-ATPase inhibitor. J Endocr 98: 221-226
13. Morgan K, Foord SM, Spurlock G, Charalambous BM, Dieguez C, Scanlon MF, Mir MA (1984) Release of an active sodium transport inhibitor (ASTI) from rat hypothalamic cells in culture. Endocrinology 115: 1642-1644
14. Morgan K, Lewis MD, Spurlock G, Collins PA, Foord SM, Southgate K, Scanlon MF, Mir MA (1985) Characterization and partial purification of the sodium-potassium-ATPase inhibitor released from cultured rat hypothalamic cells. J Biol Chem 260: 13595-13600
15. Chang JY (1981) N-terminal sequence analysis of polypeptide at the picomole level. Biochem J 199: 557-564
16. Mir MA, Chappell SP, Morgan K, Lewis MD, Scanlon MF, Lewis MJ (1986) Hypothalamic sodium transport inhibitor and vascular reactivity. J Hypertension 4 (Suppl 5): S233-S235

NEUROMEDIN U:
A NOVEL BRAIN-GUT PEPTIDE

J. Domin and S. R. Bloom

Department of Medicine, Royal Postgraduate Medical School
Hammersmith Hospital, London, U.K.

The technique of isolating novel peptides from tissue extracts based on their bioactivity was the foundation of modern peptide research. Many peptides originally isolated from the mammalian gastrointestinal tract in this way were subsequently found in the central nervous system, thus giving rise to the term "brain-gut peptides" (1). Today, peptides such as vasoactive intestinal polypeptide (VIP) and substance P are viewed primarily as neural peptides with biological roles extending beyond the gastrointestinal tract. Interestingly, however, it was not until the early part of this decade that a concerted effort was made to search systematically for novel neuropeptides in the mammalian spinal cord, when Naoto Minamino and colleagues began a systematic isolation of novel peptides from extracts of porcine cord. Purifications were monitored by bioassay, using the ability to contract smooth muscle preparations. The peptides isolated in this way were termed "neuromedins" and suffixed by a letter that was determined by amino acid sequence homology with existing peptide families. Hence, neuromedin K and neuromedin L were known as kassinin-like (2,3), neuromedin B and neuromedin C were bombesin-like (4,5), and neuromedin N was neurotensin-like (6) (Fig. 1).

Neuromedin U, the most recently isolated neuromedin, was an exception. This peptide lacked strong sequence homology with established peptide groups and was therefore suffixed by the letter "U" on the basis of its ability to contract uterine smooth muscle preparations (7).

NmK:	Asp Met His Asp Phe Phe Val Gly Leu Met-NH_2
NmL:	His Lys Thr Asp Ser Phe Val Gly Leu Met-NH_2
NmB:	Gly Asn Leu Trp Ala Thr Gly His Phe Met-NH_2
NmC:	Gly Asn His Trp Ala Val Gly His Leu Met-NH_2
NmN:	Lys Ile Pro Tyr Ile Leu-NH_2
NmU-25:	Phe Lys Val Asp Glu Glu Phe Gln Gly Pro Ile Val Ser Gln Asn Arg Arg Tyr Phe Leu Phe Arg Pro Arg Asn-NH_2
NmU-8:	Tyr Phe Leu Phe Arg Pro Arg Asn-NH_2

Figure 1. The amino acid sequences of neuromedins isolated from porcine spinal cord.

NmU-8:	Y F L F R P R N-NH_2
NmU-25:	F K V D E E F Q G P I V S Q N R R Y F L F R P R N-NH_2
Porcine PP :	... A T P E Q M A Q Y A A E L R R Y I N M L T R P R Y-NH_2
Rat PP :	... A T H E Q R A Q Y E T Q L R R Y I N T L T R P R Y-NH_2

Figure 2. Amino acid sequence comparison of neuromedin U-8, neuromedin U-25 and pancreatic polypeptide. Reprinted with permission from ref. 10, Copyright 1987, Pergamon Press plc.

Neuromedin U

Two molecular forms of neuromedin U were purified from side fractions obtained in the isolation of other neuromedins (2-6). Amino acid sequence analysis revealed a C-terminally amidated octapeptide that was termed neuromedin U-8, and an icosapentapeptide, termed neuromedin U-25, which contained the amino acid sequence of neuromedin U-8 at its C-terminus (Fig. 2).

Since neuromedin U-8 was isolated and its sequence within neuromedin U-25 was preceded by an Arg-Arg pair, a trypsin-like cleavage (8) of the larger molecular form was postulated to occur in vivo. In addition to their ability to stimulate smooth muscle contraction, both neuromedin U-8 and neuromedin U-25 exhibited hypertensive effects in the rat (7). Recent studies in dogs have shown that both molecular forms reduced mesenteric blood flow whilst slightly increasing pancreatic blood flow (9). As neither peptide altered axillary blood flow, these results suggested a selective effect on splanchnic circulation.

Using synthetic porcine neuromedin U-8, we developed a sensitive and specific radioimmunoassay that enabled us to study the distribution and to characterize the molecular forms of neuromedin U-like immunoreactivity in several species. The distribution of neuromedin U-like immunoreactivity was found to be localized and distinct from that of other regulatory peptides (10). In the rat, the highest tissue concentrations of immunoreactivity were found in the pituitary gland and the gastrointestinal tract, with lower concentrations also detected in the brain, spinal cord and urogenital tract. Neuromedin U-like immunoreactivity in the pituitary gland was primarily restricted to the anterior lobe, although low concentrations of immunoreactivity were also found in extracts of the posterior lobe (Table 1).

TABLE 1

DISTRIBUTION OF NEUROMEDIN U-LIKE IMMUNOREACTIVITY IN RAT PITUITARY

	NmU-LI (picomole/gland)
Anterior lobe	6.3 ± 0.5
Posterior lobe	0.3 ± 0.1

Radioimmunoassay analysis, however, was unable to show what proportion of the total immunoreactivity attributed to the posterior lobe was present in the adherent neurointermediate lobe (11). Immunocytochemical studies have recently resolved this issue by demonstrating the localization of neuromedin U immunoreactivity in the pituitary gland and hypothalamus (12). These results confirmed that in rat and mouse, neuromedin U immunoreactivity was predominantly concentrated in the anterior lobe

TABLE 2

DISTRIBUTION OF NEUROMEDIN U-LIKE IMMUNOREACTIVITY IN THE RAT GASTROINTESTINAL TRACT

	NmU-LI (picomole/g)		NmU-LI (picomole/g)
Esophagus	7.9 ± 1.5	Asc. Colon	74.1 ± 10.0
Stomach	1.5 ± 0.3	Trans. Colon	71.1 ± 17.7
Duodenum	131.9 ± 30.3	Desc. Colon	115.5 ± 17.2
Jejunum	178.2 ± 33.1	Rectum	57.9 ± 6.7
Ileum	181.9 ± 6.1	Liver	-
Caecum	42.8 ± 3.7	Pancreas	-

Reprinted with permission from ref. 10, Copyright 1987, Pergamon Press plc.

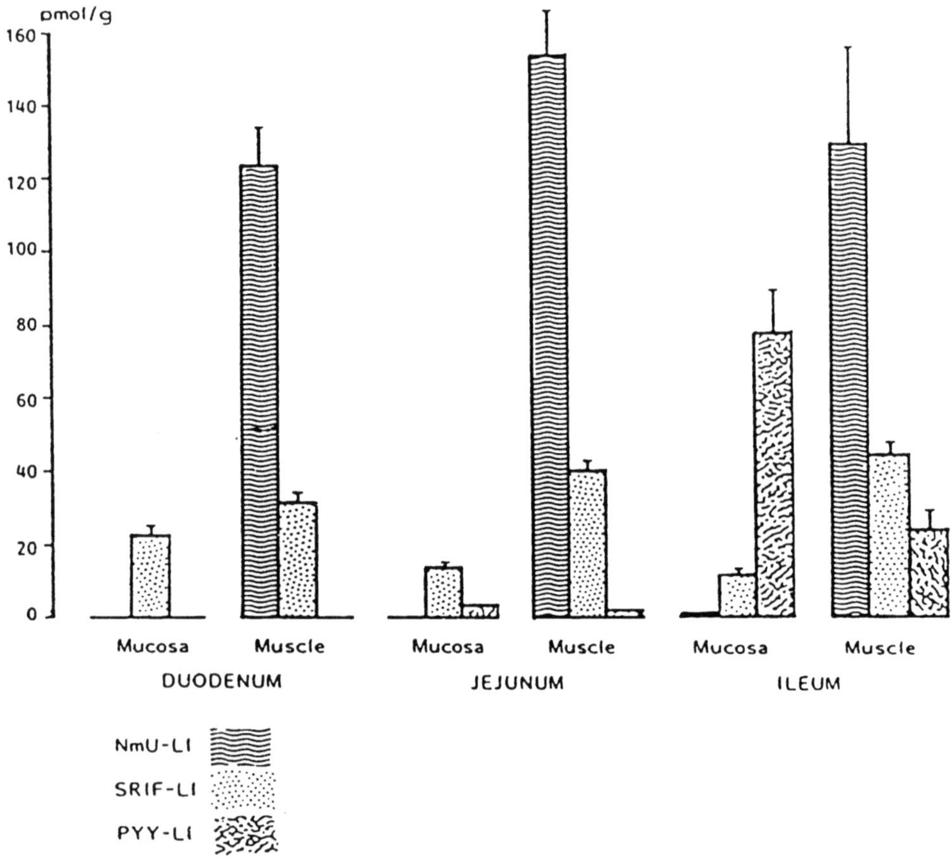

Figure 3. Concentrations of neuromedin U-, SRIF- and PYY-like immunoreactivity in extracts of small intestinal mucosal scrapings and underlying muscle layers. Reprinted with permission from ref. 10, Copyright 1987, Pergamon Press plc.

TABLE 3

DISTRIBUTION OF NEUROMEDIN U-LIKE IMMUNOREACTIVITY IN THE RAT CENTRAL NERVOUS SYSTEM

	NmU-LI (picomole/g)		NmU-LI (picomole/g)
Olfactory Bulb	-	Cervical	
Frontal Cortex	3.6 ± 1.9	Dorsal	24.9 ± 3.5
Nucleus Accumbens	40.9 ± 4.9	Ventral	9.7 ± 1.5
Septum	30.6 ± 4.3	Thoracic	
Hippocampus	-	Dorsal	39.0 ± 8.7
Thalamus	6.8 ± 2.4	Ventral	9.0 ± 1.2
Hypothalamus	31.2 ± 5.6	Lumbar	
Globus Pallidus	14.0 ± 5.5	Dorsal	31.4 ± 7.5
Amygdala	22.8 ± 8.7	Ventral	17.8 ± 5.1
Substantia Nigra/ Tegmental Ganglion	14.6 ± 5.3	Sacral	
		Dorsal	60.8 ± 9.5
Mid Brain	6.3 ± 1.5	Ventral	25.2 ± 6.5
Cerebellum Hemisphere	3.1 ± 1.5	Dorsal Root Ganglia	
Cerebellum Vermis	5.4 ± 2.0	Cervical	58.2 ± 11.2
Pons	6.8 ± 1.2	Thoracic	63.1 ± 7.7
Medulla Oblongata	15.3 ± 1.0	Lumbar	50.3 ± 5.8
Caudate Putamen	2.4 ± 0.8	Sacral	74.9 ± 14.2

Results are given as mean SEM (n = 6). - = undetectable.
Reprinted with permission from ref. 10, Copyright 1987, Pergamon Press plc.

where it co-localized with ACTH in many cells. In addition, all cells of the intermediate lobe showed intense immunostaining, but no immunoreactivity was observed in the posterior lobe. In the human pituitary, corticotrophic cells of the anterior lobe and the anterior border of the posterior lobe were stained, but again no immunoreactivity was observed in the posterior lobe. Overall, compared to results obtained in rodent species, a smaller proportion of cells were stained in the human pituitary gland. Neuromedin U immunoreactivity was also seen in ACTH immunoreactive cells of all pituitary corticotropinomas examined (n = 6).

Gastrointestinal neuromedin U-like immunoreactivity was found in highest concentrations in the small intestine (Table 2). Extraction of mucosal scrapings and the remaining muscular layers showed that in the small intestine, neuromedin U-like immunoreactivity was localized to the muscular fraction (Fig. 3). This differential distribution of immunoreactivity indicated a neuronal localization with a corresponding absence from endocrine cells. These results were consistent with our inability to detect circulating neuromedin U-like immunoreactivity in the

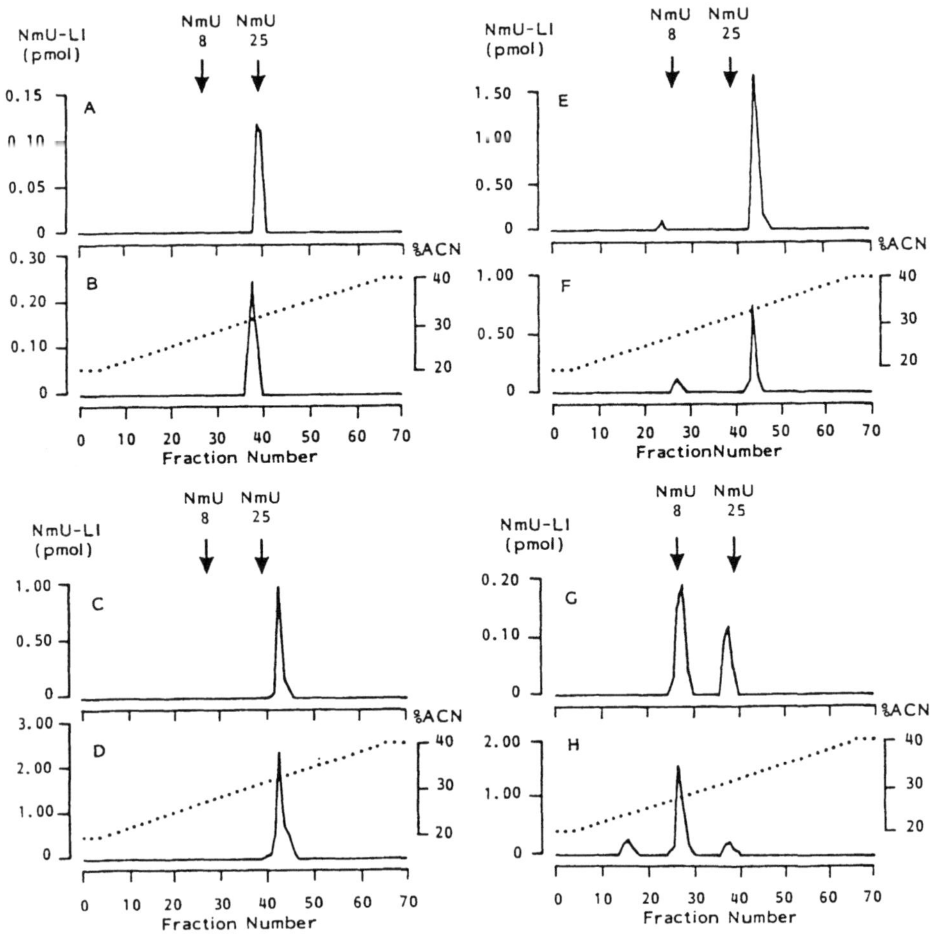

Figure 4. Representative reversed phase HPLC profiles of pig spinal cord (A) and ileum (B); rat spinal cord (C) and ileum (D); human brain (E) and ileum (F); and guinea pig spinal cord (G) and ileum (H). Reprinted from ref. 15, with permission.

rat. Subsequent immunocytochemical studies have confirmed the localization of neuromedin U immunoreactivity in the enteric nervous system, where ganglion cells in both the submucous and myenteric plexus were also found to contain immunoreactive calcitonin gene-related peptide (CGRP)(13).

Neuromedin U-like immunoreactivity was found to have widespread distribution in the rat brain (Table 3). Immunoreactive nerve fibers were widely distributed in the brain, but notably absent in the cerebellum (12). In the hypothalamus, fibers, but not perikariya, were found in the lateral, anterior and posterior sections, the paraventricular, supraoptic, suprachiasmatic and ventromedial nuclei, and the pars dorsalis and pars ventralis of the dorsomedial and periventricular nuclei. In the arcuate nucleus, both nerves and cell bodies were noted.

In the rat spinal cord, neuromedin U-like immunoreactivity displayed a dorsosacral predominance similar to that found with other neuropeptides (14). The presence of immunoreactivity in corresponding dorsal root ganglia suggested a possible sensory role for this peptide. Low concentrations of neuromedin U-like immunoreactivity were also detected in the urogenital tract.

Chromatographic analysis of pig, rat and human tissue extracts revealed the presence of only a single molecular form of neuromedin U-like immunoreactivity (15) (Fig. 4). In pig gut and spinal cord, this immunoreactivity was characterized as neuromedin U-25. In rat and in man, however, although the immunoreactivity was of a similar apparent molecular weight, it failed to coelute with the synthetic porcine neuromedin U-25 standard, thereby indicating species differences. In addition, we have recently detected the presence of neuromedin U-like immunoreactivity in the amphibian Rana temporaria (16). The distribution of immunoreactivity in this species closely paralleled that found in the rat, but the concentrations detected were approximately one order of magnitude greater. When analyzed chromatographically, neuromedin U-like immunoreactivity in this species showed marked intra-species molecular heterogeneity, with the absence of immunoreactivity coeluting with synthetic porcine neuromedin U standards.

Where such interspecies molecular variation is apparent, caution must be exercised when drawing conclusions based on direct quantitative comparisons of immunoreactivity. In addition, direct extrapolation of biological activity across species should also be treated tentatively.

In conclusion, neuromedin U-like immunoreactivity displays a discrete regional distribution in the mammal, with the highest tissue concentrations present in the pituitary gland and gastrointestinal tract. The physiological role of neuromedin U is currently unknown, although both porcine forms are clearly bioactive in pharmacological studies.

REFERENCES

1. Dockray GJ (1983) Brain-gut peptides. Viewpoints on Dig Dis 13: 5-8
2. Kangawa K, Minamino N, Fukuda A, Matsuo H (1983) Neuromedin K: A novel tachykinin identified in porcine spinal cord. Biochem Biophys Res Commun 114: 533-540
3. Minamino N, Kangawa K, Fukada A, Matsuo H (1984) Neuromedin L: A novel mammalian tachykinin identified in porcine spinal cord. Neuropeptides 4: 157-166
4. Minamino N, Kangawa K, Matsuo H (1983) Neuromedin B: A novel bombesin like peptide in porcine spinal cord. Biochem Biophys Res Commun 114: 541-548
5. Minamino N, Kangawa K, Matsuo H (1984) Neuromedin C: A bombesin-like peptide identified in porcine spinal cord. Biochem Biophys Res Commun 119: 14-20
6. Minamino N, Kangawa K, Matsuo H (1984) Neuromedin N: A novel neurotensin like peptide identified in porcine spinal cord. Biochem Biophys Res Commun 122: 542-549
7. Minamino N, Kangawa K, Matsuo H (1985) Neuromedin U-8 and U-25: Novel uterus stimulating and hypertensive peptides identified in porcine spinal cord. Biochem Biophys Res Commun 130: 1078-1085
8. Steiner DF, Docherty K, Carroll R (1984) Golgi/granule processing of peptide hormone and neuropeptide precursors: A minireview. J Cell Biochem 24: 121-130
9. Sumi S, Inoue K, Kogire M, Doi R, Takaori K, Suzuki T, Yajima H, Tobe T (1987) Effects of synthetic neuromedin U-8 and U-25, novel peptides identified in porcine spinal cord, on splanchnic circulation in dogs. Life Sci 41: 1585-1590
10. Domin J, Ghatei MA, Chohan P, Bloom SR. (1987) Neuromedin U - a study of its distribution in the rat. Peptides 8: 779-784
11. Lerner AB (1981) The intermediate lobe of the pituitary gland: introduction and background. In: Peptides of the pars intermedia. Ciba Foundation Symposium 81, London: Pitman Medical: pp. 3-12
12. Steel JH, Van Noorden S, Ballesta J, Gibson SJ, Ghatel MA, Burrin J, Leonhardt U, Domin J, Bloom SR, Polak JM. (1988) Localisation of 7B2, neuromedin B and neuromedin U in specific cell types of rat, mouse and human pituitary, in rat hypothalamus and extra-pituitary tumours. Endocrinology 122: 270-283
13. Bishop AE, Ballesta J, Carlei F, Steel JH, Gibson SJ, Domin J, Bloom SR, Polak JM (1987) Localisation of neuromedin U to enteric nerves and brain pathways. Reg Pep 18: 376
14. Gibson SJ, Polak JM, Anand P, Blank MA, Morrison JFB, Kelly JS, Bloom SR (1984) The distribution and origin of VIP in the spinal cord of six mammalian species. Peptides 5: 201-207

15. Domin J, Ghatei MA, Chohan P, Bloom SR (1986) Characterisation of neuromedin U-like immunoreactivity in rat, porcine, guinea-pig and human tissue extracts using a specific radioimmunoassay. Biochem Biophys Res Commun 140: 1127-1134
16. Domin J, Ghatei MA, Adolphus N, Bloom SR. Distribution and molecular characterisation of neuromedin U-, neuromedin B- and bombesin-like immunoreactivity in the amphibian Rana temporaria. Ann NY Acad Sci (in press)

B. THE CONTROL OF GONADOTROPIN SECRETION

INTRODUCTION

Synthesis and release of the gonadotropins are dependent on the frequent, regular, pulsatile release of hypothalamic gonadotropin releasing hormone (GnRH), which is in turn modulated by feedback effects of gonadal hormones including both steroids and peptides. Differential secretion of LH and FSH is related, at least in part, to different feedback mechanisms. The predominant feedback on LH is via gonadal steroids, whereas recent studies have demonstrated that FSH control can be exerted independently by the gonadal peptide inhibin.

Our understanding of the physiology of gonadotropin control has led to our ability to down regulate the pituitary gonadotroph. Chronic hyperstimulation using long- acting analogues of GnRH can cause a "medical orchidectomy" with castrate levels of either testosterone or estradiol. This occurs because agonistic occupation of the GnRH receptors by these long acting analogues interfaces with the intracellular synthesis of the gonadotropins. Such effects have been put to good use in the medical treatment of carcinoma of the prostate, often with dramatic results. In the future, other indications for gonadotropin down regulation - for example, as contraception, particularly in patients for whom steroid compounds are for any reason contraindicated - will be exploited. Considerable experience of their use in these circumstances has already been established.

Opioids are known to be important in the control of GnRH production. Furthermore, if GnRH pulses and the consequent pulses in LH and FSH are not normal in size and frequency, ovulation does not occur. In patients with weight-related amenorrhea, for example, gonadotropin pulsation is abnormal, a state of affairs that can be reversed by the use of the opiate antagonist naloxone. It is clear, therefore, that opiates have an important role in the control of GnRH and that normal opiate tone is essential for normal ovulatory cycles.

Although the major feedback control of LH production is from gonadal steroids, recent studies have demonstrated that inhibin is important in the independent feedback control of FSH production. Over the last few years, we have accumulated a vast amount of knowledge on the structure and function of gonadal inhibin. Inhibin is a peptide hormone and cDNA probes have enabled the cloning of the genes encoding the inhibin α and β subunits. There is a high degree of similarity between species with regard to both subunits as well as a degree of homology between the α and β subunits themselves. Furthermore, the β subunit shows a striking homology with several other peptides, including transforming growth factor β and Mullerian inhibin substance. It seems that inhibin is part of a family of related peptides with a large number of biological actions, including intragonadal paracrine effects such as direct effects on progesterone secretion in granulosa cells.

Our increased knowledge of the control of gonadotropin secretion has already had profound effects on our understanding of more general physiological mechanisms, and it seems that we are on the threshold of the discovery of further important aspects of reproduction with our knowledge of inhibin and its related peptides.

J. A. H. Wass

REGULATION OF GONADOTROPIN SECRETION: A 1987 PERSPECTIVE

H. G. Burger

Medical Research Centre and Department of Endocrinology,
Monash Medical Centre, Prince Henry's Hospital Campus,
Melbourne, Victoria, Australia

INTRODUCTION

The secretion of the pituitary gonadotropins, follicle stimulating hormone (FSH) and luteinizing hormone (LH), is under the control of gonadotropin-releasing hormone (GnRH) from the hypothalamus and is modulated by the feedback effects of both gonadal steroids and gonadal peptides. This perspective focuses on the role of ovarian inhibin in the feedback control of gonadotropin secretion, particularly the secretion of FSH. Evidence is presented in favor of three major hypotheses: (i) that FSH and LH are regulated differentially in certain circumstances, (ii) that gonadal inhibin is involved in FSH regulation, and (iii) that FSH and inhibin form a classic endocrine closed loop feedback system.

DIFFERENTIAL REGULATION OF FSH AND LH

Under a number of physiological and pathological circumstances, the secretion of FSH and LH appears to be regulated by separate mechanisms. In men with seminiferous tubule failure leading to male infertility, for example, it can be observed that FSH levels are elevated in the presence of normal values for testosterone and LH. This implies that a feedback factor other than testosterone is being produced in lower than normal amounts as a result of the seminiferous tubule damage, and that this factor particularly affects the secretion of FSH. Leydig cell reserve capacity (1) and testosterone production rates (2) have been shown to be reduced in such individuals, but the observation of a monotropic elevation of serum FSH is consistent with a deficiency of a tubule feedback factor.

Several lines of evidence suggest that under some circumstances, FSH is constitutively secreted by the pituitary and shows little short-term dependence on GnRH, while LH secretion is intimately dependent on GnRH. In ovariectomized ewes that have undergone hypothalamo-pituitary disconnection (3), the pulsatile administration of GnRH (250 ng, 2-hourly) maintains normal castrate levels of FSH and LH. If GnRH administration is stopped, LH levels become undetectable within 6 hr, whereas FSH secretion remains elevated for several days. In long-term cultures of dispersed anterior pituitary cells (4), the rate of FSH secretion remains fairly constant for at least 28 days, while the secretion of LH rapidly falls to very low levels. The addition of inhibin to such cultures results in a major inhibition of FSH secretion, whereas more inhibin is required to cause any effect on LH. These data are compatible with the concept that FSH and LH are regulated differentially in various circumstances; FSH secretion is able to continue in the absence of GnRH and is inhibited by the gonadal peptide inhibin. LH secretion depends acutely on GnRH input and falls rapidly in its absence.

It is clear that under many physiological circumstances, common factors are involved in FSH and LH regulation. Thus GnRH leads to the acute release of both gonadotropins, although the relative amounts released change under various circumstances, e.g. predominant FSH release before puberty, LH after puberty. The sex steroids clearly inhibit the production of both gonadotropins: estradiol has a biphasic effect on both, leading to initial inhibition, followed by stimulation, which affects LH much more markedly than FSH.

The differential regulation of FSH and LH implies the existence of different regulating mechanisms and different points for regulation, and is consistent with the hypothesis that there is a gonadal factor, inhibin, which preferentially affects FSH.

INVOLVEMENT OF GONADAL INHIBIN IN FSH REGULATION

Initial evidence for a role for nonsteroidal gonadal factors in FSH regulation came from the pioneering experiments of McCullagh published more than 50 years ago (5). We obtained confirmatory evidence for the effects of crude, steroid-free extracts of the testis when it was demonstrated that preparations made from bovine testes were capable of differential suppression of FSH secretion in castrated male sheep, while similarly prepared extracts of liver were ineffective (6). On the basis of these observations, a bioassay for inhibin was established using cultured pituitary cells in vitro (7). The initial assays used

GnRH-stimulated gonadotropin secretion into the medium as the assay endpoint. Subsequently, a bioassay was developed based on measurement of the cell contents of FSH and LH in cultured pituitary cells (8). This assay exhibited satisfactory characteristics in terms of precision, sensitivity and specificity, and formed the basis of subsequent studies of inhibin purification and its physiology in the immature female rat.

Our initial attempts to purify inhibin used testicular extracts. Because follicular fluid contained much larger quantities of inhibin and was readily available, we chose to use this for our more recent attempts to purify and characterize inhibin. Bovine follicular fluid was subjected to gel filtration, reversed phase high performance liquid chromatography (HPLC) and preparative polyacrylamide gel electrophoresis (PAGE) with resultant purification of bovine follicular fluid inhibin to homogeneity (9). The purification of porcine follicular fluid inhibin was achieved shortly afterwards (10,11), and we have more recently purified ovine inhibin (12). Inhibin was initially isolated as a 58 kDa protein and subsequently as a 31 kDa protein, the latter being derived by proteolytic cleavage from the former (13). Analysis of highly purified inhibins using PAGE under reducing conditions showed that the molecules are composed of two subunits, termed α and β (9-12). The α-subunit of 58 kDa bovine inhibin (a 43 kDa protein) is cleaved to give a 20 kDa α-subunit in 31 kDa inhibin, the β-subunit remaining unchanged.

The final proof for the involvement of inhibin in FSH regulation required the demonstration that highly purified preparations would selectively lower FSH levels in vivo. The in vitro selective action on FSH had already been demonstrated, as it was this characteristic in the bioassay that was used to follow the purification. We recently injected purified 31 kDa bovine inhibin into ovariectomized ewes and observed selective suppression of serum FSH levels (14).

In order to make preliminary observations on the mechanisms by which inhibin lowers the rate of FSH secretion, experiments were performed in the hypothalamo-pituitary disconnected ewe. Three injections of bovine follicular fluid (which contains predominant inhibin bioactivity as determined by bioassay) were administered over a 30-hr period, and mRNAs for the common α-glycoprotein hormone subunit and for β-LH and β-FSH subunits were measured using Northern analysis. A selective suppression of FSH β-subunit mRNA was observed (15).

An unexpected development during the purification of porcine follicular fluid inhibin was the demonstration of FSH stimulating acitivity in fractions from the chromatographic columns (16,17). This was shown to be due to the presence of β-dimers, which have FSH stimulating activity, in contrast to the α,β-

heterodimers, which show the FSH inhibitory activity characteristic of inhibin. The FSH stimulatory dimers have been termed activins.

The availability of purified inhibin allowed the establishment of the N-terminal sequence of both subunits and the subsequent construction of cDNA probes so that the genes encoding the inhibin subunits could be cloned. A single type of both α- and β-subunits was present in the bovine ovary, but two types of β-subunit, termed β_A and β_B, were found in the porcine ovary. Cloning of the genes allowed the establishment of the entire amino acid sequence of the inhibin α- and β-subunits (18-20) in various species including cow, pig, rat, sheep and human. This led to the observation that there is a high degree of interspecies similarity in the α-subunits (80%) in the β_A-subunit (almost 100%- only one amino acid substitution in the ovine β_A-subunit) and more than 95% for the β_B-subunits. There is also a degree of homology between the α- and β-subunits themselves.

Examination of the amino acid sequence of the β-subunit revealed that there are striking homologies with the sequences of transforming growth factor-β (18) and Mullerian inhibiting substance (21). More recently, the inhibin β-subunit has been found to have sequence homology with the gene product of the decapentaplegic complex in Drosophila. This protein is involved in morphogenesis (22). These amino acid sequence homologies have led to the concept that inhibin forms part of a family of related peptides with widespread biological activities. There is evidence that inhibin and inhibin-related peptides have important intragonadal actions (23-25), including effects on Leydig cell steroidogenesis and on aromatase activity and progesterone secretion in granulosa cells. The widespread effects of the inhibin-related peptide family are exemplified by the recent isolation of a protein from a human cell line that exhibits extensive differentiation inducing activity toward Friend leukemia cells, and is identical to the inhibin β-dimer (26).

On the basis of the elucidation of the inhibin structure, a recent Serono Symposium proposed that inhibin be redefined as "a glycoprotein hormone consisting of two dissimilar disulphide-linked subunits, termed α and β, which inhibit pituitary gonadotrophin production and/or secretion, preferentially FSH" (27).

Thus the hypothesis that FSH regulation involves gonadal inhibin is supported by the establishment of the structure of inhibin and the demonstration that it preferentially suppresses FSH and is capable of that effect both in vitro and in vivo.

THE FSH-INHIBIN CLOSED LOOP FEEDBACK SYSTEM

To provide evidence that FSH and inhibin form a classic endocrine closed loop feedback system, it is necessary to show not only that inhibin suppresses the secretion of FSH, but also that FSH stimulates inhibin secretion, both in vivo and in vitro. It would be expected that endogenous FSH levels should be suppressed when inhibin levels are stimulated, e.g. by exogenous FSH. It would also be expected that there should be different circumstances when FSH and inhibin are either positively or negatively correlated. Finally, selective reduction in inhibin levels (e.g. in patients with ovarian failure whose sex steroid deficiency has been corrected by appropriate supplementation) should be accompanied by a rise in the level of FSH.

FSH Stimulation of Inhibin Secretion In Vitro

Much evidence has been obtained indicating the stimulation of inhibin production in vitro from cultured granulosa cells by FSH (28,29) and by IGF-I (29,30). FSH effects can be mimicked by cyclic AMP derivatives and by activators of adenylate cyclase. Inhibin activity in the culture medium has been measured by both bioassay and radioimmunoassay (28-30).

FSH Stimulation of Inhibin Secretion In Vivo

The regulation of inhibin production by FSH has been studied extensively both in animal models and in women. FSH, but not LH (or hCG), when administered to immature female rats causes an increase in the ovarian and peripheral blood inhibin concentrations as measured by bioassay (31; also summarized in 28). Only FSH-inhibitory activity is found in the bioassay, suggesting that any activin present is not exerting a significant physiological role in FSH regulation. There is a concomitant decline in circulating FSH levels consistent with the hypothesis that inhibin production from the ovary, stimulated by exogenous FSH, suppresses endogenous FSH levels. Pregnant Mare Serum Gonadotropin (PMSG) is particularly effective in producing major increases in circulating inhibin. In animals ovariectomized 40 hr after PMSG administration, inhibin disappears from the circulation and is undetectable 8 hr after castration (32). The disappearance curve can be represented as the sum of two exponential functions, the initial half-time being approximately 18 min. Thus it is clear that the biological activity measurable in serum following FSH administration is of gonadal origin.

In order to elucidate the mechanism by which FSH stimulates inhibin secretion, mRNA was extracted from the ovaries of immature female rats following PMSG stimulation. PMSG induced an increase in mRNA for inhibin α-

subunit and there was a close correlation between ovarian mRNA levels, circulating serum inhibin, and ovarian estradiol (33,34).

The study of inhibin regulation in women required the development of a radioimmuoassay for inhibin (35,36). The first assay used an antiserum raised against bovine 58 kDa inhibin. The antiserum so obtained was able to bind [^{125}I] inhibin of both molecular sizes. The antiserum was also capable of neutralizing bovine follicular fluid inhibin biological activity in vitro, and there was a close relationship between neutralizing capacity and inhibin binding capacity. Measurement of the endogenous serum FSH levels in the immunized rabbit showed a rise in FSH coincident with the rise in antiserum binding capacity, thus providing evidence for a physiological role for endogenous inhibin in this rabbit. The evidence was also suggestive that rabbit and bovine inhibin share common immunological determinants (35). It was established that 31 kDa inhibin was stable in serum, whereas 58 kDa inhibin showed evidence of proteolytic breakdown (35). Hence 31 kDa inhibin was chosen as tracer for the routine radioimmunoassay. Subsequent antisera have been prepared against 31 kDa inhibin, and the antiserum most recently obtained has proved useful and sensitive for the measurement of inhibin in human, ovine and rat serum.

The initial data in women were obtained from patients undergoing superovulation with human menopausal gonadotropins and clomiphene citrate as part of an in vitro fertilization (IVF) program. It was found that ovarian stimulation with FSH resulted in rising levels of inhibin parallel to and closely correlated with those of estradiol (37). There was also a close correlation between serum inhibin levels and the number of follicles found at the subsequent laparoscopy.

The above evidence therefore indicates that FSH is capable of stimulating inhibin secretion both in vivo and in vitro, and the animal data indicate suppression of endogenous FSH when circulating inhibin levels are raised.

Physiological Changes in Circulating Inhibin Concentrations

The levels of inhibin have now been measured throughout the normal human menstrual cycle, in pregnancy (including placental tissues), and in puberty. Inhibin levels remain relatively constant throughout most of the follicular phase of the menstrual cycle. There is a rise at the time of the midcycle gonadotropin surge, a short-lived fall and a further rise to reach the highest levels observed in the menstrual cycle during the midluteal phase. In the luteal phase, inhibin levels are positively correlated with progesterone (36).

When inhibin, FSH, LH, estradiol and progesterone levels were compared in the late luteal phase and in the early follicular phase of the same cycles, it was noted that there was a progressive fall in serum inhibin from the end of the luteal phase into the first few days of the follicular phase, with an inverse rise in serum FSH and no significant change in LH. The changes in inhibin paralleled those of progesterone and estradiol. The data were interpreted as suggesting that the early follicular phase rise in FSH stimulates estradiol and inhibin production as the follicle develops and is itself a consequence of release from inhibition by falling inhibin levels. Estradiol secretion from the developing follicle may be sufficient in itself to suppress FSH in association with relatively stable levels of inhibin. Falling FSH in the latter half of the follicular phase prevents a further rise in inhibin and thus a feedback balance is maintained. The FSH rise at midcycle may stimulate inhibin secretion, although the levels of inhibin begin to rise just before serum FSH, suggesting that the large dominant follicle may produce sufficient inhibin to raise circulating concentrations. Inhibin levels fall rapidly during early luteinization, but inhibin and sex steroid levels subsequently rise, resulting in FSH and LH suppression until the demise of the corpus luteum.

Examination of inhibin levels during the luteal phase of hyperstimulated cycles in the IVF program confirmed the elevation of inhibin during that time (39). The luteal phase pattern of secretion raised the possibility that inhibin is a secretory product of the human corpus luteum, as had been suggested by the demonstration of the inhibin α-subunit mRNA in the rat corpus luteum (33). Strongly confirmatory evidence was provided by the demonstration of inhibin α- and β_A-subunit mRNAs in extracts of human corpora lutea and the associated demonstration of both biological and immunological activity in that tissue (38). In subjects who became pregnant in the IVF program, inhibin levels increased in parallel with increasing levels of the sex steroids and of hCG (39). The demonstration of a rise in inhibin in early pregnancy prompted an examination of the placenta as a possible additional source of the protein. Inhibin biological and immunological activity was demonstrable in the placenta, as was inhibin α mRNA (38,40,41). That the ovary was not required for the rise in inhibin in early pregnancy was demonstrated in three patients with premature ovarian failure and absence of ovarian tissue who became pregnant following oocyte donation, fertilization of the oocyte by husband's sperm, and subsequent embryo transfer. These three patients all showed a normal rise of serum inhibin during early pregnancy (42).

Inhibin levels were also measured throughout normal human puberty and rose in parallel with levels of FSH, LH, testosterone (in boys) and estradiol (in girls) (43).

Thus in a variety of physiological circumstances, inhibin levels bear a variable relationship to circulating FSH, being inversely related during the luteal phase of the cycle, during the luteal-follicular transition, and also in early pregnancy. FSH and inhibin levels rise in parallel at the end of the follicular phase of the cycle and throughout normal human puberty. Such observations are consistent with the hypothesis that inhibin and FSH form a closed loop feedback system, with FSH driving inhibin in some circumstances and being suppressed by inhibin in others.

Selective Reduction in Circulating Inhibin

A small group of patients who lack endogenous ovarian function, as referred to above, have been treated in the IVF program. In such patients, a regimen of steroid replacement is given to mimic the physiological levels of estradiol and progesterone throughout the normal menstrual cycle. These steroid levels lead to a midcycle LH surge and produce normal secretory appearances on endometrial biopsy during the luteal phase. In patients so treated, FSH levels remain elevated despite adequate steroid replacement (44). This represents a circumstance in which inhibin has been selectively removed with the preservation of normal sex steroid feedback, and suggests that the protein is normally relevant in the physiological feedback regulation of FSH.

INHIBIN LEVELS IN DISEASED STATES

Limited observations of circulating inhibin concentrations have been made in patients with polycystic ovarian syndrome. Some earlier studies of this disorder had suggested that inhibin levels might prove to be elevated (45). We have measured inhibin twice weekly for 4 weeks in five patients with polycystic ovarian syndrome and have also measured the response of circulating inhibin to a GnRH agonist and to exogenously administered human menopausal gonadotropin. Basal inhibin levels in patients with polycystic ovarian syndrome were in the lower range of concentrations found during the early to mid-follicular phase of the menstrual cycle. Inhibin levels responded normally to endogenous and exogenous gonadotropin stimulation (46). Thus we conclude that inhibin levels are not elevated in polycystic ovarian syndrome, in which the pattern of gonadotropin levels is explicable on the basis of steroid feedback effects.

POTENTIAL COMMERCIAL APPLICATIONS OF INHIBIN

One potential commercial application of inhibin would be to use it as an immunogen to increase fertility in domestic animals. Preliminary experiments have been done using an inhibin α-subunit immunogen produced by recombinant DNA techniques. Ewes immunized with this material showed a highly significant increase in ovulation rate (47).

SUMMARY

As described above, inhibin has now been purified, partially sequenced and cloned, and its structure has been established. Bioassays and radioimmunoassays have been used to establish the major tenets of the inhibin hypothesis and to provide strong evidence for its physiological role in the female. Inhibin and FSH form a classic closed loop endocrine feedback system in which FSH stimulates inhibin production and inhibin in turn restrains FSH synthesis and secretion. Inhibin has promise in animal husbandry. A whole new field of investigation - the inhibin-related peptide family - has been opened up by the finding of FSH stimulatory activity in the β-subunit dimers and by the finding of homologies between inhibin and a variety of other biologically active peptides.

ACKNOWLEDGMENTS

The work was supported by grants from the National Health and Medical Research Council of Australia and Biotechnology Australia Pty Ltd. The collaboration of numerous colleagues is acknowledged, particularly David de Kretser, Jock Findlay, Frank Morgan, Milton Hearn, Victor Lee, Ron Carson, Rob McLachlan, David Robertson, Susan Davis, Paul Farnworth, Helen Quigg, Russell Scott, Zygmund Krozowski, Rob Forage, Andy Stewart, Laura Hutchinson, Zhang Zhiwen, Adrian Herington; Joan Williams typed the manuscript.

REFERENCES

1. de Kretser DM, Burger HG, Hudson B, Keogh EJ (1975) The HCG stimulation test in men with testicular disorders. Clin Endocrinol 4: 591-596
2. Booth JD, Merriam GR, Clark RV, Loriaux L, Sherins RJ (1987) Evidence for Leydig cell dysfunction in infertile men with a selective increase in plasma follicle-stimulating hormone. J Clin Endocr Metab 64: 1194-1198
3. Clarke IJ, Cummins JT, de Kretser DM (1983) Pituitary gland function after disconnection from direct hypothalamic influences in the sheep. Neuroendocrinology 36: 376-384
4. Farnworth PG, Robertson DM, de Kretser DM, Burger HG (1988) Effects of 31 kilodalton bovine inhibin on follicle-stimulating hormone and luteinizing hormone in rat pituitary cells in vitro: actions under basal conditions. Endocrinology 122: 207-213
5. McCullagh DR (1932) Dual endocrine activity of the testes. Science 76: 19-20
6. Keogh EJ, Lee VWK, Rennie GC, Burger HG, Hudson B, de Kretser DM (1976) Selective suppression of FSH by testicular extracts. Endocrinology 98: 997-1004
7. Eddie LW, Baker HWG, Higginson RE, Hudson B (1979) A bioassay for inhibin using pituitary cell cultures. J Endocr 81: 49-60
8. Scott RS, Burger HG, Quigg H (1980) A simple and rapid in vitro bioassay for inhibin. Endocrinology 107: 1536-1542
9. Robertson DM, Foulds LM, Leversha L, Morgan FJ, Hearn MTW, Burger HG, Wettenhall REH, de Kretser DM (1985) Isolation of inhibin from bovine follicular fluid. Biochem Biophys Res Comm 126: 220-226
10. Miyamoto K, Hasegawa Y, Fukuda M, Nomura M, Igarashi M, Kaneawa K, Matsuo H (1985) Isolation of porcine follicular fluid inhibin of 32K daltons. Biochem Biophys Res Comm 129: 396-403

11. Ling N, Ying S-Y, Ueno N, Esch F, Denoroy L, Guillemin R (1985) Isolation and partial characterization of a M_r 32,000 protein with inhibin activity from porcine follicular fluid. Proc Natl Acad Sci USA 82: 7217-7221
12. Leversha LJ, Robertson DM, de Voss FL, Morgan FJ, Hearn MTW, Wettenhall REH, Findlay JK, Burger HG, de Kretser DM (1987) Isolation of inhibin from ovine follicular fluid. J Endocrinol 113: 213-221
13. Robertson DM, de Vos FL, Foulds LM, McLachlan RI, Burger HG, Morgan FJ, Hearn MTW, de Kretser DM (1987) Isolation of a 31kD form of inhibin from bovine follicular fluid. Mol Cell Endocrinol 44: 271-277
14. Findlay JK, Robertson DM, Clarke IJ (1987) Influence of dose and route of administration of bovine follicular fluid and the suppressive effect of purified 31 kilodalton bovine inhibin on plasma FSH in ovariectomized ewes. J Reprod Fert 890: 455-461
15. Mercer JE, Clements JA, Funder JW, Clarke IJ (1987) Rapid and specific lowering of pituitary FSH β mRNA levels by inhibin. Mol Cell Endocrinol 53: 251-254
16. Ling N, Ying S-Y, Ueno N, Shimasaki S, Esch F, Hotta M, Guillemin R (1986a) Pituitary FSH is released by a heterodimer of the β-subunits from the two forms of inhibin. Nature 321: 779-782
17. Vale W, Rivier J, Vaughan J, McClintock R, Corrigan A, Woo W, Karr D, Spiess J (1986) Purification and characterization of an FSH releasing protein from porcine ovarian follicular fluid. Nature 321: 776-779
18. Mason AJ, Hayflick JS, Ling N, Esch F, Ueno N, Ying S-Y, Guillemin R, Niall H, Seeburg PH (1985) Complementary DNA sequences of ovarian follicular fluid inhibin show precursor structure and homology with transforming growth factor-B. Nature 318: 659-663
19. Forage RG, Ring JM, Brown RW, McInerney BV, Cobon GS, Gregson RP, Robertson DR, Morgan FR, Hearn MTW, Findlay JK, Wettenhall REH, Burger HG, de Kretser DM (1986) Cloning and sequence analysis of cDNA species coding for the two subunits of inhibin from bovine follicular fluid. Proc Natl Acad Sci USA 83: 3091-3095
20. Mason AJ, Niall HD, Seeburg PH (1986) Structure of two human ovarian inhibins. Biochem Biophys Res Comm 135: 957-964
21. Cate RL, Mattaliano RJ, Hession C, Tizard R, Farber NM, Cheung A, Ninfa EG, Frey AZ, Gash DJ, Chow EP, Fisher RA, Bertonis JM, Torres G, Wallner BP, Ramachandran KL, Ragin RC, Manganaro TF, MacLaughlin DT, Donahoe PK (1986) Isolation of the bovine and human genes for mullerian inhibiting substance and expression of the human gene in animal cells. Cell 45: 685-698
22. Padgett RW, St Johnston RD, Gelbart WM (1987) A transcript from a Drosophila pattern gene predicts a protein homologous to the transforming growth factor-B family. Nature 325: 81-84
23. Hsueh AJW, Dahl KD, Vaughan J, Tucker E, Rivier J, Bardin CW, Vale W (1987) Heterodimers and homodimers of inhibin subunits have different paracrine action in the modulation of luteinizing hormone-stimulated androgen biosynthesis. Proc Natl Acad Sci USA 84: 5082-5086
24. Ying S-Y, Becker A, Ling N, Ueno N, Guillemin R (1986) Inhibin and beta type transforming growth factor (TGF_B) have opposite modulation effects on the follicle stimulating hormone (FSH)-induced aromatase activity of cultured rat granulosa cells. Biochem Biophys Res Comm 136: 969-975
25. Hutchinson LA, Findlay JK, de Vos FL, Robertson DM (1987) Effects of bovine inhibin, transforming growth factor-β and bovine activin-A on granulosa cell differentiation. Biochem Biophys Res Comm 146: 1405-1412
26. Eto Y, Tsuji T, Takezawa M, Takano S, Yokogawa Y, Shibai H (1987) Purification and characterization of erythroid differentiation factor (EDF) isolated from human leukemia cell line THP-1. Biochem Biophys Res Comm 142: 1095-1103
27. Burger H, Findlay J, de Kretser D, Igarashi M (eds) (1987) Inhibin - Non Steroidal Regulation of Follicle Stimulating Hormone. New York: Raven Press
28. Burger HG, Carson RS, Davis S, Zhiwen Z (1987) Regulation of inhibin secretion from the rat ovary. In: Burger H, Findlay J, de Kretser D, Igarashi M (eds) Inhibin - Non Steroidal Regulation of Follicle Stimulating Hormone. New York: Raven Press: pp 191-202

29. Bicsak TA, Tucker EM, Cappel S, Vaughan J, Rivier J, Vale W, Hsueh AJ (1986) Hormonal regulation of granulosa cell inhibin biosynthesis. Endocrinology 119: 2711-2719
30. Zhiwen Z, Carson RS, Herington AC, Lee VWK, Burger HG (1987) FSH and somatomedin-C stimulate inhibin production by rat granulosa cells in vitro. Endocrinology 120: 1633-1638
31. Lee VWK, McMaster J, Quigg H, Findlay JK, Leversha L (1981) Ovarian and peripheral blood inhibin concentrations increase with gonadotrophin treatment in immature rats. Endocrinology 108: 2403-2405
32. Lee VWK (1983) In: Greenwald GS, Terranova PF (eds) Factors Regulating Ovarian Function. New York: Raven Press: pp 157-161
33. Davis SR, Dench F, Nikolaidis I, Clements JA, Forage RG, Burger HG (1986) Inhibin A-subunit gene expression in the ovaries of immature female rats is stimulated by pregnant mare serum gonadotrophin. Biochem Biophys Res Comm 138: 1196-1202
34. Davis SR, Krozowski Z, Carson R, Burger HG (1987) Inhibin messenger RNA levels are correlated with oestradiol in PMSG stimulated rat ovaries. In: Proceedings of the 19th Annual Meeting of the Australian Society for Reproductive Biology, Abstract No. 151
35. McLachlan RI, Robertson DM, Burger HG, de Kretser DM (1986) The radioimmunoassay of bovine and human follicular fluid and serum inhibin. Mol Cell Endocrinol 46: 175-185
36. McLachlan RI, Robertson DM, Healy DL, Burger HG, de Kretser DM (1987) Circulating immunoreactive inhibin levels during the normal human menstrual cycle. J Clin Endocrinol Metab 65: 954-961
37. McLachlan RI, Robertson DM, Healy DL, de Kretser DM, Burger HG (1986) Plasma inhibin levels during gonadotropin-induced ovarian hyperstimulation for IVF: a new index of follicular function? Lancet i: 1233-1234
38. Davis SR, Krozowski Z, McLachlan RI, Burger HG (1987) Inhibin gene expression in the human corpus luteum. J Endocrinol 115: R21-R23
39. McLachlan RI, Healy DL, Robertson DM, Burger, HG, de Kretser DM (1987) Circulating immunoreactive inhibin in the luteal phase and early gestation of women undergoing ovulation induction. Fertil Steril 48: 1001-1005
40. McLachlan RI, Healy DL, Robertson DM, Burger HG, de Kretser DM (1986) The human placenta: a novel source of inhibin. Biochem Biophys Res Comm 48: 51-58
41. Mayo KE, Cerelli GM, Spiess J, Rivier J, Rosenfeld MG, Evans RM, Vale W (1986) Inhibin A-subunit cDNAs from porcine ovary and human placenta. Proc Natl Acad Sci USA 83: 5849-5853
42. McLachlan RI, Healy DL, Lutjen PJ, Findlay JK, de Kretser DM, Burger HG (1987) The maternal ovary is not the source of circulating inhibin levels during human pregnancy. Clin Endocrinol 27: 663-668
43. Burger H, McLachlan R, Bangah M, Quigg H, Findlay J, Robertson D, de Kretser D, Warne G, Werther G, Cook J, Greco S, Yong A, Smith P (1987) Serum inhibin concentrations throughout normal male and female puberty. Proceedings of the 30th Annual Meeting of the Endocrine Society of Australia, Abstract No. 134
44. Lutjen PJ, Findlay JK, Trounson AO, Leeton J, Chan LK (1986) The effect of plasma gonadotropins of cyclic steroid replacement in women with premature ovarian failure. J Clin Endocr Metab 62: 419-423
45. Tanabe K, Kowaguchi Y, Kido S, Fukuda K, Konishi Y, Shiraishi S, Nakamura Y, Iizuka R (1987) The physiology and pathophysiology of inhibin in women. In: Burger HG, Findlay JK, de Kretser DM, Igarashi M (eds) Inhibin - Non Steroidal Regulation of Follicle Stimulating Hormone. New York: Raven Press: pp 233-238
46. Buckler HM, McLachlan RI, MacLachlan VB, Healy DL, Burger HG (1988) Serum inhibin levels in polycystic ovary syndrome: basal levels and response to LHRH agonist and exogenous gonadotropin administration. J Clin Endocrinol Metab. (in press)
47. Forage RG, Brown RW, Oliver KJ, Atrache BT, Devine PL, Hudson GC, Goss NH, Bertram KC, Tolstoshev P, Robertson DM, de Kretser DM, Doughton B, Burger HG, Findlay JK (1987) Immunisation against an inhibin subunit produced by recombinant DNA techniques results in increased ovulation rate in sheep. J Endocrinol 114: R1-R4

OPIOID PEPTIDES INHIBIT THE NORADRENERGIC INPUT TO THE ROSTRAL HYPOTHALAMUS TO REDUCE LH SECRETION IN FASTED RATS

S. Hollingsworth, S. Mansfield and R. G. Dyer

Department of Neuroendocrinology
AFRC Institute of Animal Physiology and Genetics Research
Cambridge, England, U.K.

There is considerable evidence that opioids inhibit luteinizing hormone (LH) secretion via influences upon the noradrenergic system (1,2). The noradrenergic input to the hypothalamus plays an important role in triggering the preovulatory LH surge (3) and drugs that block noradrenaline receptors, or deplete the hypothalamic content of noradrenaline, prevent naloxone-induced LH release (4). By contrast, intraventricular administration of noradrenaline can override morphine-induced suppression of LH (5). This indirect evidence for opioid control of noradrenergic transmission in the hypothalamus has been confirmed by recent in vitro experiments that have demonstrated that noradrenaline release in the preoptic area is inhibited by opioids (6). Luteinizing hormone secretion and the preovulatory LH surge are controlled by neurons in the preoptic and anterior hypothalamic area (PO/AH), and noradrenergic terminals have been found in the proximity of the luteinizing hormone-releasing hormone (LHRH) cell bodies (7,8). The major source of noradrenaline in the hypothalamus is provided by the ventral noradrenergic tract (VNAT)(9). Electrical stimulation of the VNAT increases hypothalamic noradrenaline turnover and evokes LH release (10). Thus this pathway from the brainstem is a likely target for the endogenous opioid influence on LH secretion.

In rats, fasting impairs LH secretion and causes reduced fertility (11,12). Fasting does not reduce the responsiveness of the pituitary to

LHRH (13), nor does it deplete the hypothalamic content of LHRH (11). Hence the impairment of LH secretion appears to be primarily caused by a failure of the hypothalamus to trigger LHRH release. The fasting-induced inhibition of LH can be reversed by systemic administration of the opioid antagonist naloxone (14,15). This suggests that fasting activates an endogenous opioid pathway to inhibit LHRH neurons, either directly or indirectly, by acting upon inputs to them.

We have used the fasted rat to test the hypothesis that endogenous opioids can inhibit LH secretion by inhibiting the noradrenergic input to the PO/AH. In particular, we have examined the effect of naloxone upon LH release evoked by stimulation of both the VNAT and the PO/AH in normally-fed rats and in rats fasted for 72 hr.

METHODS

Experiments were performed on adult female Wistar rats (250-350 g body wt), maintained in conditions of controlled temperature (21-23° C) and lighting (14 hr light, 10 hr dark; lights on at 05 00 hr). Each animal was anesthetized (ip) with a mixture of tribromoethanol and amylene hydrate (16) and a bilateral ovariectomy was performed. Four to six weeks later the rats were given two midday injections of estradiol benzoate (2 mg/kg body wt, im; Intervet, Cambridge, UK) - one on the day of the experiment and one 72 hr earlier (to mimic the events occurring before the LH surge). The animals were divided into two groups: one group had unlimited access to food pellets (normally-fed rats); the other was deprived of food for 72 hr following the first estradiol injection (fasted rats).

On the day of the experiment, each rat was anesthetized with urethane (ethyl carbamate, 1.15 g/kg body wt, ip), a jugular vein cannulated, and the rat placed in a stereotaxic frame. A concentric bipolar electrode (SNEX 100, Rhodes Medical Instruments) was positioned stereotaxically in the PO/AH or the VNAT to allow these structures to be stimulated electrically (0.5 mA, 1 ms matched biphasic pulses; 25 Hz: 30 sec on, 30 sec off for 15 min). At the end of each experiment, the brain was removed for histological verification of electrode positions in the PO/AH or VNAT.

Blood samples (0.3 ml, replaced with heparinized saline) were taken before and after administration of naloxone hydrochloride (2 mg/kg body wt, iv; Du Pont, UK) or an equivalent volume of saline. Stimulation was applied 15 min after the injection of naloxone or saline. Blood samples were taken immediately before the onset of stimulation, immediately after stimulation ceased (time 0), and at 5, 10 and 15 min later. Plasma samples were subsequently assayed for LH in duplicate (20 μl samples) by a radioimmunoassay system supplied in kit form (NIADDK) using the RP-2 standard.

Post-stimulation LH increments were calculated by subtraction of the prestimulus value and the four increments were added to give a measure of total evoked LH release. Two groups of normally-fed rats were subjected to PO/AH stimulation after treatment with naloxone (n=10) or saline (n=9), and another two groups (8 animals in each) were similarly treated except no pulses were delivered through the stimulating electrode (sham-stimulated controls). Groups of fasted rats were treated similarly (both stimulated groups n=11; both sham-stimulated groups n=8). This experimental protocol was also followed for the VNAT stimulation study (Normally-fed rats: naloxone + stimulation n=12, saline + stimulation n=16. Fasted rats: naloxone + stimulation n=12, saline + stimulation n=14. All four sham-stimulated groups n=6.).

RESULTS

Evoked LH Release Following PO/AH Stimulation

Fasting did not significantly alter the amount of LH released following PO/AH stimulation ($P > 0.05$; t test), confirming that the fasting-induced impairment of LH release is a consequence neither of an inability of the anterior pituitary to release LH in fasted rats, nor of an impaired capacity of the PO/AH to release adequate LHRH. Pretreatment with naloxone had no significant effects on evoked LH release in either fasted or normally-fed control animals (Fig. 1). Thus the data provided no evidence for endogenous opioids influencing LH secretion by an action on either the cell bodies or terminals of LHRH neurons, or on the pituitary itself.

Evoked LH Release Following Stimulation of the Noradrenergic Pathway to the Hypothalamus (VNAT)

Stimulation of the VNAT evoked substantially less LH release compared to PO/AH stimulation. Fasting significantly reduced evoked LH release ($P < 0.05$; t test), and this suppression was reversed by administration of naloxone ($P < 0.05$; fasted saline-treated rats vs. fasted naloxone-treated rats). The opioid antagonist brought LH levels back to those seen in saline treated normally-fed rats (Fig. 1). In normally-fed rats, naloxone

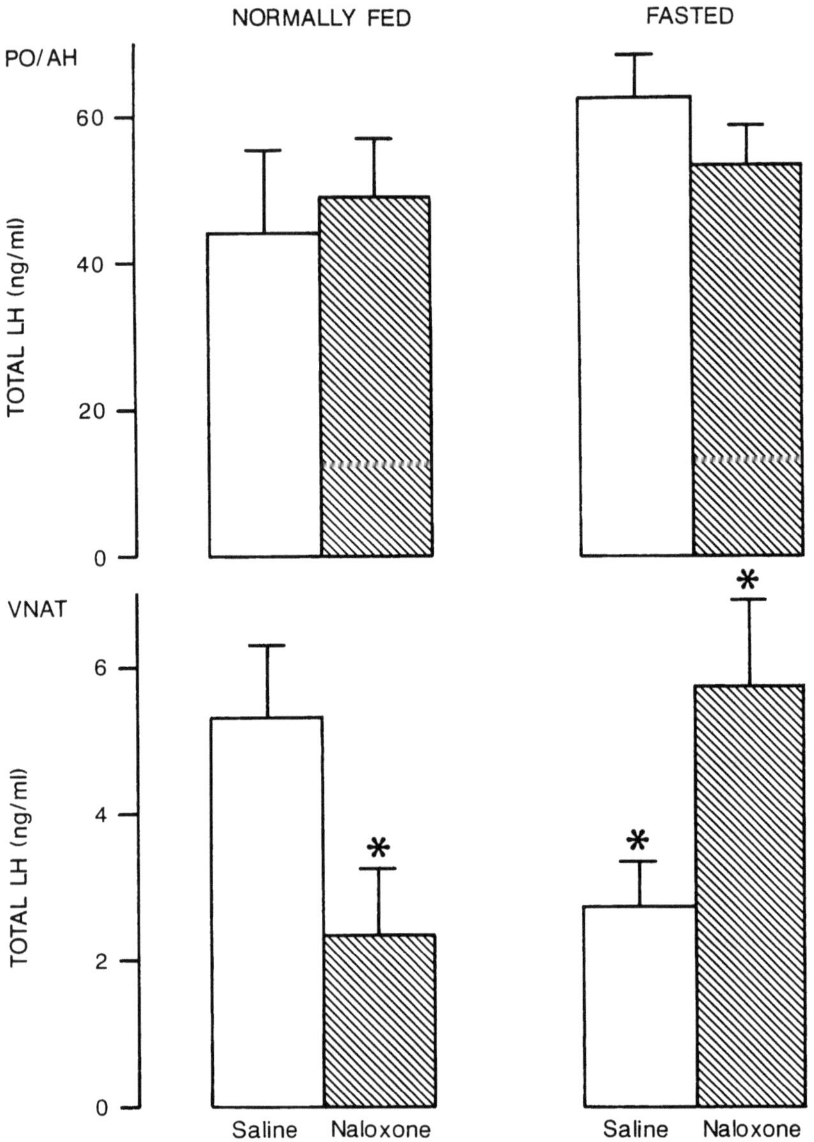

Figure 1. Total LH (ng/ml; RP2 standard) release evoked by electrical stimulation of the PO/AH (top) or VNAT (bottom) in normally-fed rats (left) and in rats fasted for 72 hr (right) following iv injection of either saline (open bars) or naloxone (shaded bars). Stimulation of the PO/AH was at least as effective in evoking LH release in fasted rats as in normally-fed rats, and evoked release was not influenced by naloxone. By contrast, VNAT stimulation was less effective in evoking LH release in fasted rats than in normally-fed rats, but this difference was reversed by administration of the opioid antagonist naloxone.
* $P < 0.05$; t test.

administration significantly reduced LH release ($P < 0.05$), i.e. the effects of naloxone upon LH release evoked by VNAT stimulation were different in normally-fed and fasted rats.

Naloxone administration had no significant effects on LH secretion in any group of sham-stimulated rats.

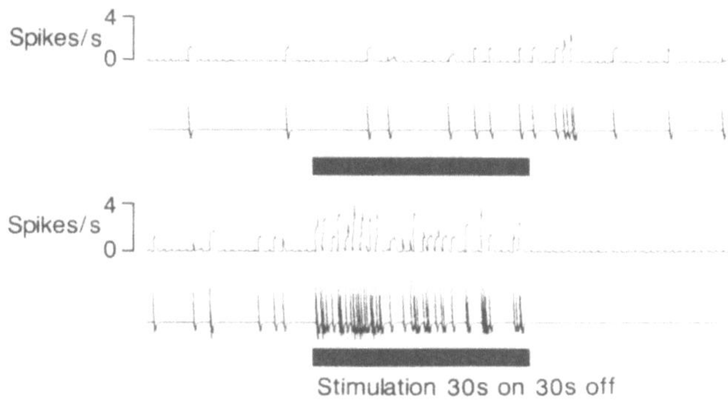

Figure 2. Polygraph records showing the response of a PO/AH neuron from a fasted rat to electrical stimulation of the VNAT (solid bars) before (top trace) and after (lower trace) iv injection of naloxone (2 mg/kg). The parameters for stimulation were identical to those used for the LH release study. The neuron was identified antidromically as projecting to the mediobasal hypothalamus, and hence is a putative LHRH neuron. Naloxone administration enhanced the excitatory response of this neuron to VNAT stimulation.

DISCUSSION

Our data indicate that the inhibitory effect of fasting upon LH secretion is mediated, in part at least, by opioid mechanisms acting upon an excitatory noradrenergic input to the neural circuitry controlling LH secretion, and not by a failure of the LHRH neurons located in the rostral hypothalamus. This conclusion is reinforced by results obtained from parallel electrophysiological studies in which we have shown that fasting alters the responsiveness of PO/AH neurons to VNAT stimulation (17) and that systemic naloxone administration increases the excitatory effects of VNAT stimulation upon PO/AH neurons recorded in fasted rats (Fig. 2). Therefore we have good evidence to support the hypothesis (15) that the impairment of LH release following a 72 hr fast is a consequence of the activation of an endogenous opioid pathway preventing noradrenergic stimulation of LHRH neurons.

Although there is evidence that opioid peptides can act on LHRH nerve terminals in the median eminence (18), our data exclude this as a major site of opioid action in the inhibition of LH secretion in fasted rats. The question to be answered next concerns the identification of the specific class of opioids responsible for modulation of the VNAT input into the rostral hypothalamus. We have shown in previous in vitro experiments (6), that β-endorphin and met-enkephalin, but not dynorphin A (1-3,6,10,13,14,18), inhibit the stimulated release of tritiated noradrenaline in the preoptic area. It follows, therefore, that μ or δ receptors are likely to be involved in the effects reported in this paper.

ACKNOWLEDGMENT

S. Hollingsworth thanks the MRC for a studentship.

REFERENCES

1. Barraclough CA, Wise PM (1982) The role of catecholamines in the regulation of pituitary luteinising hormone and follicle-stimulating hormone secretion. Endocr Rev 3: 91-119
2. Bicknell RJ (1985) Endogenous opioid peptides and hypothalamic neuroendocrine neurones. J Endocr 107: 437-446
3. Dyer RG (1985) Neural signals for oxytocin and LH release. In: Clarke JR (ed) Oxford Review of Reproductive Biology, Vol 7. Oxford: Oxford University Press: pp 223-260
4. Kalra SP, Simpkins JW (1981) Evidence for noradrenergic mediation of opioid effects on luteinising hormone secretion. Endocrinology 109: 776-786
5. Kalra SP, Gallo RV (1983) Effects of intraventricular administration of catecholamines on luteinizing hormone release in morphine-treated rats. Endocrinology 109: 776-782
6. Diez-Guerra FJ, Augood S, Emson PC, Dyer RG (1987) Opioid peptides inhibit the release of noradrenaline from slices of rat medial preoptic area. Exp Brain Res 66: 378-384
7. Hoffman GE (1985) Organisation of LHRH cells: differential apposition of neurotensin, substance P and catecholamine axons. Peptides 6: 439-461
8. Jennes L, Beckman WC, Stumpf WE, Grzanna R (1982) Anatomical relationships of serotonergic and noradrenergic projections with the GnRH system in septum and hypothalamus. Exp Brain Res 46: 331-338
9. Ungerstedt U (1971) Stereotaxic mapping of the monoaminergic pathways in the brainstem. Acta Physiol Scand Suppl 367: 1-48
10. Dean AD, Mansfield S, Dyer RG (1985) Electrical stimulation of the ventral noradrenergic tract (VNAT) releases hypothalamic noradrenaline and stimulates prolactin secretion in ovariectomised female rats treated with oestrogen. Neurosci Lett 57: 73-78

11. Dyer RG, McClure TJ (1981) Fasting inhibits oestrogen-stimulated secretion of luteinising hormone in female rats. J Physiol 319: 91-92P
12. McClure TJ, Saunders J (1985) Effects of withholding food for 0-72 h on mating, pregnancy rate and pituitary function in female rats. J Reprod Fert 74: 57-64
13. Campbell GA, Kurcz M, Marshall S, Meites J (1977) Effects of starvation in rats on serum levels of follicle stimulating hormone, luteinising hormone, thyrotropin, growth hormone and prolactin; response to LH-releasing hormone and thyrotropin releasing hormone. Endocrinology 100: 580-587
14. Corbet HJ, Dyer RG, Mansfield S (1982) Inhibition of oestrogen-stimulated secretion of luteinising hormone (LH) in fasted rats is reversed by naloxone. J Physiol 327: 37P
15. Dyer RG, Mansfield S, Corbet HJ, Dean ADP (1985) Fasting impairs LH secretion in female rats by activating an inhibitory opioid pathway. J Endocr 105: 91-97
16. Dyer RG, Weick RG, Mansfield S, Corbet H (1981) Secretion of luteinising hormone in ovariectomised adult rats treated neonatally with monosodium glutamate. J Endocr 91: 341-346
17. Dyer RG, Hollingsworth S (1985) The effect of fasting in female rats on the electrical activity of preoptic and anterior hypothalamic (PO/AH) neurones and their response to ventral noradrenergic tract (VNAT) stimulation. J Physiol 371: 178P
18. Drouva SV, Epelbaum J, Tapia-Arancibia L, Laplante E, Kordon C (1981) Opiate receptors modulate LHRH and SIRF release from mediobasal hypothalamic neurones. Neuroendocrinology 32: 163-167

ATTENUATION OF THE INHIBITORY ACTION OF CORTICOTROPIN-RELEASING HORMONE ON LUTEINIZING HORMONE SECRETION AFTER CHRONIC OPIOID RECEPTOR OCCUPATION

K. E. Nikolarakis, O. F. X. Almeida* and A. Herz

Department of Neuropharmacology, Max-Planck Institut fur Psychiatrie
Planegg-Martinsried, F.R.G.
*Institute for Pharmacology, Faculty of Veterinary Medicine
Ludwig-Maximilians University, Munich, F.R.G.

INTRODUCTION

The rat hypothalamus is rich in opioid peptides, such as β-endorphin, dynorphin, and the enkephalins (1), which contribute to the neural control of the secretion of adenohypophyseal hormones, including that of luteinizing hormone (LH) (2). The acute administration of endogenous opioid peptides results in a fall in plasma levels of LH; a similar response follows treatment with synthetic opioid peptide analogues and the opiate alkaloid morphine. Opioid receptor antagonists, such as naloxone, reverse these effects, and in some instances, produce an increase in LH secretion when given alone (3,4). Recent studies have shown that the secretion of hypothalamic opioids is under the stimulatory influence of corticotropin-releasing hormone (CRH), a 41-residue peptide synthesized in the parvocellular region of the paraventricular nucleus of the hypothalamus (5,6). CRH itself has been shown to be involved in the regulation of reproductive processes, ranging from sexual behavior and gonadotropin secretion to pregnancy. We will review some of this evidence as a prelude to our description of new data from an experiment with opioid-tolerant rats. The new data support the view that endogenous opioids provide the link between CRH and gonadotropin-releasing hormone (GnRH).

CRH-OPIOID INTERACTIONS IN THE CONTROL OF GnRH AND LH SECRETIONAND REPRODUCTIVE BEHAVIOR

There are now several lines of evidence indicating that CRH disrupts reproductive function by activating endogenous opioidergic patways. An interesting finding that emerges from studies on this subject is that CRH blocks reproduction in stressful and nonstressful conditions (CRH plays a key role in activating the pituitary-adrenal axis).

Infusion of CRH into the arcuate-ventromedial area of the hypothalamus, or into the mesencephalic central gray, abolishes GnRH-dependent lordosis behavior in estrogen-primed ovariectomized rats (7); infusion of CRH into the third cerebral ventricle suppresses masculine sexual behavior of male rats (8). The suppressive effects of CRH upon both male and female sexual behavior have been shown to be mediated through opioid receptors, since they are reduced in the presence of either antibodies against opioids or the opioid antagonist naloxone (7,8).

Exposure to stressful conditions, or disorders associated with increased activity of the hypothalamic-pituitary-adrenal axis, e.g. depression and anorexia nervosa, frequently result in dysfunction of the reproductive system. CRH synthesis (9) and release (10) increase, and GnRH and LH secretion decrease, upon exposure to stress (11); at the same time, hypothalamic opioid peptide synthesis (9) and release (12-14) also increase.

Central administration of the CRH-receptor antagonist α-helical CRF_{9-41} reverses the inhibitory effects of stress on LH, a finding that suggests a role for CRH in the modulation of LH during stress (15). Since naloxone and antisera against endogenous opioids also block the stress-induced reduction in LH secretion (3,14), it may be inferred that the stress-related CRH suppression of LH is mediated through opioidergic pathways.

The central application of CRH itself significantly decreases plasma LH levels in gonadectomized female (16) and male (17) rats. CRH also inhibits GnRH release from hypothalamic slices in vitro (18,19), while at the same time it stimulates the release of β-endorphin, dynorphin and met-enkephalin (5,6). The in vivo actions of CRH are significantly attenuated after acute or chronic pretreatment with naloxone (17), other opioid antagonists, or with antisera raised against endogenous opioid peptides (14), thus supporting the idea that opioidergic mechanisms underlie the suppression of LH by CRH. In addition, the in vitro superfusion of hypothalamic slices, or the in vivo perfusion of the mediobasal hypothalamus with the CRH receptor antagonist α-helical CRF_{9-41}, results in concomitant decreases in the release of GnRH and increases in that of opioid peptides (17).

Further evidence for CRH-opioid interactions in the control of LH was obtained in the experiment described below, using a different approach: the effectiveness of CRH in inhibiting LH secretion was tested in rats made tolerant to morphine.

EFFECTS OF CRH UPON LH SECRETION IN MORPHINE-TOLERANT RATS

We reasoned that if opioids are the link between CRH and the GnRH-LH releasing system, CRH should be less effective in suppressing LH release in rats made tolerant to an exogenous opioid receptor agonist, e.g. morphine, because the function of opioid receptors seems to be impaired during such a tolerant state (17).

Five days after having been bilaterally castrated, adult male Wistar rats (250-300 g) were implanted subcutaneously with 1 or 2 pellets of morphine. The pellets, each containing 75 mg morphine base and prepared according to the methods of Gibson and Tingstad (2), were implanted in the rats under light ether anaesthesia; control rats received placebo pellets. Each treatment group consisted of 6-8 rats. At various intervals after pellet implantation (24, 48 and 72 hr), the animals were given a subcutaneous injection of either morphine-HCl (5 or 10 mg/kg) or saline (0.5 ml), and decapitated 30 min later. Trunk blood was collected and the plasma examined for LH concentrations.

LH levels under basal conditions and following the acute injection of morphine varied among the different groups. The criteria used for determining which rats had become tolerant to chronic treatment with morphine was the presence of baseline LH levels similar to those found in placebo-treated controls and the failure of acute administration of morphine to suppress plasma LH levels further. According to these criteria, rats implanted with two pellets of morphine for 48 hr were found to be tolerant.

For the study with CRH, rats were gonadectomized and implanted with a stainless steel cannula into the lateral ventricle and a catheter into the jugular vein 5 days prior to the implantation of morphine or placebo pellets. To insure a high degree of tolerance, two pellets were implanted for 48 hr and a third pellet was implanted during the last 24 hr of the experiment, i.e. between 48 and 72 hr. The effectiveness of CRH (0.2 nmol administered via the icv cannula) in inhibiting LH secretion was tested 72 hr after implantation of the first morphine pellets (controls received 5

μl of water icv); blood was withdrawn through the jugular catheter for determination of LH levels, which were then compared for the morphine- and placebo-treated groups.

Basal plasma levels of LH were not significantly different between placebo- and morphine-treated animals, indicating that tolerance had developed in the latter group. Within 30 min following the administration of 0.2 nmol CRH, control rats showed a clear-cut suppression of baseline plasma levels ($P < 0.01$), whereas suppression was of smaller magnitude but still present ($P < 0.05$) in morphine-tolerant rats (Fig. 1). In both tolerant and non-tolerant rats, LH secretion after CRH treatment remained significantly lower than after water injections, for a period of 2 hr, in accordance with previous findings (16,17).

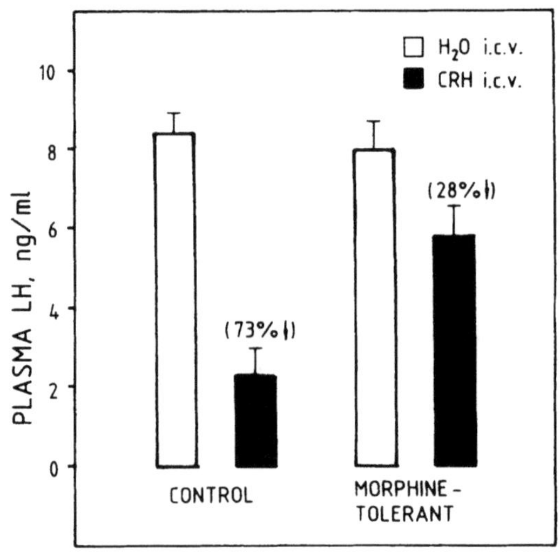

Figure 1. CRH-induced suppression in plasma LH levels in castrate male rats that were pretreated with either placebo pellets (control) or morphine pellets (morphine-tolerant). The data shown were obtained from groups of 6-8 rats, 30 min after the injection of either 5 μl of water or 0.2 nmol CRH (icv). Mean $\pm$ SEM are shown.

These data point to an opioid involvement in the mediation of suppressive effects of CRH upon LH secretion, as has also been suggested by other investigations (14,17). As other studies have also indicated (17), however, it appears that not all of the LH-suppressing actions of CRH are opioid-mediated, since in our study, some suppression (approximately 28%) was observed in morphine-tolerant rats. Thus CRH may act directly upon GnRH neurons to suppress LH secretion, or indirectly via other, nonopioid neuronal systems that impinge upon GnRH neurons. It also cannot be ruled out that some of the effects of CRH upon LH secretion may be secondary to activation of the pituitary-adrenal axis; in this regard, it is worth noting that glucocorticoids have been shown to have an inhibitory effect upon LH secretion (21,22).

Finally, two brief remarks on the validity of the tolerant castrate model used in this study are necessary. While Hahn and Fishman (23) have reported increases in whole brain opioid binding sites following castration, other authors (24) have failed to observe any changes in opioid binding in the hypothalami of castrate rats. Secondly, castration does not alter the metabolism or disposition of morphine in the rat brain (24). Thus use of morphine-tolerant castrate rats in these studies is likely to reflect the situation present in intact rats. The choice of 5-day castrate rats was made to produce elevated plasma LH levels and thus to magnify the CRH-induced suppression of plasma LH levels.

The results of these experiments provide further support of the view that endogenous opioid peptides play a role in mediating inhibitory CRH signals to GnRH-secreting neurons. The contribution of opioids in this cascade of events seems to be considerable, and thus treatments with opioid receptor antagonists may represent a useful approach for alleviating reproductive disorders due to hypersecretion of CRH.

ACKNOWLEDGMENT

Studies in the authors' laboratory were supported by SFB 220 from the Deutsche Forschungsgemeinschaft, Bonn.

REFERENCES

1. Watson SJ, Akil H, Khachaturian H, Young E, Lewis ME (1984) Opioid systems: anatomical physiological and clinical perspectives. In: Hughes J, Collier HDJ, Rance MJ, Tyers MB (eds) Opioids: Past, Present and Future. Philadelphia: Tylor and Francis: pp 145-178

2. Millan MJ, Herz A (1985) The endocrinology of the opioids. Int Rev Neurobiol 26: 1-84
3. Meites J (1984) Effects of opiates on neuroendocrine functions in animals: overview. In: Delitala G, Motta M, Serio M (eds) Opioid Modulation of Endocrine Function. New York: Raven Press: pp 53-64
4. Almeida OFX, Schulz R, Herz A (1986) Paradoxical LH and prolactin responses to naloxone after chronic treatment with morphine. J Endocrinol 108: 181-189
5. Nikolarakis KE, Almeida OFX, Herz A (1986a) Stimulation of hypothalamic -endorphin and dynorphin release by corticotropin-releasing factor (in vitro). Brain Res 399: 152-155
6. Nikolarakis KE, Almeida OFX, Herz A (1987) Factors affecting hypothalamic met-enkephalin content and release. Neuroscience 22 (Suppl.) Abstr. 197P
7. Sirinathsinghji DJS (1985) Modulation of lordosis behavior in the female rat by corticotropin releasing factor, β-endorphin and gonadotropin releasing hormone in the mesencephalic central gray. Brain Res 336: 45-55
8. Sirinathsinghji DJS (1987) Inhibitory influence of corticotropin releasing factor on components of sexual behavior in the male rat. Brain Res 407: 185-190
9. Lightman SL, Scott-Young W (1987) Changes in hypothalamic mRNA following stress and opiate withdrawal. Nature 328: 643-645
10. Plotsky P, Vale W (1984) Hemorrhage-induced secretion of corticotropin releasing factor-like immunoreactivity into the rat portal circulation and its inhibition by glucocorticoids. Endocrinology 114: 164
11. Petraglia F, Sutton S, Vale W, Plotsky P (1987) Corticotropin-releasing factor decreases plasma luteinizing hormone levels in female rats by inhibiting gonadotropin-releasing hormone release into hypophysial portal circulation. Endocrinology 1083-1088
12. Rossier J, French ED, Rivier C, Ling N, Guillemin R, Bloom FE (1977) Foot-shock induced stress increases B-endorphin levels in brain. Nature 270: 618-620
13. Millan MJ, Tsang YF, Przewlocki R, Hollt V, Herz A (1981) The influence of foot-shock stress upon brain, pituitary and spinal cord pools of immunoreactive dynorphin in rats. Neurosci Lett 211: 75
14. Petraglia F, Vale W, Rivier C (1986) Opioids act centrally to modulate stress-induced decrease in luteinizing hormone in the rat. Endocrinology 2445-2450
15. Rivier C, Rivier J, Vale W (1986) Stress-induced inhibition of reproductive functions: role of endogenous corticotropin releasing factor. Science 231: 607-609
16. Rivier C, Vale W (1984) Influence of corticotropin-releasing factor on reproductive functions in the rat. Endocrinology 914-921
17. Almeida OFX, Nikolarakis KE, Sirinathsinghji DJS, Herz A (1988) Opioid-mediated inhibition of sexual behavior and luteinising hormone secretion by corticotropin-releasing hormone. In: Dyer RG, Bicknell RJ (eds) Brain Opioid Systems and Reproduction. Oxford: Oxford University Press: (in press)
18. Nikolarakis KE, Almeida OFX, Herz A (1986) Corticotropin releasing factor inhibits gonadotropin-releasing hormone release from superfused rat hypothalami in vitro. Brain Res 377: 388-390
19. Gambacciani M, Yen SSC, Rasmussen DD (1986) GnRH release from the mediobasal hypothalamus: in vitro inhibition by corticotropin-releasing factor. Neuroendocrinology 43: 533-536
20. Gibson RD, Tingstad JF (1970) Formulation of morphine implantation pellet suitable for tolerance - physical dependence studies in mice. J Pharmaceut Sci 59: 428-437
21. Mann D, Jackson G, Blank M (1982) Influence of adrenocorticotropin and adrenalectomy on gonadotropin secretion in immature rats. Neuroendocrinology 34: 20-26

22. Rivier C, Vale W (1985) Effect of the long-term administration of corticotropin-releasing factor on the pituitary-adrenal and pituitary-gonadal axis in the male rat. J Clin Invest 75: 689-694
23. Hahn EF, Fishman J ((1985) Castration affects male rat brain opiate receptor content. Neurondocrinology 41: 60-63
24. Cicero T, O'Connor L, Bell R (1987) Reevaluation of the effects of castration on naloxone-sensitive opiate receptors in the male rat brain. Neuroendocrinology 46: 176-184

A SPECTRUM OF DISORDERS OF LH PULSATILITY IN UNDERWEIGHT WOMEN WITH MULTIFOLLICULAR OVARIES

H. D. Mason, M. Sagle, D. W. Polson, D. Kiddy,
J. Adams* and S. Franks

Department of Obstetrics and Gynaecology
St Mary's Hospital Medical School
*Department of Ultrasound, The Middlesex Hospital
London, U.K.

INTRODUCTION

In 1974 Boyar reported various abnormalities of pulsatile LH secretion in women with anorexia nervosa (1). In these and subsequent studies, it has been shown that the degree of disturbance of LH secretion is proportional to the extent of weight loss. The most severely underweight patients regress to a "prepubertal" state, with low gonadotrophin levels throughout the day and no detectable pulses (1). As these patients regain weight, the gonadotrophin levels show sleep-related rises, similar to the pattern seen in puberty.

In our initial studies, we have identified a group of women with secondary amenorrhea who were either moderately underweight at presentation or had a recent history of weight loss, but whose ovaries showed some evidence of follicular activity on ultrasound. We described this appearance as multifollicular (2). This differs from a polycystic ovary in that there is no increase in stroma and the ovaries are of normal size (2). These patients were found to have normal concentrations of serum LH. Testosterone measurements were in the normal range (2). Serum estradiol concentrations were in the normal early-mid follicular phase range, but negative progestagen withdrawal tests and a thin endometrium on ultrasound examination of the uterus provided biological evidence of estrogen deficiency. A transverse ultrasound scan of the pelvis of a woman with multifollicular ovaries can be seen in Figure 1. The ovaries are of normal size and filled with six or more cysts of 4-10 mm in diameter, scattered throughout the stroma which, in contrast to women with polycystic ovaries, is not thickened.

We postulated that these women had a hypothalamic abnormality of gonadotrophin regulation and the purpose of our present study was to investigate LH pulsatility in a group of women with multifollicular ovaries. We included assessment of diurnal changes of gonadotrophin secretion.

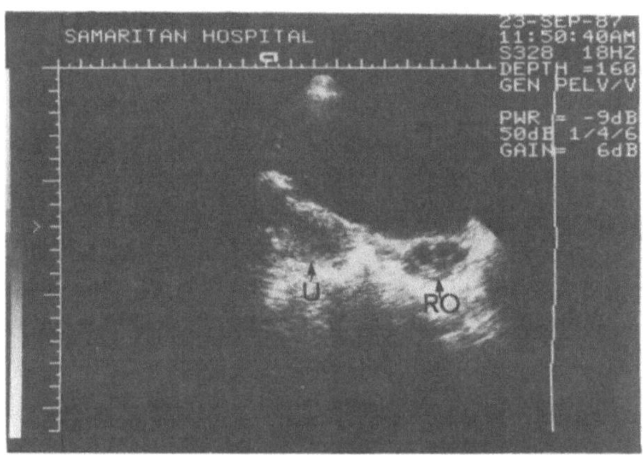

Figure 1. Transverse ultrasound scan of the pelvis of a woman with multifollicular ovaries showing right ovary (RO) and uterus (U). The ovary is of normal size and filled with follicles with no increase in stroma.

PATIENTS AND METHODS

Ten patients were studied who presented with secondary amenorrhea and had multifollicular ovaries on ultrasound. Three of the patients were of normal weight for height at presentation, but all had a recent history of weight loss. The mean body mass index (BMI) of the patients at the time of the study was 18.7 kg/m^2, the normal range being 19-25. Seven patients were studied for 8 hr, an additional three for 24 hr.

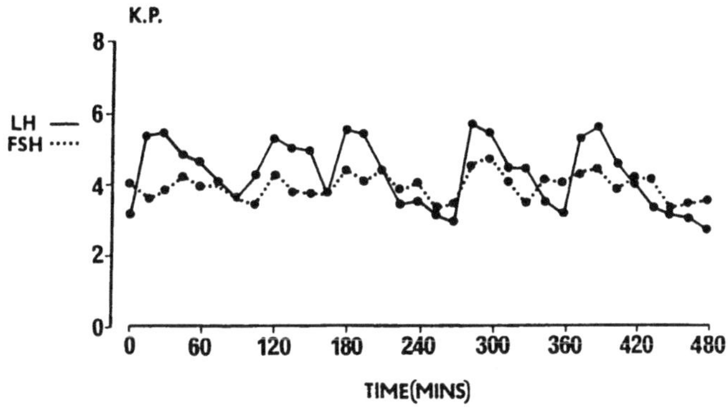

Figure 2. Serum gonadotrophin concentrations in samples taken at 15 min intervals for 8 hr in a normal woman in the early follicular phase. From ref. 3, with permission.

Five normal volunteers were studied in the early follicular phase. All were of normal weight for height and had no history of weight loss. They had regular ovulatory cycles and the ovaries were normal on ultrasound. The patients and normal volunteers were admitted to a quiet observation unit and blood samples were taken every fifteen minutes from an intravenous forearm cannula. Sleep was recorded by direct observation. Serum was stored at -20°C until the time of hormone measurement by RIA (2) and all samples from each patient were measured in the same assay.

Pulse Analysis

LH pulses were identified using a modification of the method of Santen and Bardin as described in our previous studies (4). A pulse was defined as an increase in LH from nadir to peak that was greater than 20% of the mean of all samples in the profile. This was in effect not less than 3 and up to 5 times the coefficient of variation of the assay. Because of the long half-life of FSH in serum, pulses are more difficult to detect and were not analyzed in this study. The 8 hr mean FSH, however, was calculated.

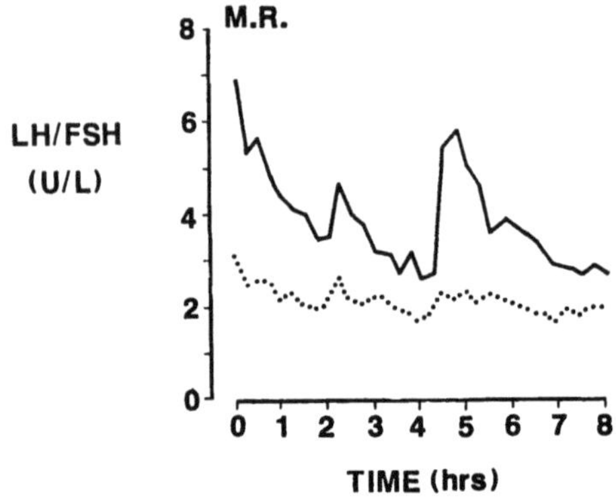

Figure 3. Gonadotrophin concentrations in serum taken at 15 min intervals for 8 hr in a woman with multifollicular ovaries, LH (—), FSH(··). Note reduced pulse frequency.

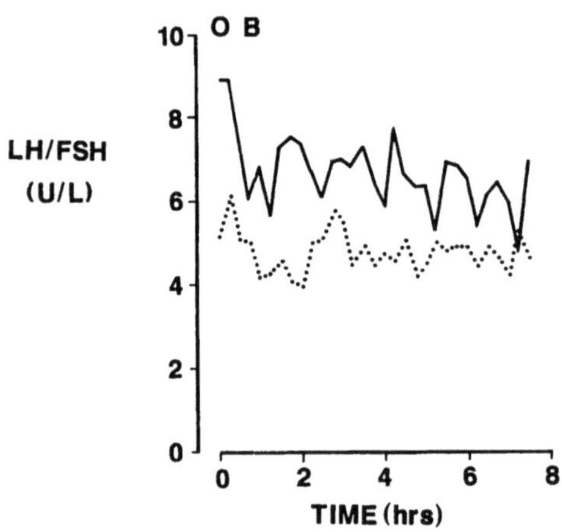

Figure 4. Gonadotrophin profile in a woman with multifollicular ovaries who later developed a normal pattern of follicles in the ovary. Note similarity to normal follicular phase profile. From ref. 3, with permission.

RESULTS

The pattern of gonadotrophin secretion in a normal woman during the early follicular phase is shown in Figure 2. There are five LH pulses across the 8 hr study period.

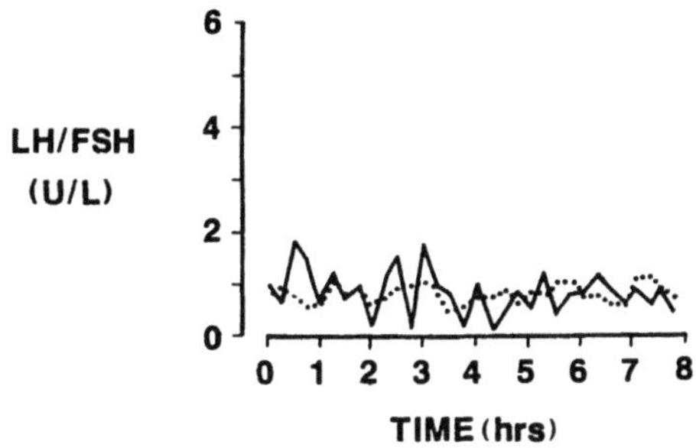

Figure 5. Gonadotrophin profile in women with multifollicular ovaries, LH (—), FSH (··). Note persistently low gonadotrophins throughout.

A typical example of the most common profile found in women with multifollicular ovaries is shown in Figure 3. Seven of the ten patients with multifollicular ovaries studied had between 1 and 3 pulses of normal amplitude during the 8 hr study period, compared to 4 or 5 in the normal group (Fig. 3). The mean FSH levels were normal.

One patient had an LH profile that was indistinguishable from that of a normal woman in the follicular phase (Fig. 4). This was despite the fact that the patient had amenorrhea at the time of the study with a BMI of only 17.5 and a multifollicular appearance on ultrasound. As in the normal profile shown in Figure 2, there are 5 LH pulses of normal amplitude in the 8 hr study period. This patient developed a normal pattern of follicles in the ovary shortly after this study was completed. In contrast, two patients were found to have persistently low gonadotrophins (Fig. 5). The mean LH level was less than 1 IU/L throughout and the peak pulse amplitude did not exceed 2 IU/L at any time. One of these patients was studied for 24 hr and was found to have low amplitude pulses throughout with no sleep-related changes in LH (Fig. 6). This pattern of gonadotrophin secretion is usually associated with small, unstimulated ovaries.

In the two other patients upon whom 24 hr studies were performed, the LH pulse frequency was normal. In contrast to normal, however, there was an increase in LH concentrations with sleep (Fig. 7). This is accounted for by an increase in pulse amplitude without a change in frequency. This is similar to the pattern described by Boyar in women gaining weight after anorexia nervosa (1) and is indicative of a stage of recovery of the hypothalamus.

The results in all patients are summarized in Table 1. There was no significant difference between the two groups in either the 8 hr mean LH

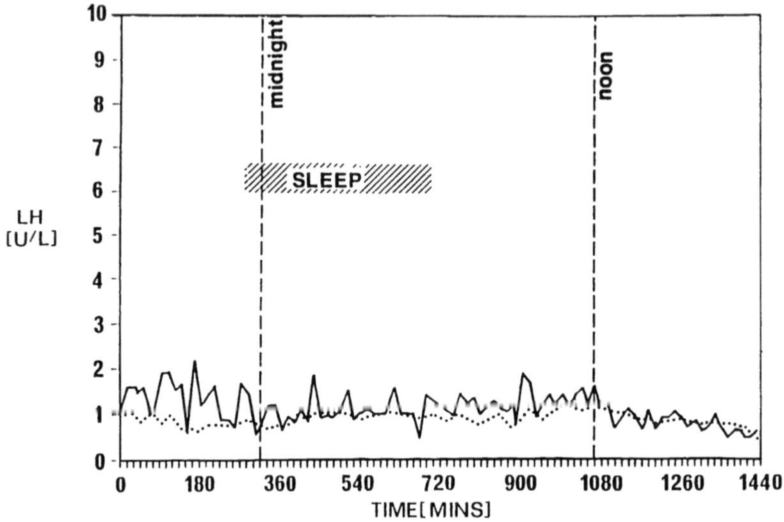

Figure 6. Serum gonadotrophin concentrations in samples taken at 15 min intervals for 24 hr from a woman with multifollicular ovaries. Note persistently low gonadotrophins throughout.

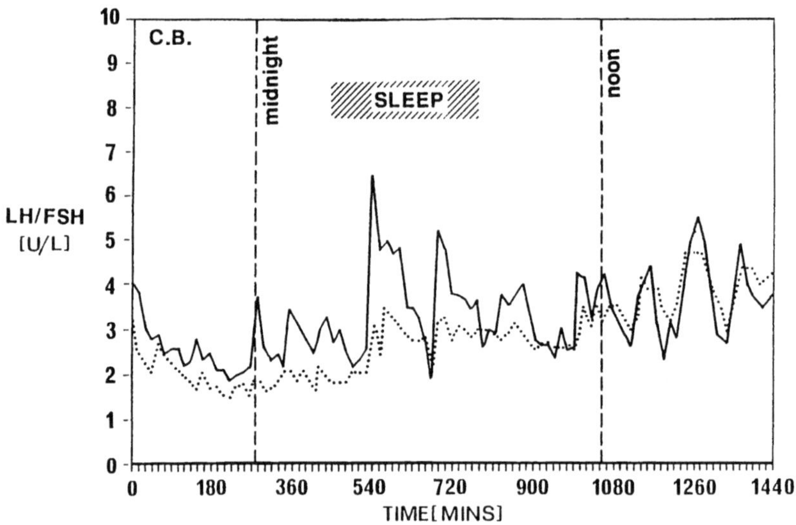

Figure 7. 24 hr gonadotrophin profile in a woman with multifollicular ovaries. Note normal LH pulse frequency, but increased LH pulse amplitude with sleep. From ref. 3, with permission.

TABLE 1

LH PULSE FREQUENCY IN WOMEN WITH NORMAL AND WITH MULTIFOLLICULAR OVARIES*

	8 hr Mean LH ± SD IU/L	Mean Frequency LH pulses/8hrs	Mean LH Amplitude IU/L	8 hr Mean FSH IU/L
Normal	5.7 ± 2.5	4.8 ± 1.5	2.2 ± 1.3	3.8 ± 1.2
Multi-follicular	5.0 ± 0.9	2.8 ± 1.6	2.1 ± 0.5	3.5 ± 1.9

*From ref. 3, with permission.

or the pulse amplitude. There was, however, a significant reduction in the mean pulse frequency in women with multifollicular ovaries. The 8 hr mean FSH levels were not different between the two groups.

DISCUSSION

Our results show that most patients with multifollicular ovaries had a gonadotrophin profile that showed LH pulses of normal amplitude, but reduced frequency. This hypothalamic abnormality of LH pulse regulation can be corrected by weight gain or by providing exogenous LHRH pulses at a physiological frequency (1,2).

It is, however, evident that there was a spectrum of disordered LH patterns that included two women with persistently low gonadotrophins and one with a profile that was similar to normal. One possible explanation for the heterogenous pattern of LH secretion in these women is that it reflects the varying degrees of weight loss, but we found no correlation between LH pulse frequency or amplitude and BMI at presentation in this group of women with mild to moderate weight loss.

The reduced LH pulse frequency with normal amplitude appears to be a stage in the recovery of the hypothalamus which follows weight gain after an episode of weight loss. There appears to be a differing sensitivity between individual patients in both the response of the hypothalamus to the degree of weight loss and that of the ovary to the LH signal.

In conclusion, our data suggest that a normal LH pulse frequency is critical for the development of a dominant follicle in the ovary.

ACKNOWLEDGMENT

Table 1 and Figures 2, 4 and 7 are reproduced from H. D. Mason, M. Sagle, D. Polson, D. Dobriansky, J. Adams, S. Franks (1988) Reduced frequency of luteinising hormone pulses in women with weight loss-related amenorrhea and multifollicular ovaries. *Clin End,* in press, published by Blackwell Scientific Publishers, Ltd., Oxford, England.

REFERENCES

1. Boyar RM, Katz J, Finkelstein JW, Kapen S, Weiner H, Weitzman ED, Hellman L (1974) Immaturity of the 24-hour luteinizing hormone secretory pattern in anorexia nervosa. New Eng J Med 291: 861-865
2. Adams J, Franks S, Polson DW, et al. (1985) Multifollicular ovaries: clinical and endocrine features and response to pulsatile luteinising hormone releasing hormone. Lancet ii, 1375-1378
3. Mason HD, Sagle M, Polson D, Dobriansky D, Adams, J, Franks S (1988) Reduced frequency of luteinising hormone pulses in women with weight loss-related amenorrhea and multifollicular ovaries. Clin End, in press
4. Mason HD, Polson DW, Jacobs HS, Adams J, Franks S (1985) Pulsatile gonadotrophin secretion in women with polycystic or multicystic ovaries. J Endocrinol 104 (suppl): 39

CHANGES IN PROLACTIN AND PROOPIOMELANOCORTIN MESSENGER RNA IN RAT PITUITARY AS SHOWN BY IN SITU HYBRIDIZATION

J. H. Steel, Q. Hamid, S. Van Noorden, L. Chandrachud+,
P. Jones**, P. Denny*, J. Burrin**, A. M. McNicol+, S. Legon*,
S.R. Bloom** and J.M. Polak

Departments of Histochemistry, *Chemical Pathology and **Medicine
Royal Postgraduate Medical School, London, U.K.
+Department of Pathology, Royal Infirmary, Glasgow, U.K.

INTRODUCTION

Although the peptide content of endocrine cells is reliably demonstrated by immunocytochemistry, it may be unrelated to the immediate physiological state of the cell and the turnover rate of cellular products. By studying the mRNA present in the cytoplasm by in situ hybridization, it is possible to evaluate the synthetic activity of a cell. Although the cytoplasmic mRNA content is dependent on factors such as the rate of post-transcriptional processing and mRNA stability as well as on the rate of formation of primary gene transcripts (1), in situ hybridization of mRNA is still a useful technique for comparing changes in physiological activity. Our hypothesis was that the combination of immunocytochemistry and in situ hybridization might provide a more complete picture of the dynamic endocrinology of a tissue. Prolactin and POMC gene expression in rat pituitary were used as examples. We studied prolactin synthesis and secretion during pregnancy and lactation, and also after ovariectomy, because estrogen is known to stimulate the synthesis and secretion of prolactin in vivo and in vitro (2,3), such regulation taking place at the level of prolactin gene transcription (4,5). Prolactin mRNA levels increase in pituitaries of normal rats or hypogonadal mice given estrogen treatment (6,7). The technique of in situ hybridization has already been used to demonstrate increased prolactin mRNA in sections of pituitary from rats given chronic diethylstilboestrol treatment (8) or estrogen (9).

Northern blot analysis of total RNA extracts was used to confirm our results and plasma prolactin levels were measured in order to compare prolactin secretion with pituitary prolactin mRNA and stored peptide levels.

Since POMC gene expression in the anterior pituitary gland is regulated by circulating levels of glucocorticoids (10,11), we studied POMC-mRNA after adrenalectomy and corticosterone treatment using in situ hybridization, which has been used to study POMC in the rat pituitary (12) and hypothalamus (13).

MATERIALS AND METHODS

Animals and Tissues

For the prolactin study, pregnant (n = 15), lactating (n = 15), ovariectomized (n = 10) and control non-pregnant (n = 15) female Wistar rats were obtained. The ovariectomized rats were killed 17 days postoperatively.

For the POMC study, normal male Sprague Dawley rats were adrenalectomized or sham-operated (n = 6); half of each group was killed after two days, the remainder after one month. Additional animals received 14 daily intraperitoneal injections of either corticosterone (500 μg/kg) (n = 7) or saline (n = 7).

For in situ hybridization, 5 rats from each group of females and all the males were anesthetized with ether and perfused intracardially with phosphate-buffered saline (PBS) followed by 4% paraformaldehyde in PBS. Pituitaries were removed and snap-frozen. Five rats from each group of females were perfused with Bouin's fluid; the pituitaries from these animals were processed to wax for immunocytochemistry. Before perfusion, about 5 ml of blood were taken by cardiac puncture from each of the pregnant, lactating and control rats under light ether anesthesia, for assay of serum prolactin levels by RIA (14). For Northern blots analysis pregnant, lactating and control rats (5 of each) were killed with ether, their pituitaries removed fresh and immediately snap-frozen in liquid nitrogen.

In Situ Hybridization

Complementary DNA clones corresponding to rat prolactin or POMC genes were inserted into SP6 plasmids, which were linearized with Bam HI and Sal64, respectively, and transcribed in vitro as described previously (15) to yield $[^{32}P]$-labeled complementary RNA (cRNA) probes. Cryostat sections (10 μm) of pituitaries from each group were cut on the same day and mounted on poly-L-lysine-coated (16) glass slides. Hybridization was carried out on all sections simultaneously, as described previously (15), using 2-3 ng per section of the appropriate cRNA probe dissolved in 20 μl of hybridization buffer. Sections were incubated at 42°C for 16-20 hr and then washed in standard saline citrate (SSC), dried and dipped in Ilford K5 emulsion for autoradiography, then exposed in light-tight boxes at 4°C for 72 hr. As negative controls one section from each pituitary was hybridized with a sense probe for prolactin or POMC whose sequence was identical to that of the mRNA, and one was pretreated with RNase before application of the cRNA probe. Also an unrelated probe was applied to some sections (a cRNA probe for NPY), and the cRNA probes were applied to sections of rat skin. The absence of positive hybridization with these controls was taken as confirmation of the specificity of the cRNA probe binding in the pituitary.

Immunocytochemistry

Cryostat sections (10 μm) of paraformaldehyde-fixed pituitaries adjacent to those used for in situ hybridization and paraffin sections (5 μm) of the Bouin's-fixed female rat pituitaries were immunostained using antiserum to rat prolactin or POMC and the peroxidase-antiperoxidase technique (17). The prolactin antiserum was used at optimal dilution and at three serial supra-optimal dilutions. Higher than optimal dilutions were used in order to detect any small differences in antigen concentration (18-20).

Isolation of Total RNA and Northern Blot Analysis

Each group of 5 fresh, frozen pituitaries from pregnant, lactating and control rats was pooled and total RNA extracted as described previously (21). Total RNA was separated by electrophoresis and transferred to a nitrocellulose filter, which was hybridized with the [^{32}P]-labeled cRNA prolactin probe. After washing and drying, the filters were exposed to photographic film at room temperature for 90 min.

RESULTS

Positive hybridization occurred in the anterior pituitary cells of all groups when the appropriate cRNA probe was used. The frequency of cells giving a positive signal for prolactin mRNA in control rat pituitaries was similar to the number of cells immunostained for prolactin. The frequency was much reduced and the density of silver grains over positive cells was much lower in pregnant rat pituitaries than in controls (Figs. 1 and 2). Sections of lactating rat pituitaries were so densely covered with silver grains that individual positive cells could not easily be identified (Fig. 3). In ovariectomized rat pituitaries, the frequency of cells that were positive for prolactin mRNA was lower than in intact controls (Fig. 4). The method control slides were all negative. No differences in prolactin immunoreactivity could be detected between pituitaries of control and pregnant or lactating rats, even at supra-optimal dilutions (Figs. 5-7). In ovariectomized rat pituitaries, the frequency of prolactin-immunoreactive cells (Fig. 8) was similar to positive cells in autoradiographs, i.e. less than in controls. Northern blot analysis revealed that the total prolactin mRNA level was below normal in pregnant and above normal in lactating rat pituitaries, confirming the results obtained using in situ hybridization (Fig. 9). Serum prolactin levels, measured by RIA, were as follows: controls 51.7 ± 10.48 ng/ml, pregnant 6.4 ± 0.65 ng/ml, lactating 51.1 ± 7.23 ng/ml (mean ± SEM, n = 5, $P < 0.001$ for difference between control and pregnant rats); and thus there was a significant decrease in serum prolactin in pregnant rats, but no change in lactating rats.

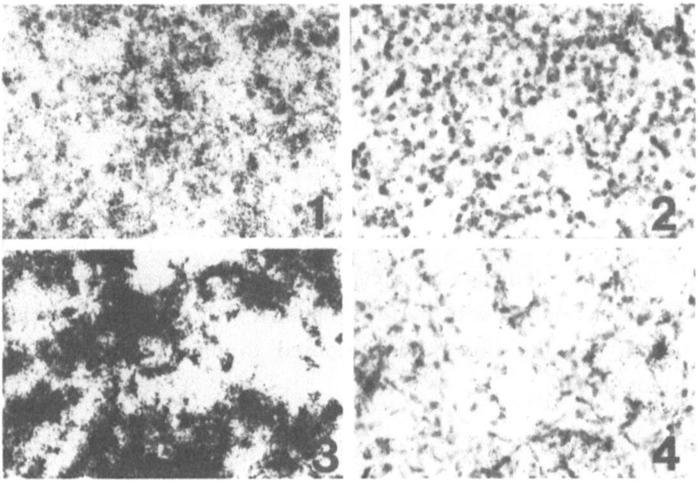

Figures 1-4. In situ hybridization of prolactin mRNA in pituitaries of (1) control, (2) pregnant, (3) lactating and (4) ovariectomized rats using a [^{32}P]-labeled cRNA probe. Tissues fixed in 4% paraformaldehyde.

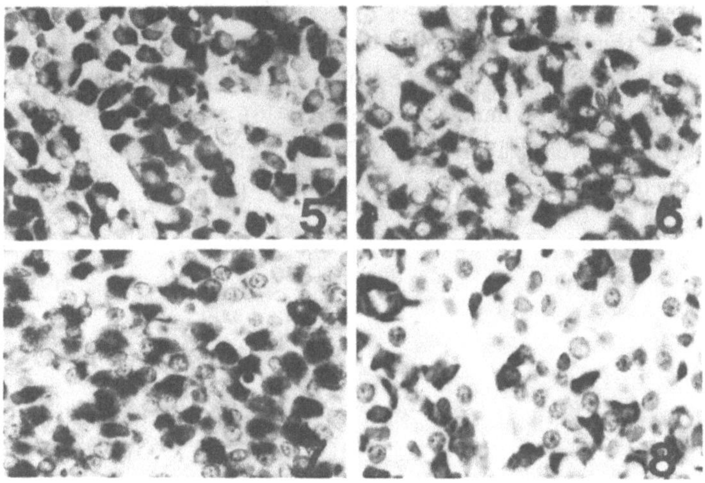

Figures 5-8. Sections of Bouin's-fixed rat pituitaries immunostained for prolactin: (7) control, (8) pregnant, (9) lactating, (10) ovariectomized.

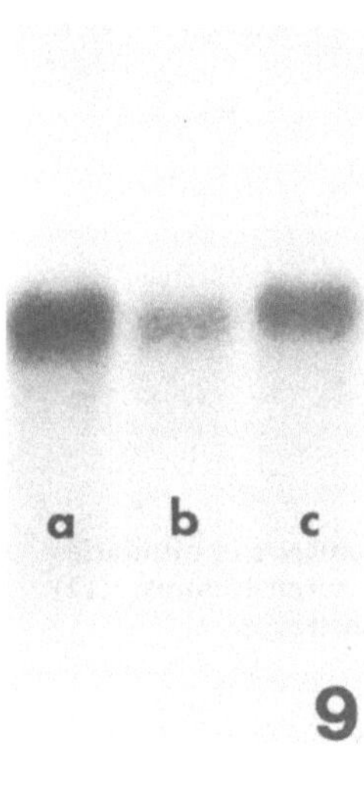

Figure 9. Northern blot analysis of total RNA from rat pituitaries. Lane a: lactating rats, lane b: pregnant rats and lane c: control rats.

In the POMC study, adrenalectomy produced a large increase in POMC mRNA signal in the anterior pituitary after 2 days (Fig. 10), which was further increased after one month (Fig. 11), compared with sham operated rats and saline controls (Fig. 12). Corticosterone-treated rats showed decreased POMC-mRNA labeling (Fig. 13). Immunostaining for POMC was reduced 2 days after adrenalectomy (Fig. 14) or sham-operation, but increased enormously one month after adrenalectomy (Fig. 15). One-month sham-operated controls resembled saline controls for POMC immunostaining (Fig. 16), as did corticosterone-treated rats (Fig. 17). The hybridization of POMC-mRNA and the POMC immunostaining in the intermediate lobe of the pituitary were similar in all rats.

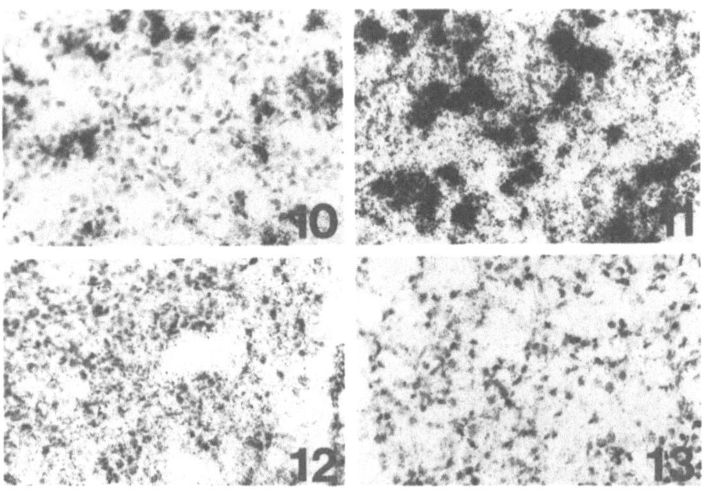

Figures 10-13. In situ hybridization of POMC-mRNA in pituitaries of rats: (10) 2 days and (11) 1 month after adrenalectomy, (12) treated with saline, and (13) treated with corticosterone.

DISCUSSION

The technique of in situ hybridization has been used to demonstrate differences in the autoradiographic signal for prolactin mRNA between pituitaries

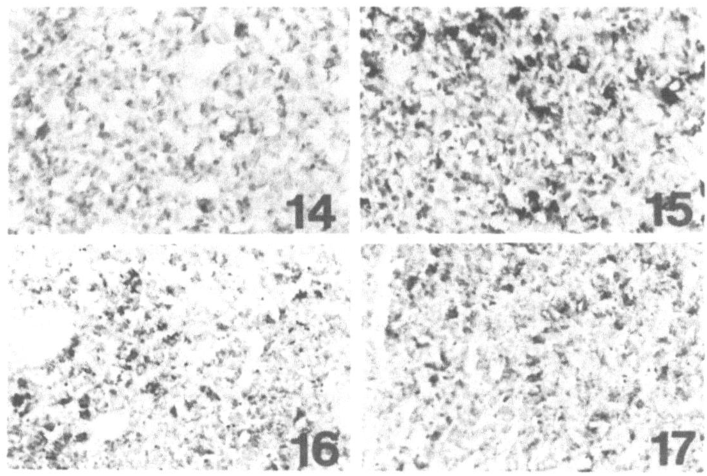

Figures 14-17. Sections of 4% paraformaldehyde-fixed rat pituitaries immunostained for POMC: (14) 2 days and (15) 1 month post-adrenalectomy, (16) saline control, (17) corticosterone-treated.

of lactating and pregnant rats and those of controls, even though no differences in prolactin immunoreactivity can be detected. In pregnant rats, the very low prolactin mRNA signal indicated that the lactotrophs were relatively inactive, but prolactin immunoreactivity was high suggesting storage of prolactin due to decreased secretion (21). Lactating rat pituitaries had very high levels of prolactin mRNA, as expected, corresponding to a raised level of prolactin synthesis; however, they had relatively normal prolactin immunoreactivity suggesting that the prolactin was being released. In ovariectomized rats, similar numbers of anterior pituitary cells were positive for prolactin and prolactin mRNA; both were less than in controls, as a result of the removal of estrogen (4).

Northern blot analysis showed that our prolactin cRNA probe detected the specific mRNA intended, and indicated that pituitary prolactin mRNA was indeed lowered in pregnancy and raised in lactation. Serum prolactin levels in pregnant rats were significantly lower than in controls, since in rats secretion as well as synthesis of prolactin is inhibited by dopamine during pregnancy (22). However, in our lactating rats, serum prolactin was not elevated as expected, despite an increased level of prolactin mRNA. Possible cessation of suckling, and thus the stimulus to prolactin secretion due to stress before the animals were killed, allowed serum prolactin levels to return to normal and cellular prolactin to be reduced by crinophagy (23,24).

Although cytoplasmic mRNA levels do reflect the rate of specific gene transcription, post-transcriptional processing can mean that there is some discrepancy between the rate of primary transcript formation and the rate of resulting mRNA synthesis. Primary transcription rates can be evaluated only by nuclear runoff assays (25) or by hybridization with specific probes to the primary gene transcript (26). A delay of several hours between changes in the rates of gene transcription and the resulting mature mRNA formation is not uncommon (25,27), and hence cessation of transcription of the prolactin gene may not have been reflected in the mRNA content of pituitary cells.

By two days following adrenalectomy, the expression of the POMC gene in the anterior pituitary was already elevated, and after one month the characteristic, raised POMC gene expression had developed, along with an increase in POMC immunostaining. However, immunocytochemistry revealed decreased POMC staining two days postoperatively, in sham-operated as well as in adrenalectomized rats, and therefore the decrease in pituitary POMC staining was not a specific effect of adrenalectomy. After one month, the sham-operated rats gave hybridization and immunocytochemistry results similar to normal unoperated rats, indicating that the sham-operated animals had returned to normal. Corticosterone treatment produced a decrease in POMC gene expression in the anterior pituitary, although immunostaining for POMC was very similar to that for saline controls. Again, only in situ hybridization revealed the reduced level of POMC gene expression after corticosterone treatment.

The changes in POMC-mRNA took place only in the anterior pituitary, since POMC in the intermediate lobe is not regulated by glucocorticoids (10,11).

In this study, using well known changes in prolactin and POMC synthesis and secretion as a model, we have shown that in situ hybridization combined with immunocytochemistry can reveal information about gene transcription and peptide storage, resulting in a more complete picture of the cellular function than could be obtained from immunocytochemistry alone.

SUMMARY

We have investigated the combined use of in situ hybridization and immunocytochemistry for the study of endocrine cell function, using as examples the expression of prolactin and proopiomelancortin (POMC) messenger RNAs (mRNAs) in rat pituitaries. In situ hybridization using a [^{32}P]-labeled cRNA probe for rat prolactin was carried out on sections of pituitaries from pregnant, lactating, ovariectomized and control rats; sections adjacent to the pituitaries were immunostained for prolactin as well. Sections from pituitaries of adrenalectomized or corticosterone-treated rats were compared with those from controls using in situ hybridization with a [^{32}P]-labeled POMC probe and by immunostaining for POMC. No differences in prolactin immunostaining were seen between pregnant or lactating rats and controls, even when the supra-optimal dilution technique was used; however, using in situ hybridization, the prolactin mRNA signal was reduced in pregnant rats and much increased in lactating rats. Northern blot analysis confirmed these changes. In ovariectomized rat pituitaries, both prolactin immunoreactivity and prolactin mRNA were decreased. Adrenalectomy produced large increases in POMC-mRNA and immunostaining after one month, whereas corticosterone treatment decreased POMC-mRNA, but not immunostaining. The combined use of in situ hybridization and immunocytochemistry can thus provide dynamic information in morphological terms.

ACKNOWLEDGMENTS

We thank: Dr. R. A. Maurer of the Department of Physiology and Biophysics, University of Iowa, USA for the cloned prolactin probes used in this study; Dr. J. L. Roberts of the College of Physicians and Surgeons of Columbia University, New York, NY, USA for the POMC probes; Dr. A. F. Parlow and the National Institute of Arthritis, Metabolism and Digestive Diseases (NIAMDD, Bethesda, MD, USA) for the supply of reagent for the rat prolactin assay and the rat prolactin antiserum for immunocytochemistry; Dr. M. Chretién of the Department of Biochemical and Molecular Neuroendocrinology, Clinical Research Institute of Montreal, Quebec, Canada, for the POMC antiserum; and Dr. J. E. Dixon of Purdue University, Indiana, USA for the riboprobe for rat NPY mRNA, used as a negative control.

REFERENCES

1. Potter E, Nicholaisen AK, Ong ES, Evans RM, Rosenfeld MG (1981) Thyrotropin-releasing hormone exerts rapid nuclear effects to increase proudction of the primary prolactin mRNA transcript. Proc Natl Acad Sci USA 11: 6662-6666
2. MacLeod RM, Abad A, Eidson LL (1969) In vivo effects of sex hormones on the in vitro synthesis of prolactin and growth hormone in normal and pituitary tumor-bearing rats. Endocrinology 84: 1475-1483

3. Maurer RA, Gorski J (1977) Effects of estradiol-17β and pimozide on prolactin synthesis in male and female rats. Endocrinology 101: 76-84
4. Maurer RA (1982) Estradiol regulates the transcription of the prolactin gene. J Biol Chem 257: 2133-2136
5. Shull JD, Gorski J (1985) Estrogen regulates the transcription of the rat prolactin gene in vivo through at least two independent mechanisms. Endocrinology 116: 2456-2462
6. Seo H, Refetoff S, Martino E, Vassart G, Brocas H (1979) The differential stimulatory effect of thyroid hormone on growth hormone synthesis and estrogen on prolactin synthesis due to accumulation of specific messenger ribonucleic acids. Endocrinology 104: 1083-1090
7. Stanley HF, Curtis A, Sheward WJ, Roberts JL, Fink G (1986) Prolactin messenger ribonucleic acid levels in the normal and hypogonadal mouse pituitary gland. Endocrinology 119: 2422-2426
8. Lloyd RV, Landefeld TD (1986) Detection of prolactin messenger RNA in rat anterior pituitary by in situ hybridization. Am J Pathol 125: 35-44
9. Yamamoto N, Seo H, Suganuma N, Matsui N, Nakane T, Kuwayama A, Kageyama N (1986) Effect of estrogen on prolactin mRNA in the rat pituitary. Analysis by in situ hybridization and immunohistochemistry. Neuroendocrinology 42: 494-497
10. Schachter BS, Johnson LK, Baxter JD, Roberts JL (1982) Differential regulation by glucocorticoids of proopiomelanocortin mRNA levels in the anterior and intermediate lobes of the rat pituitary. Endocrinology 110: 1442-1444
11. Birnberg NC, Lissitzky J-C, Hinman M, Herbert E (1983) Glucocorticoids regulate proopiomelanocortin gene expression in vivo at the levels of transcription and secretion. Proc Natl Acad Sci USA 80: 6982-6986
12. Lewis ME, Sherman TG, Burke S, Akil H, Davis LG, Arentzen R, Watson SJ (1986) Detection of proopiomelanocortin mRNA by in situ hybridization with an oligonucleotide probe. Proc Natl Acad Sci USA 83: 5419-5423
13. Gee CE, Chen C-LC, Roberts JL, Thompson R, Watson SJ (1983) Identification of proopiomelanocortin neurones in rat hypothalamus by in situ cDNA-mRNA hybridization. Nature 306: 374-376
14. Powell RC, Daniels M, Innes GK, Ashby MJ, Mashiter K (1983) Effects of trifluoroperazine on rat prolactin, growth hormone, thyroid stimulating hormone and adrenocorticotrophin secretion in vitro. Acta Endocrinol 103: 492-496
15. Hamid Q, Wharton J, Terenghi G, Hassall CJS, Aimi J, Taylor KH, Nakazato H, Dixon JE, Burnstock G, Polak JM (1987) Localization of atrial natriuretic peptide mRNA and immunoreactivity in the rat heart and human atrial appendage. Proc Natl Acad Sci USA 84: 6760-6764
16. Huang WM, Gibson SJ, Facer P, Gu J, Polak JM (1983) Improved section adhesion for immunocytochemistry using high molecular weight polymers of L-lysine as a slide coating. Histochemistry 77: 275-279
17. Sternberger LA (ed) (1979) Immunocytochemistry, ed 2. New York: Wiley
18. Vacca-Galloway LL (1985) Differential immunostaining for substance P in Huntington's diseased and normal spinal cord: significance of serial (optimal, supra-optimal and end-point) dilutions of primary antiserum in comparing biological specimens. Histochemistry 83: 561-569
19. Springall DR, Steel JH, Collina G, Polak JM (1987) The use of high dilutions of primary antiserum for immunocytochemistry to detect variations in antigen concentration. J Pathol 151: 38A (Abstract)
20. Steel JH, Van Noorden S, Ballesta J, Gibson SJ, Ghatei MA, Burrin J, Leonhardt U, Domin J, Bloom SR, Polak JM (1987) Localization of 7B2, neuromedin B and neuromedin U in specific cell types of rat, mouse and human pituitary, in rat hypothalamus and in 30 human pituitary and extra-pituitary tumors. Endocrinology (in press)
21. Steel JH, Hamid Q, Van Noorden S, Jones P, Denny P, Burrin J, Legon S, Bloom SR, Polak JM (1987) Combined use of in situ hybridization and immunocytochemistry for the investigation of prolactin gene expression in immature, pubertal, pregnant, lactating and ovariectomized rats. Histochemistry (in press)

22. Ben-Jonathan N, Oliver C, Weiner HJ, Mical RS, Porter JC (1977) Dopamine in hypophysial portal plasma of the rat during the estrous cycle and throughout pregnancy. Endocrinology 100: 452-458
23. Smith RE, Farquhar MG (1966) Lysosome function in the regulation of the secretory processes in cells of the anterior pituitary gland. J Cell Biol 31: 319-347
24. De Marco L, Mashiter K, Peters TJ (1982) Suckling withdrawal increases pituitary lysosomal enzyme activities and prolactin protease on lactating rats. Endocrinology 110: 1178-1182
25. Eberwine JH, Roberts JL (1984) Glucocorticoid regulation of pro-opiomelanocortin gene transcription in the rat pituitary. J Biol Chem 259: 2166-2170
26. Fremeau RT Jr, Lundblad JR, Pritchett DB, Wilcox JN, Roberts JL (1986) Regulation of pro-opiomelanocortin gene transcription in individual cell nuclei. Science 234: 1265-1269
27. Millington WR, Blum M, Knight R, Mueller GP, Roberts JL, O'Donohue TL (1986) A diurnal rhythm in proopiomelanocortin messenger ribonucleic acid that varies concomittantly with the content and secretion of β-endorphin in the intermediate lobe of the rat pituitary. Endocrinology 118: 829-834

EVIDENCE AGAINST D_2 DOPAMINERGIC INFLUENCE ON PHOSPHOINOSITIDE METABOLISM IN PITUITARY AND NEURAL CELLS

J. G. Baird, A. B. Cubitt, B. L. Brown
and P. R. M. Dobson

Department of Human Metabolism & Clinical Biochemistry
University of Sheffield Medical School
Sheffield, England, U.K.

INTRODUCTION

Prolactin secretion from the adenohypophysis is known to be under the tonic inhibitory control of the hypothalamus. Dopamine (DA) is believed to be the major inhibitory influence, being released from tuberoinfundibular dopaminergic neurons and transported to the pituitary gland via the hypophysial portal vessels. High affinity saturable and stereoselective DA receptors have been identified in adenohypophysial homogenates (see refs. 1-3); these have been characterized as the D_2 type (4). Immunostaining has shown these receptors to be located on the lactotrophs (5). While there is considerable evidence from our group and others to show that dopamine can reduce intracellular cyclic AMP (cAMP) levels in adenohypophysial cells (see refs. 6,7), this cannot fully account for the observed inhibition of prolactin secretion, nor its mechanism.

Canonico and co-workers reported that DA inhibited [^{32}P]orthophosphate incorporation into phosphatidylinositol (PI), an effect which could be blocked by haloperidol and pimozide (8). In contrast, our concurrent preliminary experiments suggested that dopamine did not affect the production of inositol phosphates, neither basal nor thyrotropin-releasing hormone (TRH)-stimulated, in anterior pituitary cells labeled with [^{3}H]inositol (9). We wished to investigate this possible action of dopamine further by using normal anterior pituitary cells and neuroblastoma hybrid cells, both of which exhibit dopamine inhibition of adenylate cyclase.

METHODS

Adenohypophysial cells from Sprague-Dawley rats were prepared and cultured in monolayers as described previously (10); GH_3 cells were cultured in the same way. NCB 20 cells were grown as described by Berry-Kravis et al. (11). The amount of inositol phosphates was estimated as described previously (12). DA agonists and antagonists were either solubilized in modified Krebs Ringer bicarbonate containing lithium (LiKRB) or in a small amount of dimethylsulphoxide (DMSO) before being diluted to the required concentration in LiKRB. Pertussis toxin (PT), usually at a concentration of 10 ng/ml (unless otherwise indicated), was added to cells in serum-free Ham's F10 containing 0.1% bovine serum albumin (BSA) and [^{3}H]inositol. Experiments designed to assess the effect of dopaminergic agents on the incorporation of inositol or [^{32}P]orthophosphate into phospholipids used the thin layer chromatography (TLC) method of Jolles and colleagues (13), or the deacylation method of Creba et al. (14).

RESULTS

Effects of Dopaminergic Agonists on [^{3}H]inositol Phosphate Accumulation

In experiments in which inositol phosphate accumulation and prolactin were measured on the same pituitary cell preparations, dopaminergic agonists inhibited prolactin secretion as expected; the approximate IC_{50} values for inhibition of TRH-stimulated prolactin secretion were: DA, 8 nM; apomorphine, 50 nM; and LY171555 (a gift from Eli Lilly Company), 16 nM. However, effects on inositol phosphate accumulation were small and variable. Figure 1 shows, as an example, a comparison of the effects of the agonist LY171555 on both parameters. The results depicted in Table 1 indicate that dopaminergic agonists had only slight effects on [^{3}H]inositol phosphate accumulation. When inhibition was observed it was small and variable; significant inhibition ($P < 0.05$) was observed only in a minority of experiments, and then only at high agonist concentrations.

Pre-incubation of the cells with PT resulted in some reduction in the degree of inhibition exerted by dopaminergic agonists on [^{3}H]inositol phosphate accumulation, but since the inhibitory effects were slight and variable in the first instance, any effects of PT were small, although the incidence of experiments showing significant inhibition was reduced. Studies on the ability of antagonists to affect dopaminergic agonist-induced effects on [^{3}H]inositol phosphate accumulation encountered similar

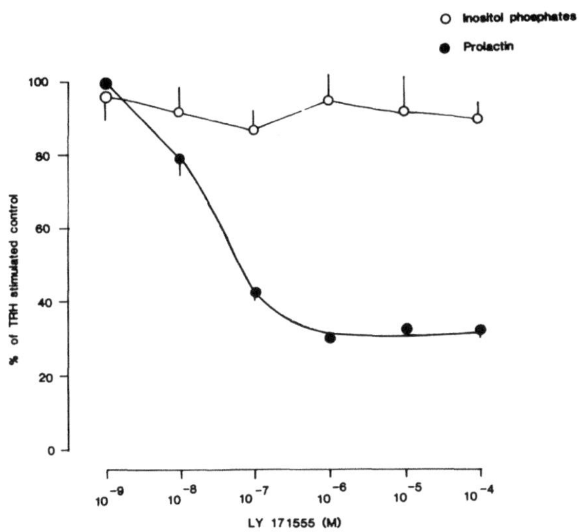

Figure 1. The effect of LY 171555 on prolactin secretion and inositol phosphate accumulation in rat anterior pituitary cells.

TABLE 1

THE EFFECTS OF DOPAMINE AGONISTS ON TRH-STIMULATED $[^3H]$INOSITOL PHOSPHATE ACCUMULATION

Agonist concentration	(μM)	Mean (±SD)	Number of observations	Number significant (P < 0.05)
Dopamine	0.1	92.7 ± 8.2	6	0
	10.0	94.7 ± 13.7	9	1
	100.0	89.1 ± 15.2	8	3
Apomorphine	0.1	96.0 ± 9.6	10	1
	0.1 + PT	93.8 ± 12.3	5	0
	10.0	81.3 ± 9.8	12	6
	10.0 + PT	89.2 ± 9.0	6	3
	100.0	84.3 ± 9.0	7	5
	100.0 + PT	84.4 ± 8.1	5	3
Bromocriptine	0.1	92.9 ± 5.0	5	3
	0.1 + PT	92.5 ± 11.0	3	1
	10.0	82.6 ± 10.3	11	8
	10.0 + PT	86.4 ± 12.6	10	4
LY 171555	0.001	95.3 ± 7.4	3	0
	0.01	91.5 ± 6.4	3	0
	0.1	86.0 ± 5.3	3	0
	1.0	94.0 ± 6.8	3	1
	10.0	90.4 ± 10.8	7	2

Data are expressed relative to the respective control in the absence of agonist (%). Triplicate wells were used in each experiment. PT (50 ng/ml) was present during the $[^3H]$inositol preincubation.

problems. Inhibition of prolactin secretion caused by the specific D_2 receptor agonist LY171555 was reversed more effectively by the antagonists or by PT preincubation, than was inhibition caused by DA, apomorphine and bromocriptine.

Effects of Dopaminergic Agonists on [^{3}H]inositol Incorporation

To investigate the possibility that dopaminergic agonists affect resynthesis of PI, the agonists were added at the same time as [^{3}H]inositol. The results (Table 2) show that there was little or no effect of DA on inositol incorporation into lipids except at very high concentrations, i.e. above 10 μM. The [^{3}H]inositol phosphates liberated were also reduced, an effect which was significant only at 100 μM and 1000 μM. High concentrations of bromocriptine also inhibited labeling and inositol phosphate accumulation. These inhibitory effects of dopaminergic agonists occurred only at levels very much higher than those needed to inhibit prolactin secretion physiologically. Moreover, these results were not reproduced in GH_3 cells.

Analysis of Dopaminergic Effects on [^{32}P]orthophosphate Labeling of Phospholipids in Rat Anterior Hemipituitary Glands

The experiments described by Canonico and colleagues showing dopaminergic inhibition of [^{32}P]orthophosphate incorporation (8,15), were repeated. The cumulated data from phospholipid separations using standard TLC plates with various test solutions in the incubations showed that the values obtained were variable and indicated no significant changes in the degree of labeling in any phospholipid (data not shown).

Effect of Dopamine in NCB20 Cells

Dopamine inhibits adenylate cyclase activity in the neuroblastoma x Chinese hamster brain explant cell line (NCB-20). There is evidence to suggest that the dopamine receptor on these cells is similar to that found on pituitary cells. We found that dopamine had no effect on inositol phosphate accumulation, or on the levels of the parent phospholipids under either basal or stimulated conditions (16).

DISCUSSION

The mechanism of action of DA in the inhibition of prolactin secretion is not clearly understood, although there is evidence for

TABLE 2

EFFECTS OF DOPAMINE PREINCUBATION ON [^{3}H]INOSITOL INCORPORATION INTO PHOSPHOLIPIDS AND THE RESULTING [^{3}H]INOSITOL PHOSPHATES PRODUCED IN NORMAL PITUITARY CELLS

Concen. dopamine (μM)	Number of experiments	Phosphoinositides		Inositol phosphate	
		Basal	TRH	Basal	TRH
0.1	3	103.4 ± 7.1	108.3 ± 4.3	90.5 ± 6.8	107.4 ± 11.6
1.0	9	97.0 ± 2.8	98.4 ± 3.2	113.9 ± 6.3	116.9 ± 3.9
10.0	14	93.6 ± 3.6	88.0 ± 2.4	110.9 ± 7.0	109.8 ± 4.0
100.0	14	91.7 ± 3.2	85.2 ± 3.0	*88.5 ± 4.3	105.2 ± 2.3
1000.0	10	81.0 ± 7.9	*82.1 ± 4.8	**60.5 ± 6.4	**67.03 ± 6.1

Rat anterior pituitary cells were preincubated with dopamine for 24 hr in Ham's F10 containing 0.1% BSA, [^{3}H]inositol and ascorbic acid (0.1 mM). After a 1 hr incubation in LiKRB in the absence or presence of TRH (55 nM), the experiment was terminated. The radioactivity in the phosphoinositides was that remaining in the organic phase, while the inositol phosphates were measured in the aqueous phase. Data is the mean ± SEM. All data is normalized with respect to the controls, i.e. in the absence of dopamine. Statistically significant inhibition is denoted as: *$P < 0.02$, **$P < 0.002$.

mediation by both cAMP or Ca^{2+}. It has been reported that DA can lower cAMP levels in pituitary glands and cells (for example, see refs. 6,7,17,18). Evidence showing that DA inhibits pituitary adenylate cyclase (19,20) and that PT can prevent the inhibitory effects of DA on prolactin secretion (21) have led to the proposal that DA inhibits prolactin secretion by working through the inhibitory (N_i) subunit of adenylate cyclase. However, it has been shown that inhibition of prolactin secretion can still occur when cAMP levels were raised by IBMX, theophylline, cholera toxin and forskolin (7,22). The importance of Ca^{2+} as a physiological regulator of prolactin secretion has been stressed (23-25) and it has been suggested that DA might act by preventing the Ca^{2+} influx required to maintain secretion. Furthermore, there is evidence that DA can lower intracellular Ca^{2+}(26,27).

The possible involvement of phosphoinositide hydrolysis in these processes becomes intriguing. There is considerable evidence that the increase in prolactin secretion caused by TRH is mediated by increased phosphoinositide hydrolysis (for review, see ref. 28), and therefore DA-mediated inhibition of TRH-induced $[Ca^{2+}]_i$ and prolactin secretion may indicate an inhibition of phosphoinositide hydrolysis. This concept was supported by the results of Canonico and colleagues (8,15), who used [^{32}P]orthophosphate incorporation into PI as a measure of PI turnover. In the present study, the dopaminergic agonists inhibited TRH-stimulated prolactin secretion markedly. While PT and antagonists could prevent most of the actions of dopaminergic agonists, some secondary effects became apparent at higher agonist concentrations, which were not manifested to the same degree by the specific D_2 agonist LY171555. In contrast, decreases in phosphoinositide hydrolysis in the same experiments were small and variable, occurring only at very high agonist concentrations, although some of these experiments did not show significant inhibition. Neither PT preincubation nor antagonists totally abolished the small inhibitions observed, although the degree of incidence was reduced slightly. Dopamine and bromocriptine did cause decreases in phosphoinositide labeling and [^{3}H]inositol phosphate accumulation, but the effects were apparent only at levels very much higher than those needed to inhibit prolactin secretion. In addition, there were no consistent effects of dopaminergic agonists on [^{32}P]orthophosphate incorporation into rat anterior hemipituitary glands. It can be argued the [^{32}P]orthophosphate incorporation into PI is a secondary event and may occur as a consequence of a primary event, such as the activation of N_i or lowered intracellular Ca^{2+} levels. The other effects on phosphoinositide metabolism could also be indirect. One possible explanation is that at these very high concentrations, some secondary effect of the dopaminergic agonist can occur, which may not be related to receptor occupation. However, the absence of any effect in GH_3 cells may suggest that the presence of DA receptors is necessary.

Simmonds and Strange reported that low concentrations of DA could inhibit stimulated [^{3}H]inositol phosphate formation in bovine anterior pituitary cells, and the effect was blocked by antagonists (29). Furthermore, Enjalbert and co-workers reported that DA could inhibit angiotensin-stimulated production of [^{3}H]inositol phosphates (30). However, our studies would suggest that dopaminergic agonists, while being able to produce marked inhibition of prolactin secretion, do not exert effects on phosphoinositide metabolism at comparable concentrations. Indeed, this view has now been confirmed by Canonico and colleagues (31), who also could not repeat the observation of Enjalbert et al. (30). In addition, the data obtained using the DA-responsive neuroblastoma glioma hybridoma line NCB-20 agree with our observations using adenohypophysial cells that phosphoinositide metabolism may not be involved in DA action.

It has been suggested that the DA receptor is coupled to the inhibitory guanine nucleotide binding protein of adenylate cyclase (21), but it is tempting to speculate that the effects of dopaminergic agonists observed on calcium levels may indicate a direct link between the receptor and an ion channel. The intracellular processes mediating the inhibition of both basal and stimulated prolactin secretion are probably more complicated than previously thought. We would conclude that secondary and, possibly, nonspecific effects may be responsible for some of the small inhibitory effects on phosphoinositide metabolism that have been observed.

REFERENCES

1. Creese I, Schneider R, Snyder SH (1977) ^{3}H-spiroperidol labels dopamine receptors in pituitary and brain. Eur J Pharmacol 46: 377-381
2. Cronin MJ, Roberts JM, Weiner RI (1978) Dopamine and dihydroergocryptine binding to the anterior pituitary and other brain areas of the rat and sheep. Endocrinology 103: 302-309
3. Cronin MJ, Weiner RI (1979) [^{3}H]spiroperidol (spiperone) binding to a putative dopamine receptor in sheep and steer pituitary and stalk median eminence. Endocrinology 104: 307-312
4 Kebabian JW, Calne DB (1979) Multiple receptors for dopamine. Nature 277: 93-96
5. Goldsmith PC, Cronin MJ, Rubin RJ, Weiner RI (1978) Immunocytochemical staining of dopamine receptors on mammotrophs. VIIIth Annual Meeting of Society for Neuroscience, St Louis, MO, Abstract 1098
6. Barnes GD, Brown BL, Gard TG, Atkinson D,. Ekins RP (1978) Effect of TRH and dopamine on cyclic AMP levels in enriched mammotroph and thyrotroph cells. Mol Cell Endocrinol 12: 273-284
7. Ray KP, Wallis M (1983) Actions of dopamine on prolactin secretion and cyclic AMP metabolism in ovine pituitary cells. Mol Cell Endocrinol 27: 139-155
8. Canonico PL, Valdenegro CA, MacLeod RM (1982) Dopamine inhibits ^{32}P incorporation into phosphatidylinositol in the anterior pituitary of rats. Endocrinology 111: 347-349
9. Brown BL, Baird JG, Quilliam LA, Merritt JE, Dobson PRM (1985) Calcium-mediated intracellular signalling in the control of prolactin secretion from rat anterior pituitary cells. In: MacLeod RM, Thorner MO, Scapagnini U (eds) Prolactin, Basic and Clinical Correlates. Fidia Research Ser. Vol. 1., Liviana Press: Padova: pp 199-204
10. Dobson PRM, Brown BL (1985) The preparation, culture and incubation of rat anterior pituitary cells for static and dynamic studies of secretion. Meth Enzymol 109: 293-298
11. Berry-Kravis E, Freedman SB, Dawson, G (1984) Specific mediated inhibition of cyclic AMP synthesis by dopamine in a neuroblastoma x brain hybrid cell line NCB-20. J Neurochem 43: 413-420
12. Baird JG, Dobson PRM, Wojcikiewicz RJH, Brown BL (1983) Thyrotropin-releasing hormone stimulates inositol phosphate production in normal anterior pituitary cells and GH_3 tumour cells in the presence of lithium. Bioscience Reports 3: 1091-1099
13. Jolles J, Zwiers H, Dekker A, Wirtz KWA, Gispen WH (1981) Corticotropin-(1-24)-tetracosapeptide affects protein phosphorylation and polyphosphoinositide metabolism in rat brain. Biochem J 194: 283-291

14. Creba JA, Downes P, Hawkins PT, Bewster G, Michell RH, Kirk CJ (1983) Rapid breakdown of phosphatidyl inositol 4-phosphate and phosphatidylinositol 4,5-bisphosphate in rat hepatocytes. Biochem J 212: 733-747
15. Canonico PL, Valdenegro CA, MacLeod RM, O'Dell SB, Harcus CT (1983) The inhibition of phosphatidylinositol turnover: a possible postreceptor mechanism for the prolactin secretion-inhibiting effect of dopamine. Endocrinology 113: 7-14
16. Cubitt AB, Brown BL, Dobson PRM (1987) Activation of dopamine receptors does not affect phosphoinositide turnover in NCB 20 cells. J Neurochem 49: 183-188
17. Ray KP, Wallis M (1981) Effects of dopamine on prolactin secretion and cyclic AMP accumulation in the rat anterior pituitary gland. Biochem J 194: 119-128
18. Swennen L, Denef C (1982) Physiological concentrations of dopamine decrease adenosine 3',5'-monophosphate levels in cultured rat anterior pituitary cells and enriched populations of lactotrophs. Evidence for a causal relationship to inhibition of prolactin release. Endocrinology 111: 398-405
19. DeCamilli P, Macconi D, Spada A (1979) Dopamine inhibits adenylate cyclase in human prolactin-secreting pituitary adenomas. Nature 278: 252-254
20. Onali P, Schwartz JP, Costa E (1981) Dopaminergic modulation of adenylate cyclase stimulation by vasoactive intestinal peptide in anterior pituitary. Proc Natl Acad Sci USA 78: 6531-6534
21. Cronin MJ, Myers, GA, MacLeod RM, Hewlett EL (1983) Pertussis toxin uncouples dopamine agonist inhibition of prolactin release. Am J Physiol 244: E499-E504
22. Delbecke D, Scammell JC, Mortinez-Campos A, Dannies PS (1986) Dopamine inhibits prolactin release when cyclic AMP levels are elevated. Endocrinology 118: 1271-1277
23. Thorner MO, Hackett JT, Murad F, MacLeod RM (1980) Calcium rather than cyclic AMP as the physiological intracellular regulator of prolactin release. Neuroendocrinology 31: 390-402
24. Merritt JE, Brown BL (1984) The possible involvement of both Ca^{2+} and cyclic AMP in the dopaminergic inhibition of prolactin secretion. Life Sci 35: 707-711
25. Schettini G, Cronin MJ, MacLeod RM (1983) cAMP and calcium-calmodulin interrelation in the control of prolactin secretion: evidence for dopamine inhibition of cAMP accumulation and prolactin release after calcium mobilization. Endocrinology 112: 1801-1807
26. Schofield JG (1983) Use of a trapped fluorescent indicator to demonstrate effects of thyroliberin and dopamine on cytoplasmic calcium concentrations in bovine anterior pituitary cells. FEBS Letts 159: 79-82
27. Malgaroli A, Vallar L, Elahi FR, Pozzan T, Spada A, Meldolesi J (1987) Dopamine inhibits cytosolic Ca^{2+} increases in rat lactotroph cells. J Biol Chem 262:13920-13927
28. Gershengorn MC (1986) Mechanism of thyrotropin releasing hormone stimulation of pituitary hormone secretion. Ann Rev Physiol 48: 515-526
29. Simmonds SH, Strange PG (1985) Inhibition of inositol phospholipid breakdown by D_2 dopamine receptors in dissociated bovine anterior pituitary cells. Neurosci Letts 60: 267-272
30. Enjalbert A, Sladeczek F, Guillon G, Bertrand P, Shu C, Epelbaum J, Garcia-Sainz A, Jard S, Lombard C, Kordon C, Bockaert J (1986) Angiotensin II and dopamine modulate both cAMP and inositol phosphate productions in anterior pituitary cells. J Biol Chem 261: 4071-4075
31. Canonio PL, Jarvis WD, Judd AM, MacLeod RM (1986) Dopamine does not attenuate phosphoinositide hydrolysis in rat anterior pituitary cells. J Endocrinol 110: 389-393

"CURED" PROLACTINOMA PATIENTS AFTER TRANSSPHENOIDAL ADENOMECTOMY: PREDICTION OF RECURRENCE

E. Ciccarelli, E. Ghigo, L. Savino, C. Miola,
A. Bertagna and F. Camanni

Dipartimento di Biomedicina Endocrino-Metabolica e Gastroenterologica,
Cattedra di Endocrinologia, Università di Torino
Torino, Italy

INTRODUCTION

A high successful rate of cure after transsphenoidal adenomectomy has been demonstrated in patients with prolactinoma, especially those with a microadenoma (1-5). However, a high rate of recurrence after apparently successful adenomectomy has been reported in some studies (2,5). This could suggest the presence of a persistent underlying abnormality in the pituitary or the hypothalamus. To identify the persistence of any underlying abnormality of hypothalmic-pituitary control after surgery and to predict relapses, we studied the long-term follow-up and the PRL secretory dynamics in a group of prolactinoma patients determined "cured" after transsphenoidal adenomectomy.

PATIENTS AND METHODS

Nineteen patients (18 females and 1 male), selected on the basis of normal PRL levels and pituitary function after transsphenoidal removal of a PRL-secreting adenoma, were studied for 12-96 months (mean 40 months) after surgery. All patients gave their informed consent to participate in the study. Prolactin responses to nomifensine (NOM), an indirect dopamine agonist, and to TRH, as well as PRL and TSH responses to domperidone (DOM), a peripheral antidopaminergic drug, were evaluated at yearly intervals.

All samples were drawn via an indwelling cannula placed in an antecubital vein and kept patent by slow isotonic saline infusion. Starting at 0830 hr following an overnight fast, the patients received, on separate occasions with a minimum interval of 2 days, NOM (200 mg, orally), DOM (10 mg iv, bolus) and TRH (200 μg iv, bolus). After two basal blood samples were obtained (-15 and 0 min), additional blood samples were taken at 15, 30, 45, 60 and 90 min during the DOM and TRH tests and at 30 min intervals for 4 hrs during the NOM test.

Data are expressed as the absolute decreases (NOM) or increases (DOM and TRH) below or above the basal values, compared with that obtained in normal subjects.

Plasma PRL and TSH were measured by highly specific radioimmunoassays, using reagents provided by CEA-IRE-SORIN (Saluggia, Italy). The intraassay coefficients of variation for PRL and TSH were 6.6 and 8.1, respectively. The upper limit of normal for PRL levels was 20 ng/ml in women and 12 ng/ml in men, and for TSH levels 5.5 μU/ml in both sexes. The results were analyzed by paired and unpaired Student's t tests.

RESULTS

At the last assessment (2-91 months) after adenomectomy, in the total group of patients, NOM did not induce any significant change in plasma PRL (basal values mean ± SEM = 9.1 ± 1.0 ng/ml, nadir 8.1 ± 1.1 ng/ml), in sharp contrast to the PRL-lowering effect of the drug in 25 normal subjects (basal values 11.7 ± 1.0 ng/ml, nadir 5.9 ± 0.6 ng/ml). DOM induced a significant rise in plasma PRL (basal values mean ± SEM = 8.7 ± 1.2 ng/ml, peak 75.5 ± 13.6 ng/ml), but its magnitude was markedly lower than that induced in normal subjects (basal values mean ± SEM = 7.8 ± 1.1 ng/ml, peak 143.8 ± 10.9 ng/ml). In fact, analysis of individual data revealed that normal increments in plasma PRL were shown only in 7/19 patients. Moreover, in 4 of these patients the normalization of the PRL response to DOM was observed only 2-4 years after surgery. A similar pattern was observed for the PRL response to TRH (basal values mean ± SEM = 8.3 ± 1.1 ng/ml, peak 39.5 ± 5.0 ng/ml in patients vs. basal values 7.9 ng/ml, peak 80.7 ± 14.3 ng/ml in normal subjects). After surgery, a persistent exaggerated TSH response to DOM was found in 7/16 patients (43.8%)(data not shown).

A recurrence of hyperprolactinemia has been observed 5-90 months (mean 46 months) after adenomectomy in 6/12 (50%) of patients, all of whom had a blunted response to DOM. On the other hand, in the 7 subjects in whom this response became normal after surgery, normo-prolactinemia has so far persisted at 12-96 months after surgery.

DISCUSSION

Our results show that in prolactinoma patients apparently cured after surgery, with regard to the persistence or re-establishment of the altered dopaminergic neuroregulation of PRL secretion, two subpopulations of subjects may be identified. A high frequency of recurrence has been documented only in patients bearing a persistent alteration in the dopaminergic PRL control, which could indicate that a primary defect in hypothalamic function may be present and responsible for the recurrences. However, the possibility of a persistent hyperprolactinemia-induced hypothalamic dysfunction (6) or a reduced pool of residual lactotrophs cannot be excluded.

Our results are similar to those observed in some (4,7,8), but not all (1,9,10), studies with different stimulating (TRH, insulin-induced hypoglycemia, various dopamine-antagonist drugs) or inhibiting (L-dopa-Carbidopa) tests. In most cases, however, the discrepancies could be explained on the basis of the different criteria adopted to define a "normal" response. In fact, a normal PRL response to a releasing pattern often has been defined to be at least 100% over baseline levels, which is a widely reduced value when compared to those in normal controls.

We therefore suggest that the dynamic study of PRL secretion may be helpful to evaluate the outcome of surgery in patients with prolactinoma and to predict long-term recurrence in patients at risk.

ACKNOWLEDGMENTS

This work was partially supported by grants from Ministero della Pubblica Istruzione and from Regione Piemonte, Progetto di Ricerca Finalizzata N.27 (1983).

REFERENCES

1. Barbarino A, De Marinis L, Anile C, Menini E, Merlini G, Maira G (1982) Dopaminergic mechanisms regulating prolactin secretion in patients with prolactin-secreting pituitary adenoma. Long-term studies after selective transsphenoidal surgery. Metabolism 31: 1100-1104
2. Serri O, Rasio E, Beauregard H, Hardy J, Somma M (1983) Recurrence of hyperprolactinemia after selective transsphenoidal adenomectomy in women with prolactinoma. N Engl J Med 309: 280-283
3. Rodman EF, Molitch ME, Post KD, Biller BJ, Reichlin S (1984) Long-term follow-up of transsphenoidal selective adenomectomy for prolactinoma. JAMA 252: 921-924
4. Arafah BM, Brodkey JS, Pearson OH (1986) Gradual recovery of lactotroph responsiveness to dynamic stimulation following surgical removal of prolactinomas: long-term follow-up studies. Metabolism 35: 905-912
5. Schlechte JA, Sherman BM, Chapler FK, von Gilder J (1986) Long-term follow-up of women with surgically treated prolactin-secreting pituitary tumors. J Clin Endocrinol Metab 62: 1296-1301
6. Sarkar DK, Gottschall PE, Meites J (1984) Decline of tuberoinfundibular dopaminergic function resulting from chronic hyperprolactinemia in rats. J Clin Endocrinol Metab 115: 1269-1275
7. Tucker HG, Lankford HV, Gardner DF, Blackard WG (1980) Persistent defect in regulation of prolactin secretion after successful pituitary tumor removal in women with amenorrhea-galactorrhea syndrome. J Clin Endocrinol Metab 51: 968-971
8. Schlechte JA, Sherman BM (1981) Abnormal regulation of prolactin secretion after successful surgery for prolactin-secreting pituitary tumours. Clin Endocrinol 15: 165-174
9. Jaquet P, Grisoli F, Guibout M, Lissitzky JC, Carayon P (1978) Prolactin-secreting tumors. Endocrine status before and after surgery in 33 women. J Clin Endocrinol Metab 46: 459-466
10. Reichlin S, Molitch M (1984) Neuroendocrine aspects of pituitary adenoma. In Camanni F, Muller P (eds) Pituitary Hyperfunction. New York: Raven Press: pp 47-70

ALTERED DOPAMINERGIC CONTROL OF THYROTROPIN SECRETION IN PATIENTS WITH IDIOPATHIC CENTRAL HYPOTHYROIDISM

P. Beck-Peccoz, L. Persani, C.-M. Rossi,
R. Elli, and G. Faglia

Department of Endocrinology, University of Milan
Milan, Italy

INTRODUCTION

Idiopathic central hypothyroidism (ICH) is a rare disease caused by impaired hypothalamic secretion of TRH (1-4). Previous studies have shown that in some patients with ICH, circulating TSH lacks biological activity (5-7) because of its impaired binding to the thyroid receptor (7), and that both receptor-binding and biological activities can be restored by chronic treatment with TRH (7). Idiopathic central hypothyroidism is usually part of plurihormonal pituitary deficiency of hypothalamic origin. In contrast with low circulating levels of GH, ACTH, LH and FSH, both serum TSH and PRL concentrations are frequently found in the upper limit of normal range or even slightly elevated (2-4,8-10). Because the secretion of both hormones is under the inhibitory control of dopamine (DA) (11-13), we studied the serum TSH and PRL responses to appropriate neuropharmacological treatment in six patients with ICH in order to assess their dopaminergic function.

PATIENTS

Six male patients with ICH (aged 19-35 yrs) gave their informed consent to the study. Six normal males (aged 22-43 yrs) volunteered as controls. All but one (no. 3) of the patients with ICH were born by breech delivery. Diagnosis of panhypopituitarism was made during childhood, and replacement therapies using L-thyroxine, cortisol and extracted GH commenced at that time.

Testosterone and/or gonadotrophin treatment was given when normal height was achieved. Three months before the present study, all replacement therapies were withdrawn and the assessment of hypothalamic-pituitary functions reevaluated (Tables 1 and 2). Diagnosis of central hypothyroidism was confirmed according to previously reported criteria (5), which include clinical features of hypothyroidism, low levels of circulating free thyroid hormones (FT4 and FT3), normal thyroidal response to bovine TSH (bTSH), and absence of antithyroid autoantibodies (Anti-Tg and Anti-Mc Abs) (Table 1). The secretion of TSH

TABLE 1

PATIENT CHARACTERISTICS

Parameter	Case no.						Normal range
	1	2	3	4	5	6	
Age/Sex	21/M	35/M	29/M	30/M	23/M	28/M	
TSH (mU/l)	8.1	4.5	6.6	4.3	11.7	3.1	<0.3-4.6
FT4 (pmol/l)	3.7	4.0	5.8	5.5	3.2	4.2	9.0-20.0
FT3 (pmol/l)	3.5	2.3	3.1	3.1	2.7	3.5	3.8-8.9
B/I ratio*	<0.2	<0.2	<0.2	---	<0.2	<0.2	2.9-5.6
24h ^{131}I uptake:							
-baseline (%)	14.0	6.0	11.3	---	23.4	27.5	25-50
-net increase after bTSH (%)	27.9	15.6	22.7	---	49.1	47.9	>15
T4 net increase after bTSH (%)	95	126	105	---	79	157	>50
Anti-Tg Ab	*neg*	*neg*	*neg*	*neg*	*neg*	*neg*	*neg*
Anti-Mc Ab	*neg*	*neg*	*neg*	*neg*	*neg*	*neg*	*neg*
PRL (μg/l)	20.2	8.1	49.0	16.1	34.5	36.3	<15
GH (μg/l)	0.9	0.7	0.7	0.3	0.7	0.4	<3.0
SmC (nmol/l)	14.5	7.0	11.3	17.2	8.0	15.7	18-74
ACTH (ng/l)	<15	<15	35	---	39	<15	<90
Cortisol (nmol/l)	80	44	153	252	58	54	140-700
Urinary 17OHCS (μmol/24 hr)	4.2	6.8	7.8	12.5	2.7	5.9	8-22
LH (U/l)	4.5	3.9	4.5	2.9	2.9	4.8	5.0-20.0
FSH (U/l)	2.6	3.3	3.7	4.4	2.9	3.0	4.0-15.0
Testo (nmol/l)	1.9	2.1	1.4	3.5	1.6	6.5	13.5-46.0
CT scan	*Normal*	*Partial empty sella*	*Normal*	*Normal*	*Partial empty sella*	*Normal*	

** Biologic (B) to immunologic (I) activity ratio of immunoaffinity purified serum TSH. B was measured as adenylate cyclase-stimulating activity in human thyroid membranes (7).*

with reduced biological activity has previously been documented (5,7) in five patients by means of both cytochemical bioassay and stimulation of adenylate cyclase activity in human thyroid membranes (Table 1). Serum basal TSH and PRL levels were above normal range in 3 and 5 cases, respectively. Serum basal

GH, ACTH, LH and FSH levels were definitely low, as were concentrations of circulating somatomedin C, cortisol and testosterone, as well as urinary excretion of 17OH-corticosteroids (except in the case of patient no. 4) (Table 1). Skull CT scans were normal in four patients, while a partially empty sella was seen in

TABLE 2

RESPONSES OF DIFFERENT PITUITARY HORMONES TO SPECIFIC RELEASING HORMONES (IN PARENTHESES), EXPRESSED AS NET INCREASE OVER THE BASAL VALUES

Case no.	TSH (TRH) mU/l	PRL (TRH) μg/l	GH (GHRH) μg/l	ACTH (CRH) ng/l	LH (GnRH) U/l	FSH (GnRH) U/l
1	40.3°	21	1.4	26	1.6	1.6
2	24.1°	37	1.0	7.4	1.0	0.2
3	12.5^	43	0.9	143	4.0	0.0
4	21.2°	32	---	---	3.0	1.2
5	32.3°	38	0.0	132	1.4	0.9
6	9.0^	50	0.4	0.0	1.2	1.0
Normal male range	3-13	20-80	6-20	20-150	15-60	2-10

° Prolonged and exaggerated, and ^ delayed TSH response to TRH.

two. The pituitary responses to specific releasing hormones are listed in Table 2. TSH response to TRH (200 ug) was delayed, prolonged or exaggerated in all patients, whereas PRL response to this tripeptide was normal. GH response to GHRH (1 μg/kg bw) as well as LH and FSH responses to GnRH (100 μg) were impaired in all patients. The ACTH response to CRH (1 μg/kg bw) was normal in three, impaired in one, and absent in another patient.

METHODS

Serum hormone concentrations were measured by commercial RIA kits, with the exception of TSH, which was assayed by a previously reported "in-house" method (5). Both serum TSH and PRL levels were measured before and after the following neuropharmacological treatment: a) injection of domperidone (10 mg iv; blood sampling at -120, -90, -60, -30, 0, 15, 30, 60, 90, 120 min) both in basal conditions and b) during a 240-minute DA infusion (0.06 μg/kg bw/min iv; sampling as above); or c) after pretreatment with carbidopa (50 mg per os every 6 hr for 24 hr + 35 mg 90 min before domperidone injection) plus L-dopa (100 mg per os 90 min before domperidone injection) (sampling as above, starting at -90 min). Cortisol replacement therapy (25-37.5 mg/day) was reinstated during the above testing.

Statistical analysis was carried out by analysis of variance (mixed ANOVA and split-plot design) and Tukey's test for multiple comparison (14). The results of TSH and PRL response to the various dopaminergic treatments were normalized as mean ± SD logarithm of the ratio between TSH or PRL values at a given time and their values at time 0, when domperidone was injected.

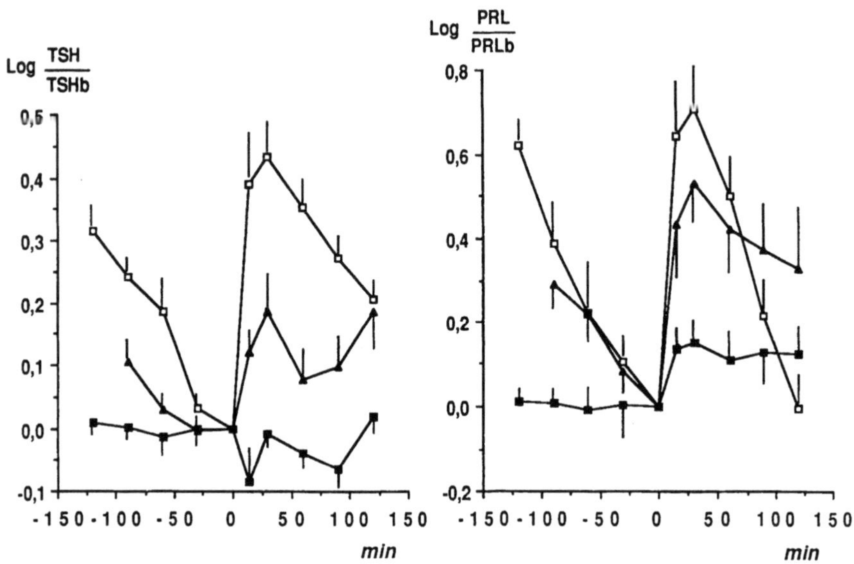

Figure 1. Serum TSH and PRL responses to domperidone (injected at time 0) both in basal conditions (■——■) and during DA administration (▲——▲) or after pretreatment with Carbidopa plus L-Dopa (□——□); here depicted are the data from the 4 responsive cases) in patients with idiopathic central hypothyroidism. Data are expressed as mean (± SD) logarithm of the ratio between serum TSH or PRL values at a given time and their basal values (time 0).

RESULTS

TSH and PRL responses to the various tests in patients with ICH are shown in Figure 1. Domperidone injection did not cause any significant variation in serum TSH levels in patients with ICH, in contrast with the significant increase ($P < 0.02$) recorded in normal controls [patients: basal = 6.1 ± 4.0 (SD), peak = 6.1 ± 3.4; controls: basal = 1.3 ± 0.9, peak = 2.4 ± 1.3 mU/l]. Serum

PRL response to domperidone in patients with ICH was significantly lower ($P < 0.01$) than that found in normal controls (patients: basal = 29.9 ± 19.6, peak = 40.2 ± 23.0; controls: basal = 8.2 ± 3.2, peak = 190.6 ± 73.0 μg/l). After DA infusion, serum TSH and PRL were lowered (TSH: from 5.1 ± 2.5 to 2.5 ± 1.1 mU/l; PRL: from 26.7 ± 15.8 to 6.1 ± 3.9 μg/l) and their response to domperidone was 7.0 ± 5.4 mU/l and 33.0 ± 21.2 μg/l, respectively. The selective enhancement of central dopaminergic tone obtained by pretreatment with carbidopa plus L-Dopa lowered both TSH and PRL concentrations, and was followed by restoration of their response to domperidone in four patients (TSH: basal = 6.4 ± 3.6, nadir = 4.4 ± 2.1, peak = 7.2 ± 2.7 mU/l; PRL: basal = 34.2 ± 17.1, nadir = 13.8 ± 8.1, peak = 56.6 ± 18.3 μg/l). For the remaining two patients, whose CT scans showed partially empty sellas, this treatment did not cause any significant variation of TSH and PRL concentrations, either before or after domperidone injection.

DISCUSSION

The present study provides evidence for an alteration of the dopaminergic control of TSH and PRL secretion in patients with idiopathic central hypothyroidism and concomitant deficiencies of other pituitary hormones (GH, ACTH, LH and FSH). In fact, the DA receptor blockade by domperidone, a drug that does not cross the blood brain barrier, did not result in an adequate increase in TSH and PRL release, thus indicating a DA deficiency at the thyrotroph and lactotroph levels (13,15). Taking into account that the hypothyroid state per se usually enhances the TSH and PRL responses to several antidopaminergic drugs, including domperidone (11), the present findings appear indicative of a DA deficiency at the pituitary cell level. Indeed the alternative explanation of these results, i.e. the existence of pituitary insensitivity to DA action, was ruled out by the finding of normal TSH and PRL inhibition during DA infusion, as well as restoration of their reactivity to domperidone administration. It is noteworthy that the restoration of normal dopaminergic control was obtained by infusion of DA in amounts (0.06 μg/kg bw/min) giving serum DA levels within the physiological range found in pituitary stalk blood (12,16). Finally, the selective enhancement of central dopaminergic tone obtained by giving L-Dopa after pretreatment with carbidopa (an agent that inhibits the pripheral, but not the central enzymatic decarboxylation of L-Dopa) (17), caused a prompt inhibition of TSH and PRL levels and restored their responses to domperidone. These results indicate that hypothalamic dopaminergic pathways are intact with respect to intrahypothalamic decarboxylation process and DA transport throughout the portal system to pituitary cells. The lack of TSH and

PRL response to this treatment, which was found in two patients, may be explained by the presence of partially empty sella that might have altered the vascular hypothalamic-pituitary connection.

These findings indicate a primary hypothalamic DA defect in patients with idiopathic central hypothyroidism and multiple pituitary trophic hormone deficiencies. This hypothalamic derangement fits well with previous evidence of multiple neurohormone deficiencies in this disease (2-4,7-10). In fact, the presence of normal or even exaggerated responses of TSH, PRL and ACTH to their specific releasing hormones rules out primary pituitary alterations. The lack of GH and gonadotropin response to GHRH and GnRH, respectively, may be explained by the long standing absence of appropriate pituitary priming by both neuro- and target hormones, which might have led to an impaired response to acute stimulation (18,19). Along this line, we may interpret the disappearance of normal gonadotropin response to GnRH previously observed in some patients.

Although little is known of the archeology of this syndrome, the fact that almost all patients were born by breech delivery suggests that traumatic or hypoxic microlesions of the hypothalamus might be responsible for the impaired release of the various neurohormones.

Finally, DA deficiency may have caused the increase in serum basal PRL levels and may contribute, along with other factors - such as low circulating levels of thyroid hormones, cortisol, and possibly somatostatin - to the finding of elevated serum basal TSH levels in patients with ICH. The possible role of DA deficiency
in the synthesis and secretion of TSH molecules with reduced biologic activity still remains to be elucidated.

ACKNOWLEDGMENTS

This work was partially supported by funds from Ministero della Pubblica Istruzione (Rome, Italy) and APREC (Milan, Italy).

REFERENCES

1. Pittman JA, Haigler ED, Hershman JM, Pittman CS (1971) Hypothalamic hypothyroidism. N Engl J Med 285: 844-845
2. Faglia G, Beck-Peccoz P, Ferrari C, Ambrosi B, Spada A, Travaglini P, Paracchi S (1973) Plasma thyrotropin response to thyrotropin-releasing hormone in patients with pituitary and hypothalamic disorders. J Clin Endocrinol Metab 37: 595-601
3. Illig R, Krawczynska M, Torresani T, Prader A (1975) Elevated plasma TSH and hypothyroidism in children with hypothalamic hypothyroidism. J Clin Endocrinol Metab 41: 722-728
4. Martin LG, Martul P, Connor TB, Wiswell JG (1972) Hypothalamic origin of idiopathic hypopituitarism. Metabolism 21: 143-149

5. Faglia G, Bitensky L, Pinchera A, Ferrari C, Paracchi S, Beck-Peccoz P, Ambrosi B, Spada A (1979) Thyrotropin secretion in patients with central hypothyroidism: evidence for reduced biological activity of immunoreactive thyrotropin. J Clin Endocrinol Metab 48: 989-998
6. Petersen VB, McGregor AM, Belchetz PE, Elkeles RS, Hall R (1978) The secretion of thyrotropin with impaired biological activity in patients with hypothalamic-pituitary disease. Clin Endocrinol 8: 397-402
7. Beck-Peccoz P, Amr S, Menezes-Ferreira MM, Faglia G, Weintraub BD (1985) Decreased receptor binding of biologically inactive thyrotropin in central hypothyroidism. Effect of treatment with TRH. N Engl J Med 312: 1085-1090
8. Costom GH, Grumbach MM, Kaplan SL (1971) Effect of thyrotropin-releasing factor on serum TSH: an approach to distinguish hypothalamic from pituitary hypothyroidism forms of idiopathic hypopituitary dwarfism. J Clin Invest 50: 2219-2225
9. Mitsuma T, Shenkman L, Suphavai A, Hollander CS (1973) Hypothalamic hypothyroidism: diminished thyroidal response to TRH. Am J Med Sci 265: 315-318
10. Sato T, Ishiguro K, Suzuki Y, Taketani T, Izumisawa A, Masuyama T, Nagaoki T, Koizumi S, Nakajima H (1976) Low setting of feed-back regulation of TSH secretion by thyroxine in pituitary dwarfism with TSH-releasing hormone deficiency. J Clin Endocrinol Metab 42: 385-389
11. Scanlon MF, Lewis M, Weightman DR, Chan V, Hall R (1980) The neuroregulation of human thyrotropin secretion. In: Martini L, Ganong WF (eds) Frontiers in Endocrinology, Volume 6. New York: Raven Press: pp 333-380
12. Connel JMC, Ball SG, Balmforth AJ, Beastall GH, DAvies DL (1985) Effect of low-dose dopamine infusion on basal and stimulated TSH and prolactin concentrations in man. Clin Endocrinol 23: 185-192
13. Delitala G, Devilla L, Lotti G (1980) Domperidone, an extracerebral inhibitor of dopamine receptors, stimulates thyrotropin and prolactin release in man. J Clin Endocrinol Metab 50: 1127-1130
14. Keppel G (1973) Design and analysis. A researcher's handbook. Engelwood Cliffs, NJ: Prentice-Hall Inc
15. Brown GM, Verhaegan H, Van Wimersma GTB, Brugmans J (1981) Endocrine effects of domperidone: a peripheral dopamine blocking agent. Clin Endocrinol 15: 275-282
16. Gibbs DM, Neill JD (1978) Dopamine levels in hypophysial stalk blood in the rat are sufficient to inhibit prolactin secretion in vitro. Endocrinology 102: 1895-1900
17. Fine SA, Frohman LA (1978) Loss of central nervous system component of dopaminergic inhibition of prolactin secretion in patients with prolactin-secreting pituitary tumors. J Clin Invest 61: 973-978
18. Borges JLC, Blizzard RM, Gelato MC (1983) Effects of human pancreatic tumor growth hormone releasing factor on growth hormone and somatomedin C levels in patients with idiopathic growth hormone deficiency. Lancet ii: 119-124
19. Mortimer CH (1973) Luteinizing hormone and follicle stimulating hormone releasing hormone test in patients with hypothalamic-pituitary-gonadal dysfunction. Br Med J 4: 73-81

C. GROWTH HORMONE NEUROREGULATION I

INTRODUCTION

There have been rapid advances in the understanding of somatotroph function in recent years. The major peptidic hypothalamic regulators of GH gene expression, synthesis and release have now been characterized. This has allowed further exploration of the interactions and intracellular message pathways involved in the control of the somatotroph. Using in vitro systems, it has also been possible to study the control of the synthesis release and gene expression of the relevant peptides themselves. For example, does the somatotroph produce IGF-I and does this or related molecules exert endocrine or paracrine actions? Using IGF-I as a model, it may be possible to understand further how functional control of anterior pituitary cell types relates to growth control in the same cell types.

In addition to neuropeptide-related mechanisms, acetylcholine has emerged as a very important hypothalamic regulator of GH function - probably by exerting a tonic inhibitory influence on the hypothalamic somatostatinergic neuron. The use of cholinergic muscarinic antagonist drugs, although of no benefit in the pathological GH hypersecretion of acromegaly, may well prove to be of benefit in the inhibition of GH release in disease states with an intact hypothalamic-pituitary axis such as diabetes mellitus.

M. F. Scanlon

LEVELS OF NEUROENDOCRINE AND PHARMACOLOGIC REGULATION OF GROWTH HORMONE SECRETION

F. F. Casanueva, B. Burguera, X. Casabiell, C. Alvarez, C. Muruais, F. Dominguez and C. Dieguez

Departments of Medicine and Physiology
Faculty of Medicine and Hospital General Galicia
Santiago de Compostela University
Santiago de Compostela, Spain

INTRODUCTION

Neuroendocrine regulation is complex. Each pituitary hormone is controlled at different levels and by a large array of messengers, in order to achieve carefully modulated release with respect to both quantity and frequency. The fine tuning of growth hormone (GH) control is exerted at pituitary, hypothalamic and suprahypothalamic levels. At each level neurotransmitters, hormones and nutrients interact to regulate GH synthesis and release. It is interesting that GH is the only pituitary hormone closely regulated by nutrients: amino acids stimulate GH secretion and both glucose and free fatty acids (FFAs) block GH release (1). However, the mechanisms underlying such actions are largely unknown.

FREE FATTY ACIDS (FFAs)

FFAs are potent regulators of GH release. Lowering of FFAs induces release of GH, a lipolytic hormone. On the other hand, FFA administration blocks GH secretion elicited by stimuli such as arginine, glucagon, L-dopa hypoglycemia, exercise, sleep, and lowered endogenous FFA levels (2-5). In order to understand the mechanism of action of FFAs, GH release was stimulated in normal volunteers by GHRH administration (1 μg/kg) with or without administration of FFAs. In order to raise FFA levels, volunteers received a lipid-heparin infusion (250 ml 10% Intralipid plus 2500 U of heparin) from -30 until 120 min. Using this approach, half the volunteers responded normally to the GHRH challenge, while in the other five GH secretion was totally blocked. Similar results were obtained by Imaki and co-workers (6). To explain this "all or none" pattern of GH blockade by FFAs, we hypothesized that fatty acids need a certain time to exert their action. In contrast, the action of GHRH is virtually immediate (7). When the lipid-heparin infusion was performed from -90 until 0 min in order to obtain an earlier (0 min) and higher FFA peak, GHRH-induced GH release was completely blocked (Fig. 1). This result confirms that FFAs require considerable time in order to exert maximum effects.

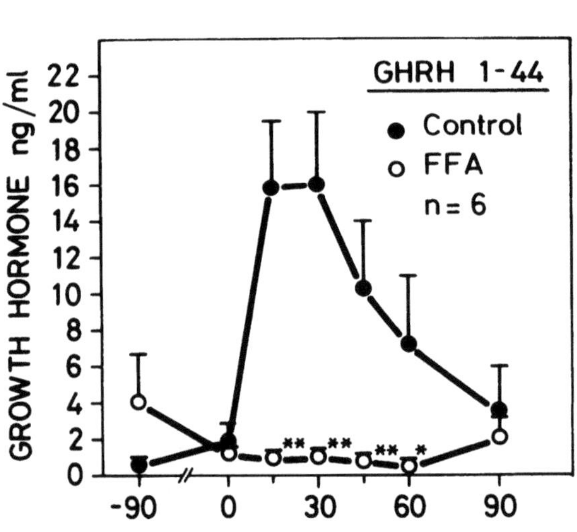

Figure 1. Plasma GH concentrations after GHRH (1 ug/kg iv) administration in six volunteers (•) and after the same stimulus (o) plus lipid-heparin infusion from -90 to 0 min. FFAs levels increased from 1.0 ± 0.1 to 11.6 ± 0.8 mMoles/l, -90 and 0 min respectively. *P < 0.05. Reprinted from ref. 8, © by The Endocrine Society, 1987, with permission.

There are two possible mechanisms for the antagonism of the GH response to exogenous GHRH by FFAs: (a) FFAs induce somatostatin release from the hypothalamus, or (b) FFAs block somatotrophs directly at the pituitary; a combination of these mechanisms may also occur. That FFAs may act through somatostatin activation is supported indirectly by our data showing that FFAs reduce the set-point of TSH secretion in normal volunteers (8). Furthermore, antisomatostatin antiserum eliminates the blocking activity of FFAs on GHRH-stimulated GH secretion in rats (9). Interestingly, in rats studied one week after complete hypothalamic ablation, FFAs are still able to block GH release. This effect may still be mediated by somatostatin (C. Dieguez, unpublished observations) since FFAs can release somatostatin from splanchnic tissues (10).

To observe if FFAs could act directly at the pituitary, rat GH secretion was studied in vitro in dispersed anterior pituitary cells incubated with or without GHRH at both 10^{-8}M and 10^{-10}M concentrations. Different concentrations of either caprylic acid (C 8) and oleic acid (C 18:1) were added to the incubation medium. Figure 2 shows that at doses close to those attained in vivo in our previous studies, FFAs were able to suppress basal or GHRH-stimulated GH secretion. At the effective dose for GH blockade, no cell toxicity was detected with any fatty acid. Nevertheless, the toxic dose of oleic acid was very close to the blocking one. A putative action of FFAs' disturbing the interaction of GHRH with its specific receptor was ruled out, because FFAs also inhibited the GH response to 10^{-5}M forskolin (8).

Although the mechanism of action of FFAs on pituitary cells is unknown, considerable data from the use of other cell models indicate that fatty acids do not act on specific receptors or by energetic transformation. In fact, after administration these compounds are not oxidized, but are immediately incorporated into cell plasma membranes. The accepted view is that FFAs alter the physical state of the lipid-bilayer, interfering with the function of integral membrane proteins, such as receptors, adenylate cyclase or others.

According to Karnowsky (11), FFAs belong to two groups: in Group A, FFAs are composed of cis-unsaturated fatty acids such as oleic acid (C 18:1); and in Group B, FFAs are composed of trans-unsaturated or cis-saturated acids such as caprylic acid (C 8). Group A FFAs partition into the fluid-phase of the membrane, i.e. more internally, and group B FFAs partition into the gel-phase lipid domain, i.e. more externally. Thus the different lipid-domain that is disrupted by each type of FFAs could account for the slight divergences in potency and the toxicity of high oleic acid

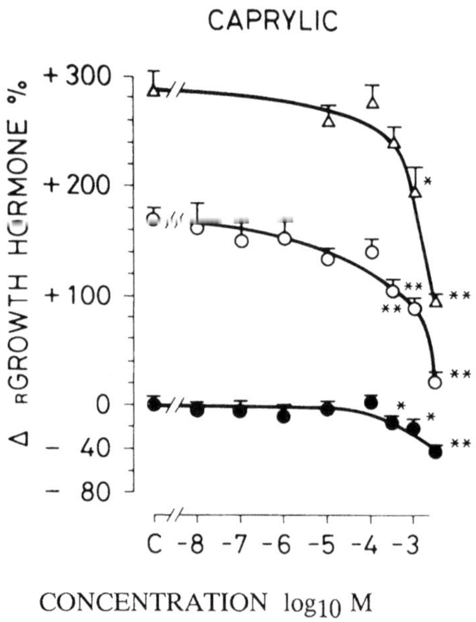

Figure 2. GH increments over basal values (%) from dispersed rat pituitary cells at different concentrations of caprylic acid after the addition to the incubation medium of GHRH 10^{-8} M (Δ), 10^{-10} M (o) or vehicle (•). C=control values, no FFAs in the medium. *$P < 0.05$. Reprinted from ref. 8, © by The Endocrine Society, 1987, with permission.

concentrations, observed in our study. This proposed mechanism of action of FFAs is reminiscent of the mechanism of action of local anesthetics on the plasma membranes (Fig. 3).

FFAs block the GH responses to all the GH stimuli tested so far and they require an appreciable amount of time in order to exert maximal effect. A combined mechanism of direct inhibition at the pituitary level and stimulation of somatostatin release from hypothalamic and/or splanchnic tissues is suggested. Although neither mechanism of regulation has been described before for the modulation of GH release by FFAs, their

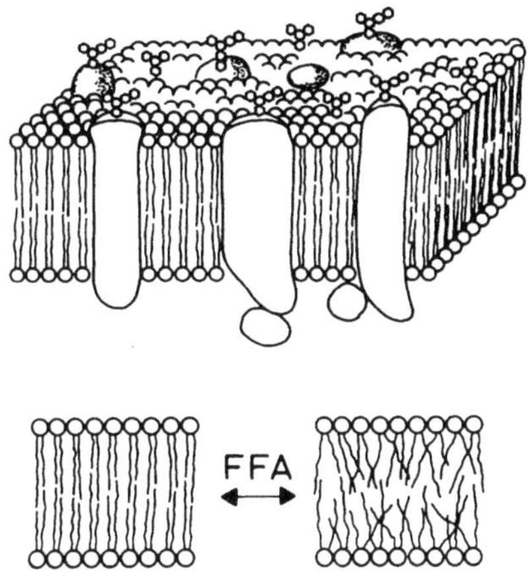

Figure 3. Model proposed for the FFA action in the plasma membrane of somatotroph cells. After incorporation into the lipid bilayer, FFAs disrupt the physical state of the integral membrane proteins, such as receptors or adenylate cyclase.

relevance if any, to the physiological regulation of GH needs to be ascertained.

ACETYLCHOLINE (Ach)

The regulation of GH secretion by acetylcholine occurs at hypothalamic levels. The final secretory state of the somatotroph is the result of the balance between GHRH and somatostatin release into the portal vessels. The release of these two neurohormones is controlled by a complex network of messengers - the hypothalamic neurotransmitters (12). Ach, the first neurotransmitter to be discovered, was neglected as a GH neuromodulator because in 1969 it was demonstrated that atropine did not prevent the reflex GH discharge after insulin hypoglycemia (13). In 1981, we demonstrated that opioid-induced GH release in the dog was completely antagonized by atropine and, surprisingly, atropine appeared more potent

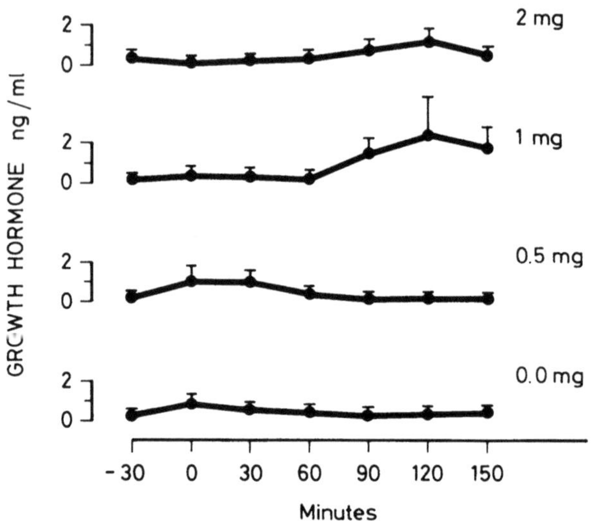

Figure 4. Administration to normal volunteers of three doses of the direct muscarinic agonist RS-86 (Sandoz) or placebo. At clinically effective doses, GH secretion was not altered, nor was cortisol secretion (data not shown).

than the specific opioid blocker naloxon (14). This finding suggested that Ach may be an important neurotransmitter in GH regulation and subsequently, other groups reported that anti-muscarinic compounds were able to block the GH responses to glucagon, arginine, clonidine and exercise (15-18). In order to understand this anticholinergic action further, a group of volunteers were challenged with GHRH (1 μg/kg iv) plus or minus atropine 1 mg sc 30 min before. Unexpectedly, the pretreatment with the antimuscarinic compound completely blocked GHRH-induced GH release and, in addition, atropine was devoid of any action on GH release by cultured anterior pituitary cells (19). This result supported the view that atropine acted through regulation of somatostatin release from the hypothalamus, and we postulated the existence of a cholinergic hypothalamic neuron exerting tonic inhibition over somatostatin release.

Recently, we have studied the action of RS-86, a direct M1 muscarinic agonist, which readily crosses the blood-brain-barrier and is devoid of side effects (20). RS-86 did not change GH basal levels in the

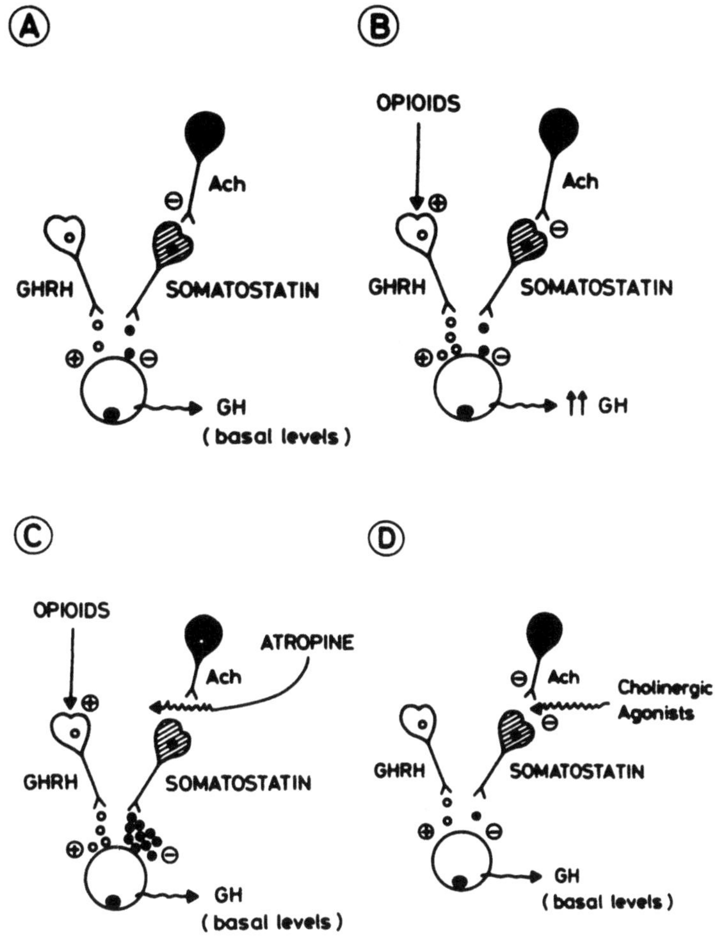

Figure 5. Working hypothesis of cholinergic regulation of growth hormone secretion. Atropine administration disinhibits the somatostatinergic hypothalamic neuron and the resulting somatostatin release is able to abolish any action of GHRH, either endogenous (e.g. by giving opioid peptides) or exogenous (by injecting GHRH).

dose-response study performed in normal volunteers (Fig. 4). This result fits well with the cholinergic hypothesis depicted in Figure 5, because increased activity at the putative inhibitory muscarinic receptor on somatostatinergic neurons would not necessarily alter plasma GH levels, depending on the level of endogenous cholinergic activity. The fact that atropine did not alter hypoglycemia-induced GH release has no explanation at the moment; it is possible that the hypoglycemia acts dually, activating endogenous GHRH release and directly inhibiting the somatostatin release, thus preventing the effect of atropine. Direct experimental evidence for the cholinergic-somatostatin hypothesis has been recently provided by Locatelli et al. (21).

The increase in our knowledge concerning the cholinergic control of GH secretion has stimulated clinical innovation. Massara and coworkers have reported that administration of the cholinesterase inhibitor piridostigmine, the net effect of which is to activate the cholinergic synapse, potentiates GHRH-stimulated GH release, and more importantly, prevents the desensitization to GHRH that occurs as a result of previous GHRH administration (22). This result suggests that combined administration of GHRH and oral piridostigmine might abolish nonresponsiveness to GHRH and the effectiveness of GHRH diagnostically and therapeutically in short stature. Another potential application came from the report that pirenzepine, an anticholinergic compound well tolerated by oral route, when administered at night prevents the sleep-associated GH release. This could provide a useful tool in situations in which a GH reduction is desirable, i.e. diabetes mellitus (23).

GLUCOCORTICOIDS

We have discussed one nutrient (FFAs) and one neurotransmitter (Ach), which regulate GH secretion at different levels and by different mechanisms. Of the peripheral hormones modulating GH release, glucocorticoids are important because of the clinical implications. Glucocorticoids are potent stimulators of GH release in rats both in vivo and in vitro, and they potentiate the action of GHRH in these situation (24,25). On the other hand, we know that in hypercortisolism, and following short-term treatment with glucocorticoids, the GH response to diverse stimuli, such as arginine, hypoglycemia or exercise, is reduced and somatic growth is blunted (26,27).

In order to understand further the mechanism of action of glucocorticoids, we performed the following GHRH tests in normal volunteers at 1200 hr. (a) control, only GHRH (1 μg/kg iv), (b) dexamethasone 4 mg was administered iv 3 hr before GHRH, and (c) dexamethasone 8 mg po was administered 12 hr before GHRH (0000 hr). GHRH induced only a moderate rise in GH levels, due to the fact that two of the six volunteers showed a very low response. Pretreatment with 8 mg dexamethasone 12 hr before, completely suppressed GHRH-induced GH release. Surprisingly, the low dexamethasone dose significantly potentiated GHRH-stimulated GH release. An interesting observation was that in the case of four volunteers, selected for nonresponse to a previous GHRH test, treatment with dexamethasone 4 mg iv 3 hr before, reestablished the normal GH response to GHRH administration.

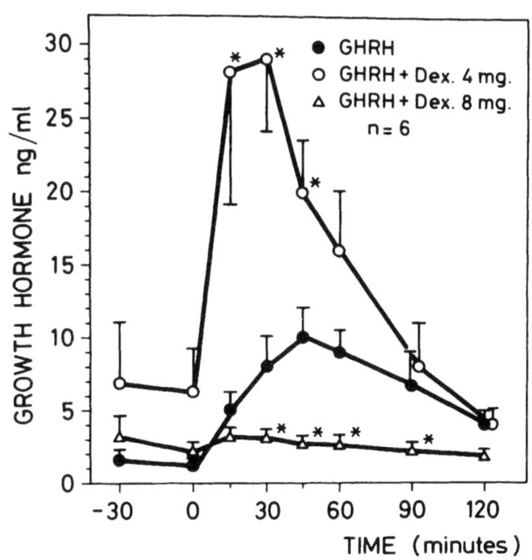

Figure 6. GHRH-induced GH secretion in three different corticosteroid situations. From ref. 28, with permission.

At present, we cannot establish the mechanism of action for the dual effects of glucocorticoid pretreatment. In rats it has been postulated that corticoids reduce somatotroph sensitivity to somatostatin while increasing GHRH receptor number. However, we cannot extrapolate from these results to man because of the known differences in the control of GH between these models. Furthermore, in rats, glucocorticoids have not been shown to inhibit GH secretion. Our hypothesis is that glucocorticoids inhibit GHRH-induced GH release through activation of somatostatin release. The short-term low-dose dexamethasone effect could be explained by a nonspecific permissive action of these steroids over GH secretion, as has been reported for insulin.

In summary, regulation of GH secretion is exerted at at least three anatomic levels. The net secretory state of the somatotrophs is determined by the complex interaction of a large variety of central and peripheral signals. The wide range of these signals is remarkable, from common metabolites such as glucose or FFAs, to complex hormones either peptidic or steroidal. The extensive research performed worldwide on these topics is expanding our knowledge about the mechanism and level of action of these compounds and will provide new tools for both diagnostic and therapeutic procedures.

ACKNOWLEDGMENTS

Supported by grants F1S 86-1033 and 87-1243, and by Xunta de Galicia 609026-1066. Figures 1 and 2 are reproduced from Casanueva et al., J Clin Endocrinol Metab (1987) Vol 65, with authorization. Figure 6 has been reproduced from Casanueva et al., Neuroendocrinology (1988) Vol 47 (i), with permission from S. Karger AG, Basel.

REFERENCES

1. Martin JB, Reichlin S (1987). Clin Neuroendocrinology 2d ed. Philadelphia: F A Davis Co
2. Quabbe JI, Bratzke HJ, Siegers U, Elban K (1972) Studies on the relationship between plasma free fatty acids and growth hormone secretion in man. J Clin Invest 51: 2388-2392
3. Blackard WG, Hull EW, Lopez A (1971) Effects of lipids on growth hormone secretion. J Clin Invest 50: 1439-1445
4. Irie M, Sakuma M, Tsushima T, Shizume K, Nakao K (1967) Effects of nicotinic acid administration on plasma growth hormone concentrations. Proc Soc Exp Biol Med 126: 708-713
5. Casanueva FF, Villanueva L, Peñalva A, Vila T, Cabezas J (1981) Free fatty acids inhibition of exercise-induced growth hormone secretion. Horm Metab Res 13: 348-352
6. Imaki T, Shibasaki T, Shizume K, Masuda A, Hotta M, Kiyosawa Y, Jibiki K, Demura H, Tsushima T, Ling N (1985) The effect of free fatty acids on growth hormone-releasing hormone-mediated GH secretion in man. J Clin Endocrinol Metab 60: 290-296

7. Frohman LA, Jansson JO (1986) Growth hormone releasing hormone. Endocrine Rev 7: 223-253
8. Casanueva FF, Villanueva L, Dieguez C, Diaz Y, Scanlon MF, Schally AV, Fernandez-Cruz A (1987) Free fatty acids block GHRH-stimulated GH secretion in man directly at the pituitary. J Clin Endocrinol Metab 65: 634-642
9. Imaki T, Shibasaki T, Masuda A, Hotta M, Yamauchi N, Demura H, Wakabayashi I, Ling N (1986) The effect of glucose and free fatty acids on growth hormone (GH)-releasing factor-mediated GH secretion in rats. Endocrinology 118: 2390-2395
10. Wasada T, Howard B, McCorkle K, Harris V, Hunger RH (1981) High plasma free fatty acid levels contribute to the hypersomatostatinemia of insulin deficiency. Diabetes 30: 358-361
11. Karnowsky MJ (1979) Lipid domains in biological membranes. Amer J Pathol 97: 212-216
12. Lightman SL, Everitt BJ (1986) Neuroendocrinology. Oxford: Blackwell Sci Pub
13. Blackard WG, Waddell CC (1969) Cholinergic blockade of growth hormone responsiveness to insulin hypoglycemia. Proc Soc Exp Biol Med 131: 192-196
14. Casaneuva FF, Betti R, Frigerio C, Cocchi D, Mantegazza P, Muller EE (1980) Growth hormone releasing effect of an enkephalin analog in the dog: evidence for a cholinergic mediation. Endocrinology 106: 1239-1245
15. Delitala G, Frulio T, Pacifico A, Maioli M (1982) Participation of cholinergic muscarinic receptors in glucagon and arginine-mediated growth hormone secretion in man. J Clin Endocrinol Metab 55: 1231-1237
16. Delitala G, Maioli M, Pacifico A, Brianda S, Palermo M, Manneli M (1983) Cholinergic receptor control mechanisms for L-dopa, apomorphine and clonidine-induced growth hormone secretion in man. J Clin Endocrinol Metab 57: 1145-1149
17. Casanueva FF, Villanueva L, Cabranes JA, Cabezas J, Fernande-Cruz A (1984) Cholinergic mediation of growth hormone secretion elicited by arginine, clonidine and physical exercise in man. J Clin Endocrinol Metab 59: 526-530
18. Casanueva FF, Villanueva L, Diza Y, Devesa J, Fernandez-Cruz A, Schally AV (1986) Atropine selectively blocks GHRH-induced GH secretion without altering LH, FSH, TSH and ACTH/cortisol secretion elicited by their specific hypothalamic releasing factors. Clin Endocrinol 25: 319-323
19. Casanueva FF, Villanueva L, Dieguez C, Scanlon MF, Schally AV, Fernandez-Cruz A (1986) Atropine blockade of GHRH-induced GH secretion in man is not exerted at pituitary level. J Clin Endocrinol Metab 62: 186-191
20. Spiegel R (1984) Effect of RS-86 an orally active cholinergic agonist on sleep in man. Psychiat Res 11: 1-13
21. Locatelli V, Torsello A, Redaelli M, Ghigo E, Massara F, Muller EE (1986) Chloinergic agonist and antagonist drugs modulate the growth hormone response to growth hormone-releasing hormone in the rat: evidence for mediation by somatostatin. J Endocrinol 111: 271-278
22. Massara F, Ghigo E, Milinatti P, Mazza E, Locatelli V, Muller EE, Camanni F (1986) Potentiation of cholinergic tone by pyridostigmine bromide reinstates and potentiates the growth hormone responsiveness to intermittent administration of growth hormone releasing factor in man. Acta Endocrinol 113: 12-16
23. Page MD, Koppeschaar H, Dieguez C, Gibbs JT, Hall R, Peters J, Scanlon MF (1987) Cholinergic muscarinic receptor blockade with pirenzepine abolishes slow wave sleep-related growth hormone release in young patients with insulin-dependent diabetes mellitus. Clin Endocrinol 26: 355-359

24. Vale W, Vaughan J, Yamamoto G, Spiess J, Rivier J (1983) Effects of synthetic human pancreatic (tumor) GH releasing factor and somatostatin, triiodothyronine and dexamethasone on GH secretion in vitro. Endocrinology 112: 1553-1556
25. Wehrenberg WB, Baird A, Ling N (1983) Potent interaction between glucocorticoids and growth hormone releasing factor in vivo. Science 221: 556-558
26. Takeda R, Nakabayashi H, Usukura U, Hosogima H, Kawato M, Uedo M (1974) Suppressive effect of glucocorticoids on arginine-induced growth hormone release in normal subjects. Steroid Lipid Res 5: 193-199
27. Von Werder K, Hane S, Forsham PH (1973) Suppression of the hypothalamo-pituitary-adrenal axis and growth hormone release with dexamethasone. Horm Metabol Res 5: 225-228
28. Casanueva FF, Burguera B, Tome M, Lima L, Tresguerres JAF, Devesa J, Dieguez C (1988) Depending on the time of administration, dexamethasone potentiates or blocks growth hormone-releasing hormone-induced growth hormone release in man. Neuroendocrinology 47: 46-49

INTRACELLULAR MECHANISMS IN GH CONTROL

G. Schettini, T. Florio, O. Meucci, E. Landolfi, M. Grimaldi,
G. Magri, B. Merola* and G. Lombardi*

Departments of Pharmacology and *Endocrinology
II School of Medicine, University of Naples
Naples, Italy

INTRODUCTION

The neuroendocrine regulation of GH secretion involves two major factors of hypothalamic origin: the recently isolated GH releasing hormone (GRF) and the inhibitory factor somatostatin (SRIF) (1-5). After having been secreted into the hypophyseal portal blood, these factors reach the anterior pituitary gland, where they interact with specific receptors (probably located on somatotrophs), thus eliciting physiological responses.

Several cellular messengers are involved in the response of secretory processes to stimulatory or inhibitory inputs. In the present chapter, we will discuss the intracellular mechanisms governing GH secretion, in particular the roles of the adenylate cyclase (AC)-cyclic AMP (cAMP), calcium (Ca^{2+}), and phosphatidylinositol-protein kinase C systems in the control of GH secretion and its responses to GRF and SRIF.

THE ADENYLATE CYCLASE-cAMP SYSTEM

It is well known that changes in cAMP accumulation, which are induced by forskolin, cholera toxin or phosphodiesterase inhibitors, as well as the use of cAMP derivatives, are associated with modifications of GH release (6-9). Cyclic AMP accumulation increases rapidly and is maintained for hours during exposure to GRF, whereas this effect is reduced by SRIF (10,11). In our experiments, we have studied the coupling of GRF and SRIF receptors with the AC-cAMP system (Fig. 1).

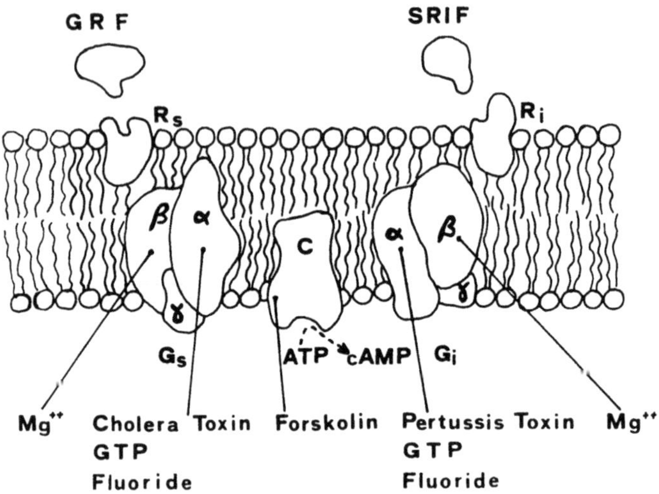

Figure 1. The adenylate cyclase system (AC) consists of three major components: the receptor; the GTP-binding protein, composed of three subunits (alpha, beta, gamma); and the catalytic unit. Both receptor and GTP-binding protein may be stimulatory and inhibitory in nature.

This figure is a schematic representation of the coupling between the GRF and SRIF receptors and the AC-cAMP system in the somatotrophs, according to the model of the enzyme proposed by Gilman. The binding of GRF to the stimulatory receptors (R_s) induces an increase in catalytic unit (C) activity through a stimulatory GTP-binding protein (G_s). Similarly the SRIF receptor-induced inhibition of C activity is mediated by an inhibitory binding protein (G_i). Both stimulatory and inhibitory GTP-binding proteins require bound GTP in order to mediate their effects on C and exert their action by dissociating their subunits into two complexes: α and β-γ. The α_s subunit, containing the GTP-binding site, interacts with C directly to stimulate cAMP formation, while the dissociation of G_i leads either to the direct inhibition by α_i on C or to the binding of the β-γ complex to α_s, preventing its stimulation of C. When GTP is hydrolized to GDP, GTP-binding proteins are inactivated. C converts ATP to cAMP, which stimulates GH release. Forskolin is a diterpene capable of activating C in the absence of both R_s and G_s. Cholera toxin inhibits GTP hydrolysis, allowing G_s to remain active in the absence of receptor stimuli; pertussis toxin by ADP rybosilation of G_i prevents its dissociation into subunits, uncoupling inhibitory receptors stimulus to C.

GRF stimulates, in a dose- (Table 1) and time-dependent manner, AC activity in membrane preparations of anterior pituitary glands (12). The stimulatory effect of GRF occurs within 30 seconds and is long-lasting. Conversely SRIF dose-dependently inhibits AC activity in membrane preparations of anterior pituitary cells (Table 1) (12,13) as well as cAMP accumulation in, and GH release by, primary cultures of anterior pituitary cells (11). Pretreatment of anterior pituitary cells with pertussis toxin, which by ADP ribosilating the α

TABLE 1

EFFECT OF VARIOUS SRIF CONCENTRATIONS ON GRF-STIMULATED ANTERIOR PITUITARY ADENYLATE CYCLASE ACTIVITY BOTH IN FEMALE AND MALE RATS

GRF (nM)		0	3		10		30		100	
SRIF (μM) ♀	0	6.08 ± 0.2	7.05 ± 0.4		9.72 ± 0.6*		12.64 ± 0.8**		21.40 ± 1.5**	
	0.1	5.92 ± 0.2			12.16 ± 0.8	+2.44	14.10 ± 1.1	+1.46	20.67 ± 0.9	-0.73
	1.0	5.39 ± 0.2			9.24 ± 0.6	-0.48	11.92 ± 0.5	-0.72	18.71 ± 1.2	-2.67
	10.0	5.41 ± 0.3	6.57 ± 0.2	-0.48	7.53 ± 0.4+	-2.19	9.97 ± 0.5+	-2.67	17.26 ± 1.1+	-4.14
	100.0	4.62 ± 0.2**	5.22 ± 0.3+	-1.83	6.86 ± 0.3+	-2.86	8.31 ± 0.7++	-4.33	10.34 ± 0.7++	-11.06
SRIF (μM) ♂	0	3.91 ± 0.2	4.44 ± 0.5		5.47 ± 0.3*		13.29 ± 0.9**		27.37 ± 0.8**	
	0.1	2.90 ± 0.2*			6.41 ± 0.6	+0.94	12.19 ± 1.3	-1.1	25.08 ± 1.4	-2.29
	1.0	3.37 ± 0.4			4.69 ± 0.5	-0.68	10.00 ± 0.4+	-3.29	24.39 ± 0.5+	-2.98
	10.0	2.66 ± 0.3*			4.22 ± 0.2+	-1.25	9.53 ± 0.6+	-3.76	21.40 ± 0.6++	-5.97
	100.0	1.22 ± 0.3**			3.28 ± 0.3++	-2.19	4.69 ± 0.2++	-8.6	9.85 ± 1.1++	-17.52

GRF dose dependently enhances enzyme activity with a greater effectiveness in male than in female tissue, the maximal stimulation is +700% in male versus +352% in female. SRIF inhibition is more pronounced in male tissue (-69% versus -24%), is dose-dependent and is dependent upon GRF concentration.

Data are expressed as picomoles of cAMP/mg prot/min ± S.E.M. and as picomoles changes versus respective GRF-stimulated control value.

*$p < 0.5$ versus respective basal value.
**$p < 0.1$ versus respective basal value.
+$p < 0.05$ versus respective GRF stimulation.
++$p < 0.01$ versus respective GRF stimulation.

Reprinted from ref. 13, with permission.

subunit of GTP-binding protein uncouples the inhibitory receptor from the catalyst, nearly abolishes SRIF inhibition of both basal and forskolin-stimulated AC activity (Fig. 2) (13) according to a previous report by Cronin et al. (14). These results confirm that the SRIF receptor is coupled with the AC enzyme.

The interaction between GRF and SRIF in modulating GH secretion is also reflected in the interaction of these two factors with the anterior pituitary AC enzyme. Indeed, by testing increasing concentrations of SRIF with those of of GRF, an enhancement of SRIF inhibition of GRF-stimulated AC activity is observed (Table 1). These results suggest that GRF and SRIF receptors are possibly coupled with a common entity of the AC enzyme, likely located on the same pituitary cell (Fig. 1).

Previous studies both in vivo and in vitro have shown a different sensitivity on the part of GH secretion between male and female animals in their responses to GRF and SRIF (15). Here we report a different sensitivity to GRF or SRIF action shown by male and female anterior pituitary AC enzyme activity. Indeed, in agreement with the secretion data, the AC enzyme derived from membranes of male anterior pituitary glands is more responsive to GRF stimulation and SRIF inhibition respectively (13). These results cannot be interpreted on the basis of a different percentage of somatotrophs between male and female alone because, although there is a higher percentage of somatotrophs in the male gland, the female pituitary contains a larger number of cells, thus the total number of GH-secreting cells is approximately equivalent in the two sexes. The results may therefore be ascribed to an intrinsically higher sensitivity to GRF and SRIF by the AC enzyme in male tissue (Table 1). The nature of this difference needs to be clarified.

These data show that changes in cAMP formation can induce modifications in GH release. However Barzeau et al. demonstrated that Ca^{2+} is necessary for GRF stimulation of both GH release and cAMP formation (6), whereas Bilezykjian et al. showed that verapamil, a Ca^{2+} channel blocker, inhibits GRF-stimulated GH secretion without affecting GRF-induced cAMP accumulation (11). All these findings suggest that an interrelationship between cAMP and Ca^{2+} in the control of GH release exists.

THE Ca^{2+} SYSTEM

Although several stimuli affect GH secretion by modifying intracellular cAMP content, they fail to elicit this response in the absence of Ca^{2+}. Thus Ca^{2+} seems central to the GH secretory process. The GRF receptor is clearly coupled with the AC-cAMP system, but little evidence is available about the relationship between GRF and Ca^{2+} homeostasis. GRF does not modify

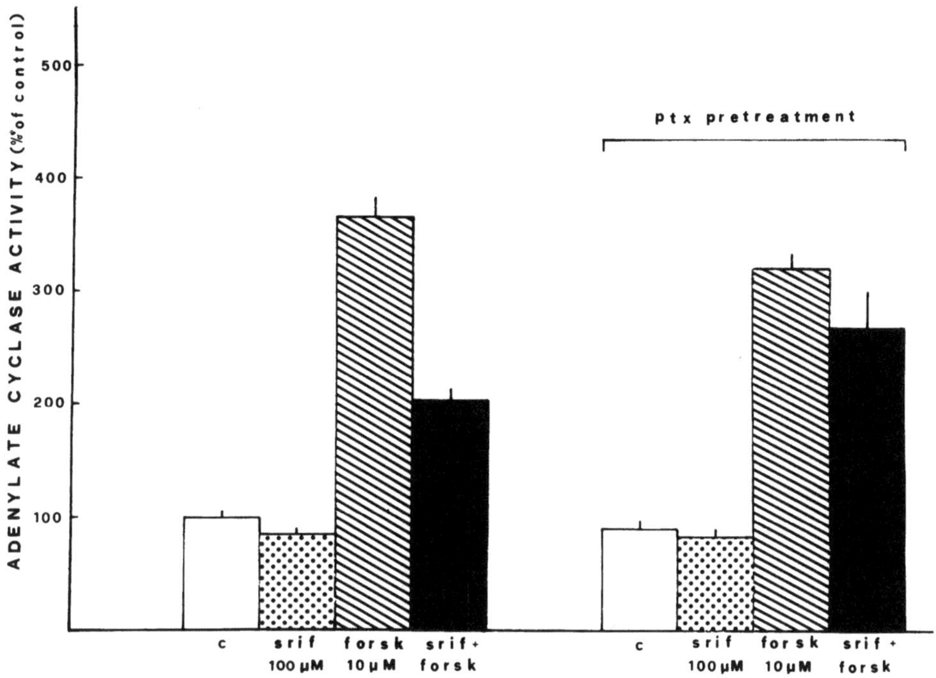

Figure 2. Effect of pertussis toxin pretreatment (100 ng/ml for 18 hr) on SRIF (100 μM) inhibition of basal and forskolin-stimulated (10 μM) AC activity in dispersed anterior pituitary cells. The toxin nearly abolished the SRIF inhibition in both conditions (from -14% to -6% in basal and from -42% to -18% in forskolin-stimulated enzyme activity). Reprinted from ref. 13, with permission.

intracellular free Ca^{2+} concentrations, as measured by the Quin 2 fluorescent probe, in anterior pituitary cells (unreported data). However, Login et al. (16) in a discussion of perifused anterior pituitary cells preloaded with $^{45}Ca^{2+}$, reported a stimulatory effect of GRF on $^{45}Ca^{2+}$ efflux that was parallel to that on GH secretion. Therefore the role of Ca^{2+} as intracellular messenger for GRF action requires further confirmation.

The mechanism of action of SRIF occurs not only by interacting with the AC-cAMP system, but also by interfering with Ca^{2+} metabolism. Indeed, several lines of evidence show the ability of SRIF to reduce intracellular Ca^{2+} concentrations in normal anterior pituitary cells, GH_3 cells, AtT20 ACTH secreting clonal cells, and β-pancreatic islets (17-20). Moreover, studies by Schlegel et al. and by Luini and colleagues show that pretreatment with pertussis toxin abolishes the SRIF inhibition of Ca^{2+} mobilization (19,21), suggesting that the coupling of SRIF receptors with Ca^{2+} channels is also modulated by a GTP-binding protein (Go). Furthermore, at least in AtT20 clonal cells, using the patch clamp technique, it has been shown that SRIF receptors are coupled with voltage-dependent Ca^{2+} channels (19). The data provided by Login et al. also support regulation of Ca^{2+} flux by SRIF. Indeed they show that SRIF is able to inhibit fractional Ca^{2+} efflux from $^{45}Ca^{2+}$-prelabeled dispersed pituitary cells, an effect that coincides temporarily with the effect on GH release (22). While an involvement for Ca^{2+} in GRF action is questionable, Ca^{2+} ions clearly seem to modulate SRIF action.

The Ca^{2+} involvement in the GH secretory process has also been studied using various methods aimed at modifying Ca^{2+} mobilization, but the nature of the regulation of GH secretion by Ca^{2+} channels has not been completely addressed. For this purpose, we used maitotoxin (MTX), a marine toxin (23) that reportedly stimulates norepinephrine release and $^{45}Ca^{2+}$ uptake in the pheochromocytoma cell line (PC12) and activates voltage-sensitive Ca^{2+} channels in a neuroblastoma cell line (25). MTX stimulates $^{45}Ca^{2+}$ uptake in dispersed anterior pituitary cells in suspension (Fig. 3) as well as GH release by primary cultures of anterior pituitary cells (Fig. 4) (26). Both these effects are rapid (30 sec and 1.5 min, respectively), long-lasting and blocked by Ca^{2+} antagonists.

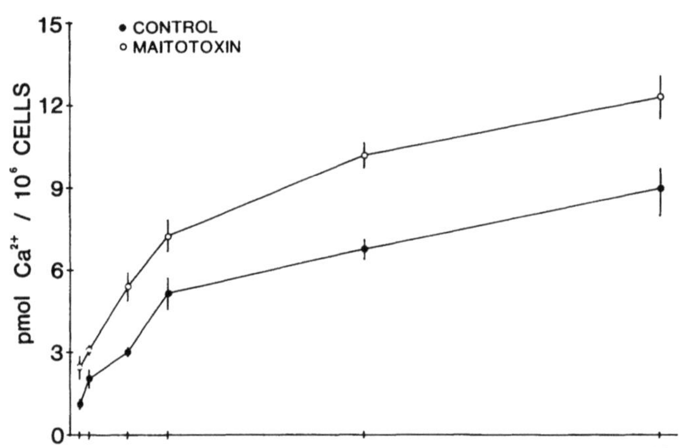

Figure 3. Effect of MTX on $^{45}Ca^{2+}$ exchange by dispersed anterior pituitary cells. Incubation of cells with $^{45}Ca^{2+}$ in the presence of 40 ng/ml MTX for indicated times caused an increase in the amount of $^{45}Ca^{2+}$ associated with cells in comparison with that of untreated cells. The effect of MTX was evident at 0.5 min ($P < 0.05$) and continued over the time course studied ($P < 0.05$). Reprinted from ref. 26, with permission.

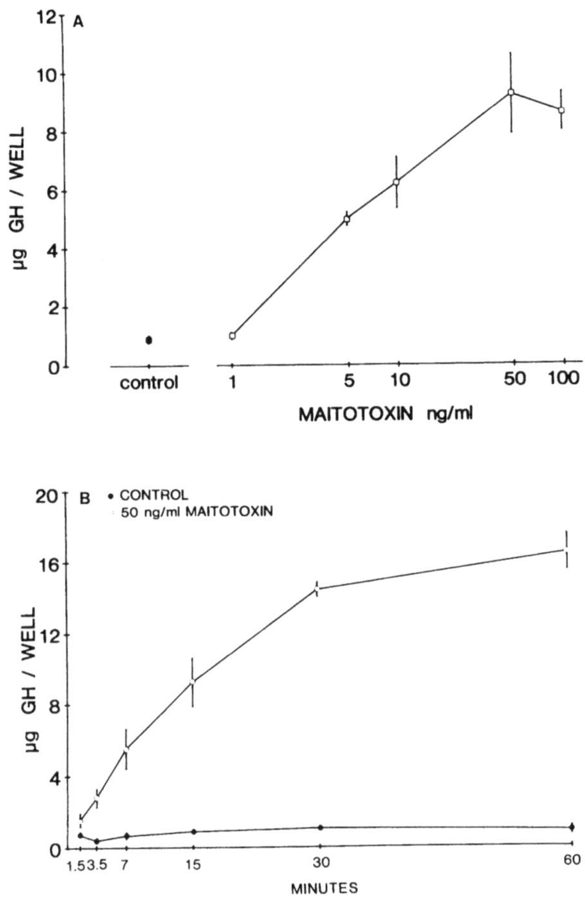

Figure 4. A) Dose response of MTX (30 min of incubation) on GH release from primary cultures of anterior pituitary cells. B) Time course of MTX stimulation of GH release from primary cultures of anterior pituitary cells. MTX (50 ng/ml) significantly ($P < 0.05$) stimulated GH release within 1.5 min of exposure, continuing throughout the next 60 min of incubation. Reprinted from ref. 26, with permission.

Thus changes in Ca^{2+} mobilization are able to modify GH secretion. By increasing intracellular Ca^{2+} influx, it is possible to elicit GH secretion.

PHOSPHATIDYLINOSITOL-PROTEIN KINASE C SYSTEM

The recent characterization of the phosphatidylinositol (PI) cycle and protein kinase C (PKC) system led to the identification of a variety of extracellular signals that activate cellular functions which operate through membrane receptors that promote the breakdown of phosphatidylinositol bisphosphate (PIP_2) (27,28). Inositol trisphosphate and diacyglycerol, which are formed during this process, mobilize intracellular Ca^{2+} stores and activate PKC, respectively; the latter effect can also be mimicked by the tumor-promoting phorbol esters, such as tetradecanoylphorbol acetate (TPA) (27,28). The evidence present in the literature does not support a clear involvement of the PI cycle in the mechanism of GRF action. Despite the findings of GRF stimulation of [^{32}P]orthophosphate incorporation into PI (29), no report shows that GRF acts through the subsequent PI-PKC system. However, the direct activation of PKC by TPA, as well as by synthetic diacylglycerol (OAG), increases the secretion of GH, an effect blocked by Ca^{2+} antagonists and SRIF (30,31). Moreover, TPA also enhances GRF-stimulated cAMP accumulation and GH release in pituitary cells (31,32). To date, no clear evidence has been presented that supports a role for a PI-PKC pathway in SRIF mechanism of action.

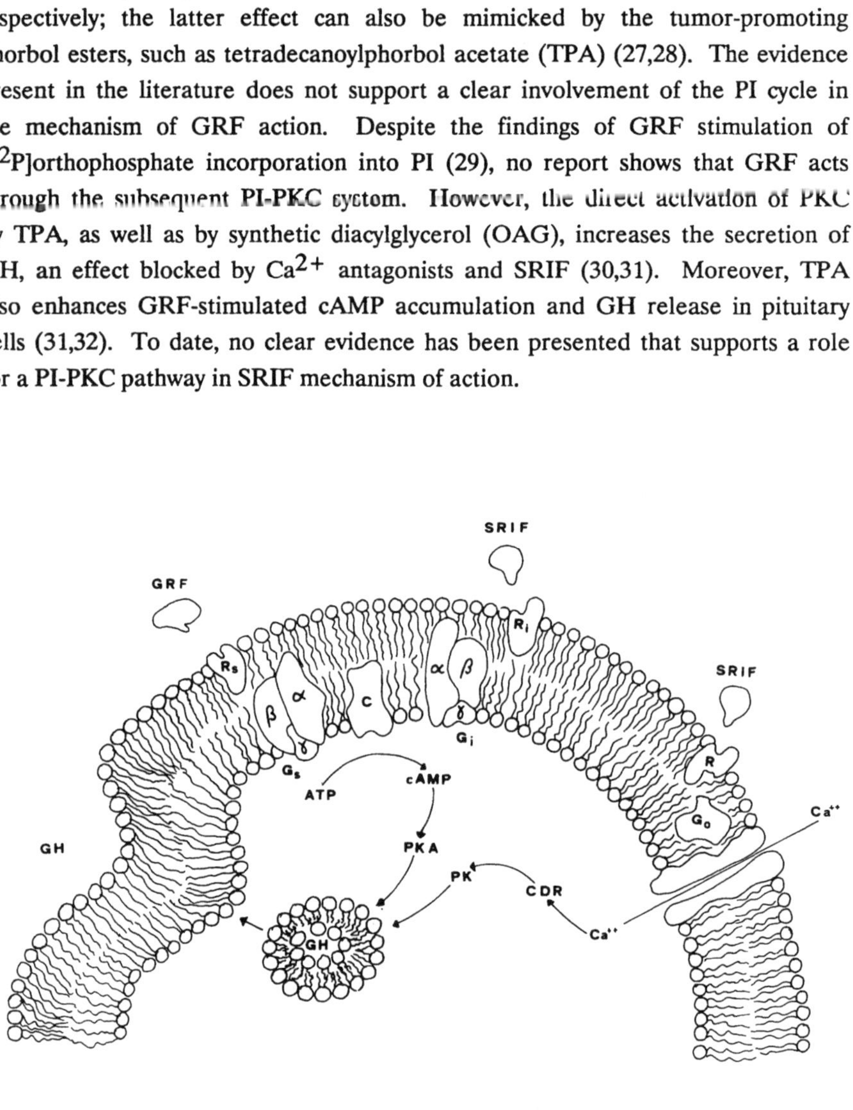

Figure 5. Schematic representation of the known intracellular mechanisms involved in GH control.

These results show that, although no direct coupling of GRF or SRIF receptors with the PI-PKC system is reported, changes in this messenger system caused by phorbol esters or OAG may modify basal or stimulated GH secretion, probably through an interaction with the cAMP generating system.

CONCLUSION

The intracellular control of GH secretion occurs primarily through the AC-cAMP and Ca^{2+} systems (Fig. 5). Cyclic AMP is the major second messenger responsible for GRF stimulation of GH release, whereas SRIF exerts its inhibitory action either by affecting the AC-cAMP generating system or by reducing Ca^{2+} mobilization.

REFERENCES

1. Brazeau P, Vale W, Burgess R, Ling N, Buccern RN, Rivier J (1973) Hypothalamic polypeptide that inhibits the secretion of immunoreactive pituitary growth hormone. Science 179: 77-79
2. Esch FS, Bohlen P, Ling N, Brazeau P, Wehremberg WV, Thorner MO, Cronin MJ, Guillemin R (1982) Characterization of a 40 residue peptide from human pancreatic tumor with growth hormone releasing activity. Biochem Biophys Res Commun 109: 152-158
3. Spiess J, Rivier J, Vale W (1983) Characterization of a rat hypothalamic growth hormone releasing factor. Nature 303: 532-535
4. Spiess J, Rivier J, Thorner MO, Vale W (1982) Sequence analysis of a growth hormone releasing factor from a human pancreatic islet tumor. Biochemistry 21: 6037-6040
5. Brazeau P, Bohlen P, Guillemin R (1982) Mechanism of the antagonism between growth hormone releasing factor (GRF) and somatostatin on the release of pituitary growth hormone. 64th Annual Meeting of the Endocrine Society Abs. 538.
6. Brazeau P, Ling N, Esch FS, Bohlen P, Mougin C Guillemin R (1982) Somatocrinine (growth hormone releasing factor) in vitro bioactivity; Ca^{2+} involvement, cAMP mediated action and additivity of effect with PGE_2. Biochem Biophys Res Commun 109: 588-595
7. Schramm M, Sellinger Z (1984) Message transmission receptor controlled adenylate cyclase system. Science 225: 1350-1356
8. Vale W, Brazeau P, Rivier C, Brown M, Boss B, Rivier LD, Burgus N, Ling N, Guillemin R (1975) Somatostatin. Recent Prog Horm Res 31: 265-279
9. MacLeod RM, Lehmeyer JE (1970) Release of pituitary growth hormone by prostaglandins and dibutyryl adenosine 3'5'-monophosphate in absence of protein synthesis. Proc Natl Acad Sci USA 67: 1172-1175
10. Cronin MJ, Rogol AD, McLeod RM, Keefer DA, Login IS, Borges JCL, Thorner MO (1983) Biological activity of a growth hormone releasing factor secreting by a human tumor. Am J Physiol 244: E346-E355
11. Bilezykjian LM, Vale W (1983) Stimulation of adenosine 3'5'-monophosphate production by growth hormone releasing factor and its inhibition by somatostatin in anterior pituitary cells in vitro. Endocrinology 113: 1726-1731
12. Schettini G, Cronin MJ, Hewelett EL, Thorner MO, MacLeod RM (1984) Human pancreatic tumor growth hormone releasing factor stimulates anterior pituitary adenylate cyclase activity, adenosine 3'5'-monophosphate accumulation and growth hormone release in a calmodulin dependent manner. Endocrinology 115: 1308-1314

13. Schettini G, Florio T, Meucci O, Landolfi E, Lombardi G, Marino A (1988) Somatostatin inhibition of anterior pituitary adenylate cyclase activity: different sensitivity between male and female rats. Brain Res 439: 322-329
14. Cronin MJ, Rogol AD, Myers GA, Hewelett EM (1983) Pertussis toxin blocks the somatostatin inhibition of growth hormone release and adenosine 3'5'-monophosphate accumulation. Endocrinology 113: 209-215
15. Cronin MJ, Rogol AD (1984) Sex difference in the cyclic adenosine 3'5'-monophosphate and growth hormone response to growth hormone releasing factor in vitro. Biol Reproduction 31: 984-988
16. Login IS, Judd AM, MacLeod RM (1986) Association of $^{45}Ca^{2+}$ mobilization with stimulation of growth hormone release by GH-releasing factor in dispersed normal anterior pituitary cells. Endocrinology 118: 239-243
17. Pandiella A, Reza Elahi F, Vallar L, Spada A (1987) Alpha-1 adrenergic stimulation of cytosolic Ca^{++} concentration and growth hormone secretion in rat somatotrophs. J Endocrinol Invest 10 (Suppl.3): 35
18. Schlegel W, Wuarin F, Wollheim CB, Zahnd GR (1984) Somatostatin lowers the cytosolic free Ca^{++} concentration in clonal rat pituitary cells (GH_3 cells). Cell Calcium 5: 223-236
19. Lewis DL, Weight F, Luini A (1986) A guanine nucleotide binding protein mediates inhibition of voltage dependent calcium current by somatostatin in a pituitary cell line. Proc Natl Acad Sci USA 83: 9035-9039
20. Sussman KE, Leitner JW, Draznin B (1987) Cytosolic free calcium concentrations in normal pancreatic islet cells: effect of secretagogues and somatostatin. Diabetes 36: 571-577
21. Schlegel W, Wuarin F, Zbaren C, Wollheim CB Zahnd GR (1985) Pertussis toxin selectively abolishes hormone induced lowering of cytosolic calcium in GH_3 cells. FEBS Lett. 189: 27-32
22. Login IS, Judd AM (1986) Trophic effects of somatostatin on calcium flux: dynamic analysis and correlation with pituitary hormone release Endocrinoloy 119: 1703-1707
23. Yasumoto T, Nakajima I, Oshima Y, Bagnis R (1979) A new toxic dinoflagellate found in association with ciguatera. In Taylo DL and Selinger H (eds) Toxic Dinoflagellate Blooms. New York: Elsevier: pp 65-70.
24. Takahashi M, Tatsumi M, Okizumi Y, Yasumoto T (1983) Ca^{2+} channel activating function of maitotoxin, the most potent marine toxin known, in clonal rat pheochromocytoma cells. J Biol Chem 258: 10944-10949
25. Freedman SB, Miller RJ, Miller DM, Tindall Dr (1984) Interaction of maitotoxin with voltage sensitive calcium channels in cultured neuronal cells. Proc Natl Acad Sci USA 81: 4582-4585
26. Schettini G, Koike K, Login IS, Judd AM, Cronin MJ, Yasumoto T, MacLeod RM (1984) Maitotoxin stimulates hormonal release and calcium flux in rat anterior pituitary cells in vitro. Am J Physiol 247: E520-E525
27. Berridge MJ (1984) Inositol trisphosphate and diacylglycerol as second messengers. Biochem J 220: 345-360
28. Nishizuka Y (1984) The role of protein kinase C in cell surface signal and tumor promotion. Nature 308: 693-698
29. Canonico PL, Cronin MJ, Thorner MO, MacLeod RM (1983) Human pancreatic GRF stimulates phosphatidylinositol labelling in cultured anterior pituitary cells. Am J Physiol 245: E587-E590
30. Ohmura E, Tsushima T, Murakami H, Wakai K, Shizume K (1984) Effect of phorbol esters on the release of growth hormone and prolactin from rat pituitary cells cultured in monolayer. Acta Endocrinol 107: 185-190
31. Summers ST, Canonico PL, MacLeod RM, Rogol AD, Cronin MJ (1985) Phorbol esters affect pituitary growth hormone and prolactin release: interaction with GH-releasing factor, somatostatin and bromocriptine. Eur J Pharmacol 111: 371-376
32. Cronin MJ, Canonico PL (1985) Tumor promoters enhance basal and growth hormone releasing factor stimuated cyclic AMP levels in anterior pituitary cells. Biochem Biophys Res Commun 129: 404-410

ALPHA-1-ADRENERGIC STIMULATION OF CYTOSOLIC Ca^{2+} CONCENTRATION AND GROWTH HORMONE (GH) IN RAT SOMATOTROPHS

A. Pandiella, F. Reza Elahi, L. Vallar and A. Spada*

CNR Center of Cytopharmacology, Department of Pharmacology and
*Department of Endocrinology, University of Milan
Milan, Italy

INTRODUCTION

It is well established that the α-adrenergic system plays a primary role in the regulation of growth hormone (GH) release in humans and animals (1-3). As far as the site of action of α-adrenergic agents is concerned, the high levels of epinephrine in the portal circulation (4) as well as the identification of α_1-adrenergic receptors in the pituitary gland, point to the possible direct effect of adrenergic agents on hormone release. However, studies with isolated anterior pituitary cells have yielded rather conflicting results: α-adrenergic drugs have been reported either to inhibit, stimulate or not to influence GH release from pituitary cells in culture (3). With respect to the mechanisms of action of α-adrenergic agents, in all the systems so far studied, such as vascular smooth muscle cells, hepatocytes and myocardial cells, α_1 adrenergic agents have been demonstrated to increase cytosolic free Ca^{2+} concentrations (5-7).

In the present study, the effects of the α_1-adrenergic agent phenylephrine (PHE) on in vitro GH release and cytosolic free Ca^{2+} concentration ($[Ca^{2+}]_i$) were investigated in highly purified rat somatotroph preparations using a selective indicator of $[Ca^{2+}]_i$.

MATERIALS AND METHODS

Preparation of Somatotrophs

Anterior pituitaries from 20 male Wistar rats weighing 250-300 g were dissociated enzymatically, applied over an albumin (0.3-2.4%) gradient (prepared according to a slight modification of ref. 8) and sedimented at 1 g for 3.2 hr. Nine 100 ml fractions were collected. Immunofluorescence analysis using specific anti-GH and anti-PRL antibodies revealed that in the three bottom fractions, the vast majority (75-80%) of cells were somatotrophs. The three fractions were combined and the cells either allowed to attach in plastic multiwells over 3 days for secretion studies or maintained in suspension in bacteriological dishes for $[Ca^{2+}]_i$ measurements after 1 day in culture.

Secretion Studies

After 3 days in culture, cells were washed twice with DMEM without serum. The cells were incubated in 1 ml serum-free DMEM containing test substances (added in 100 μl aliquots) and 0.1% ascorbic acid. Four replicate wells were used per treatment. The cells were then placed in a water-saturated atmosphere of 5% CO_2 - 95% air at 37°C for 15 min. At the end of the incubation, media were removed and stored at -20°C until radioimmunoassay (RIA). GH was measured with materials provided by the National Institute of Diabetes and Digestive Diseases and Kidney Rat (NIDDK) Pituitary Hormone Distribution Program. The intrassay and interassay coefficients of variations were 5% and 9.4%, respectively.

Measurement of $[Ca^{2+}]_i$

After 1 day in culture, the somatotroph-enriched fractions of anterior pituitary cells were collected from bacteriological dishes, washed and resuspended at 4-5 x 10^6 cells/ml in Krebs-Ringer HEPES incubation medium (KRH) that contained (in mmol/liter): NaCl, 125; KCl, 5; KH_2PO_4 and $MgSO_4$, 1.2; $CaCl_2$, 2; HEPES-NaOH, pH 7.4, 25; glucose, 6. Somatotrophs were loaded with the Ca^{2+}-indicator fura 2 (9) by incubating the cells with 5 μM fura-2-acetoxymethylester for 30 min at 37°C. Thereafter, the cell suspension was diluted with warm KRH, washed and resuspended in KRH at 5 x 10^5/mml. Fluorescence recordings were carried out in a Perkin-Elmer LS5 spectrofluorimeter and calculation of $[Ca^{2+}]_i$ was performed as recommended by Grynkiewicz et al. (9).

Materials

The Fura 2-acetoxymethylester was purchased from Molecular Probes (Junction City, OR). Anti-GH and anti-PRL antibodies were a kind gift from Dr. A. Zanini, phentolamine from Dr. E. Sher. Phenylephrine, isoproterenol, propranolol and clonidine were purchased from Sigma (St. Louis, MO). Culture Media were from Flow Laboratories (Mackenheim, West Germany); all other chemicals were reagent grade.

RESULTS

Effect of PHE on In Vitro GH Release

Incubation of somatotroph preparations with the α_1-agonist PHE caused an increase in GH release, while clonidine, a predominantly α_2-adrenergic agonist, was ineffective (Table 1). The α-adrenergic antagonist phentolamine completely abolished the effect of PHE, while the β-antagonist propranolol only partially counteracted its action (Table 1).

TABLE 1

EFFECT OF ADRENERGIC AGENTS ON GH RELEASE FROM SOMATOTROPHS IN MONOLAYER CULTURE

Drugs	GH release (ng/well/30 min)
0	105 ± 10
PHE 0.1 μM	210 ± 12
Clonidine 0.1 μM	110 ± 9
PHE 0.1 μM + Propranolol 0.1 μM	160 ± 12
PHE 0.1 μM + Phentolamine 0.1 μM	100 ± 10

Values are the means ± SE of quadruplicate determinations.

Effect of PHE on $[Ca^{2+}]_i$

Resting $[Ca^{2+}]_i$ levels of somatotrophs were variable, ranging from 106 to 248 nM. The addition of PHE caused a $[Ca^{2+}]_i$ rise that was complete within 20 sec and was maintained for at least 5-10 min. (Fig. 1). In Ca^{2+}-containing medium, $[Ca^{2+}]_i$ transients induced by 1 μM PHE were variable, the peak values ranging from 280 to 410 nM (mean ± SE: 370 ± 55 nM). When the antagonist agent phentolamine was applied after PHE in equimolar concentrations, the PHE-induced $[Ca^{2+}]_i$ rise was quickly interrupted and $[Ca^{2+}]_i$ levels rapidly declined to a steady state indistinguishable from the original basal value (Fig. 1). Moreover, preexposure of somatotrophs to phentolamine completely prevented the PHE effect on $[Ca^{2+}]_i$, without modifying resting $[Ca^{2+}]_i$. The specific β-receptor agonist isoproterenol caused an increase in $[Ca^{2+}]_i$ that was smaller and more transient than that induced by PHE. Moreover, the addition of PHE after isoproternol caused an additional marked rise in $[Ca^{2+}]_i$ (data not shown).

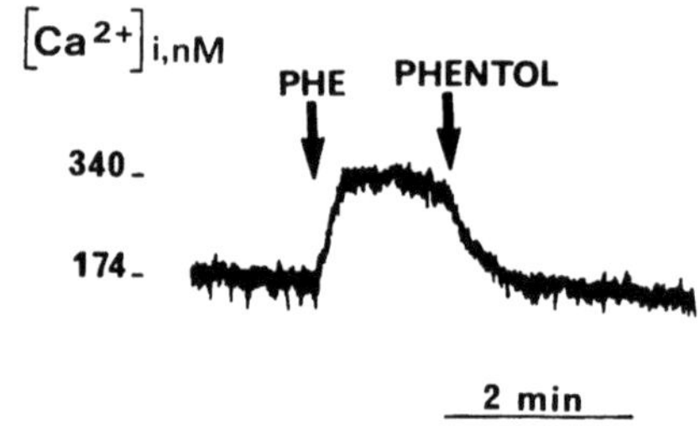

Figure 1. Fura 2 loaded somatotrophs were inserted in a thermostatically controlled cuvette. Where indicated, phenylephrine (PHE, 1 μM) and phentolamine (PHENTOL, μM) were added to the cell suspension.

Effect of SRIF on $[Ca^{2+}]_i$

Upon addition of SRIF to resting somatotrophs, $[Ca^{2+}]_i$ decreased (about 10-20% of resting $[Ca^{2+}]_i$), and a lowered steady state of $[Ca^{2+}]_i$ was reached after 1-2 min and maintained for at least 10 min. The effect of SRIF completely depended on extracellular Ca^{2+}. The exposure to SRIF dramatically affected the $[Ca^{2+}]_i$ rise induced by adrenergic activation. SRIF, added 4 min after PHE, rapidly reduced $[Ca^{2+}]_i$ to basal values, with an effect indistinguishable in terms of kinetics from that brought about by the specific receptor antagonist (Fig. 2).

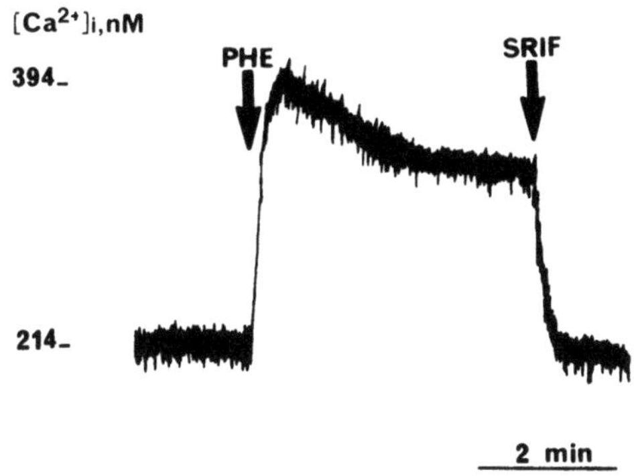

Figure 2. Effect of SRIF (500 nM) on $[Ca^{2+}]_i$ rise induced by 1 μM PHE.

DISCUSSION

In the present study, the effectiveness of PHE in stimulating GH release from rat somatotrophs in culture is clearly demonstrated. This effect appears to be mediated by α_1-adrenergic receptors since GH response to PHE was completely blocked by an α-adrenergic antagonist and only partially reversed by the specific β-adrenergic antagonist propranolol. As far as the efficacy and potency of PHE on GH release are concerned, the effectiveness of PHE was similar to that caused by the activation of β-adrenergic receptors in somatotrophs (8). It is worth noting that PHE was effective at concentrations that, although elevated (in the 10-1000 nM range), approach the concentrations of epinephrine physiologically present in the portal circulation.

In view of these results, it is tempting to speculate that the GH response to α-adrenergic stimulation may result from the sum of the direct effects of adrenergic agents at the somatotroph level and the central modulation of GRF discharge. However, it remains completely unknown whether somatotroph responsiveness to direct α-adrenergic stimulation occurs also in humans.

While GRF and β-adrenergic agents stimulate GH release by increasing cAMP levels (10,11), PHE markedly affected resting $[Ca^{2+}]_i$ in somatotrophs. The effect was mediated by activation of α_1-adrenergic receptors as shown by the use of specific α- and β-adrenergic blockers. As expected, the β-adrenergic agonist isoproterenol showed a very small effect on $[Ca^{2+}]_i$, possibly due to the slight action on α-receptors displayed by the β-adrenergic agent.

As already observed in pituitary tumor cell lines, SRIF reduced steady state $[Ca^{2+}]_i$ in normal rat somatotrophs (12). As characterized in detail in GH_3 cells, the inhibition of Ca^{2+} influx seems to occur as a consequence of SRIF-induced hyperpolarization (12). Not only $[Ca^{2+}]_i$ resting levels, which probably result from the average of a large number of spontaneous fluctuations in $[Ca^{2+}]_i$ occurring randomly, but also $[Ca^{2+}]_i$ rises due to the Ca^{2+} influx stimulated by α_1-adrenergic agents were dramatically reduced by SRIF.

In conclusion, GH release from rat somatotrophs seems to be under a complex control brought about by several stimulatory and inhibitory agents involving at least two second messenger (Ca^{2+} and cAMP) pathways. Here is reported the first evidence that in somatotrophs, $[Ca^{2+}]_i$ levels are stimulated by α_1-adrenergic agents and that this triggers $[Ca^{2+}]_i$ rise results in a stimulation of GH release.

REFERENCES

1. Martin JB (1976) Brain regulation of growth hormone secretion. In: Martini L, Ganong WF (eds) Frontiers in Neuroendocrinology. New York: Raven Press: pp 129-168
2. Arimura A, Culler MD (1985) Regulation of growth hormone secretion. In: Imura H (ed) The Pituitary Gland. New York: Raven Press: pp 221-259
3. Toumisto J, Mannisto P (1985) Neurotransmitter regulation of anterior pituitary hormones. Pharmacol Rev 37: 249-332
4. Johnston CA, Gibbs DM, Negro-Vilar A (1983) High concentrations of epinephrine derived from a central source and 5-hydroxyindole-3-acetic acid in hypophysial portal plasma. Endocrinology 113: 819-821
5. Exton JH (1985) Mechanisms involved in α- adrenergic phenomena. Am J Physiol 248: E633-E647
6. Morgan NG, Shuman EA, Exton JH, Blackmore PF (1983) Stimulation of the α_1-adrenergic control of hepatocyte redistribution by increases in cyclic AMP. J Biol Chem 258: 5110-5116
7. Blackmore PF, Waynick LE, Blackman GE, Graham LW, Sherry RS (1984) α and β-Adrenergic stimulation of parencymal cell Ca^{2+} influx. Influence of extracellular pH. J Biol Chem 259: 12322-12325
8. Swennen L, Baes M, Schramme C, Denef C (1985) β-Adrenergic stimulation of adenosine-3',5'-monophosphate (cAMP) in primary cultures of rat anterior pituitary cell populations separated by unit gravity sedimentation. Relationship to growth hormone and prolactin release and to non-secreting cells. Neuroendocrinology 40: 78-83
9. Grynkiewicz G, Poenie M, Tsien RY (1985) A new generation of Ca^{2+} indicators with greatly improved fluorescence properties. J Biol Chem 260: 3440-3450
10. Spada A, Vallar L, Giannattasio G (1984) Presence of an adenylate cyclase dually regulated by somatostatin and hormone pancreatic growth hormone releasing factor in GH secreting cells. Endocrinology 115: 1203-1209
11. Rivier J, Spiess J, Thorner M, Vale W (1982) Characterization of growth hormone-releasing factor from a pancreatic islet tumour. Nature 300: 276-278
12. Schlegel W, Wuarin F, Zbaren C, Wollheim CB, Zahnd GR (1985) Pertussis toxin selectively abolishes hormone induced lowering of cytosolic calcium in GH_3 cells. FEBS Lett 189: 27-32

L-DOPA RELEASES GH VIA A GRF-DEPENDENT MECHANISM IN NORMAL HUMAN SUBJECTS, WHEREAS ARGININE, CLONIDINE AND ADRENALIN PLUS PROPRANOLOL DO NOT

C. Dieguez*, R. Valcavi**, M. D. Page, M. Zini**,
P. Casoli**, C. A. Edwards, I. Portioli** and M. F. Scanlon

Neuroendocrine Unit, Department of Medicine, University of Wales
College of Medicine, Cardiff, Wales, U. K.
*Department of Physiology, School of Medicine,
Santiago de Compostela, Spain
**Sezione Endocrino Metabolica, Arcispedale S. Maria Nuova,
Reggio Emilia, Italy

INTRODUCTION

Growth hormone (GH) secretion by the anterior pituitary (AP) gland is regulated by several neurotransmitter pathways and a variety of peripheral feedback signals that act either directly at the AP level or by modulating GRF and somatostatin (SRIF) release from the hypothalamus (1). Until recently, clinical investigation of GH reserve was based on the use of several different tests including insulin-induced hypoglycemia, arginine, L-dopa and clonidine, although the mechanisms by which these secretagogues elicit GH secretion is poorly understood. In an attempt to answer this question, we have used the experimental model first suggested by the findings of Shibasaki et al.: administration of 200 μg of GRF to normal human subjects completely abolishes the GH responses to GRF given two hr later, while GH responses to hypoglycemia are maintained (2). This indicates that hypoglycemia releases GH through a non-GRF-dependent mechanism, and in the studies reported here we have used α-adrenergic agonism, arginine and L-dopa as the GH secretagogues rather than hypoglycemia.

SUBJECTS AND METHODS

During the in vivo studies, tests were performed in random order, at least one week apart using five groups of normal subjects (aged 18-42). The subjects fasted from midnight and remained recumbent throughout the studies, which commenced at 0830 hr. Thirty min after the insertion of forearm intraveneous cannulae, six subjects received GRF 1-44 (200 μg iv Sanofi) at time 0 min and again at 120 min. Another six subjects received arginine (30 gm iv over 30 min), and on a separate occasion GRF 1-44 (200 μg iv at 0 min), followed by arginine from 120 min to 150 min. An additional six subjects received L-dopa (500 mg po), and on a separate occasion GRF (200 μg iv at 0 min) followed by L-dopa at 120 min. Five subjects underwent a propranolol (1.5 mg/min iv from 110 min to 120 min) plus adrenalin (10 μg/min iv from 120 min to 150 min) infusion after either GRF (200 μg iv at 0 min) or placebo administration. Finally, seven subjects received clonidine (150 μg bolus iv) at 120 min after prior GRF (200 μg iv at 0 min) or placebo administration. In vitro studies were carried out using monolayer cultures of rat AP cells as described previously (3).

RESULTS

In Vivo Studies

GRF administration at 0 min led to a clear rise in plasma GH, which did not occur when GRF administration was repeated at 120 min (Fig. 1). GH levels peaked at 45 min after arginine administration on both occasions (Fig. 2). However, when arginine was administered following GRF pretreatment, the peak GH level attained was higher than when arginine was administered alone (55.3 ± 16.5 vs. 19.3 ± 6.4, SEM, $p < 0.02$) as was the area under curve (mean ± SEM 1404 ± 412 vs. 523 ± 233, $p < 0.02$). Following GRF pretreatment, L-dopa administration failed to increase plasma GH, whereas L-dopa alone led to a clear increase in GH (peak GH, 30.8 ± 13 mU/l) (Fig. 2). Propranolol-adrenalin infusion (Fig. 3) or clonidine administration (data not shown) led to similar increases in plasma GH levels irrespective of whether GRF or placebo was administered previously at time 0 min.

In Vitro Studies

As expected, GRF increased and SRIF decreased GH secretion by rat AP cells in monolayer culture (Fig. 4). Neither arginine nor L-dopa affected basal GH secretion or GH responses to GRF and SRIF.

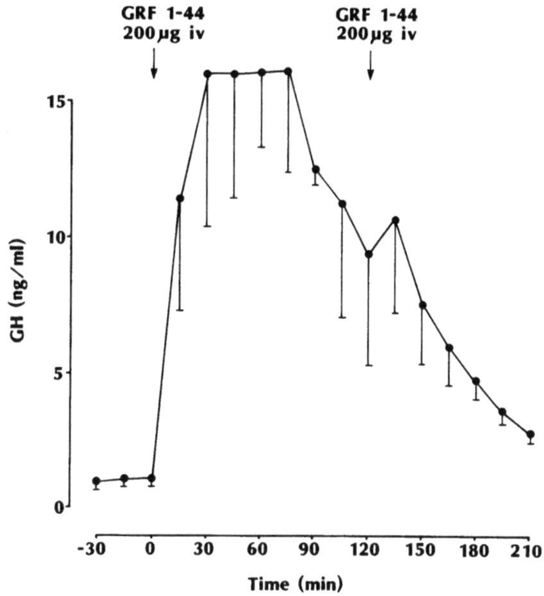

Figure 1. GH responses to GRF 1-44 administration at time 0 and 120 min in normal subjects (n = 6). Values are mean ± SEM.

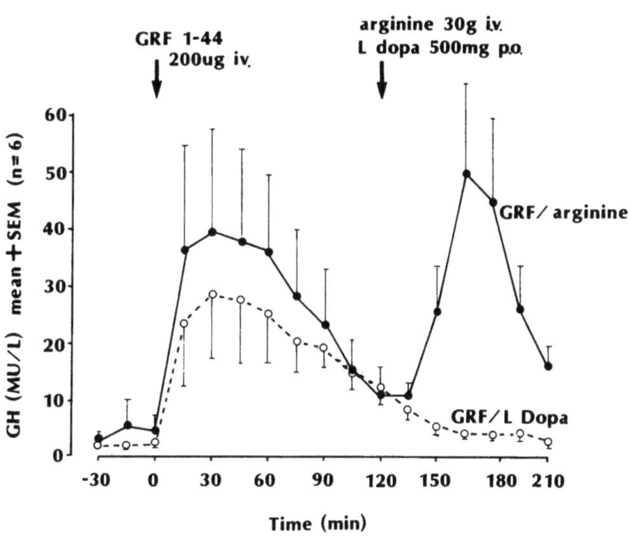

Figure 2. GH response to L-dopa and arginine following GRF pretreatment in normal male subjects (n = 6). Values are mean ± SEM.

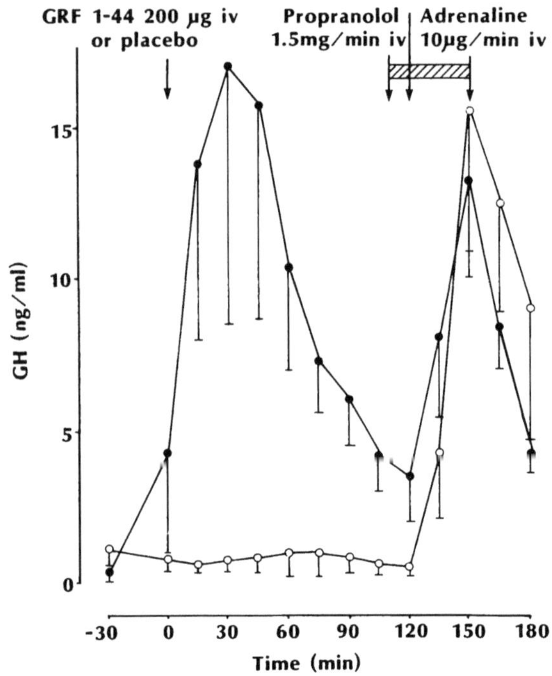

Figure 3. Effect of GRF 1-44 (●—●) or placebo (o—o) administration given at time 0 min on the GH responses to propranolol-adrenalin infusion, in normal subjects (n = 5). Values are mean ± SEM.

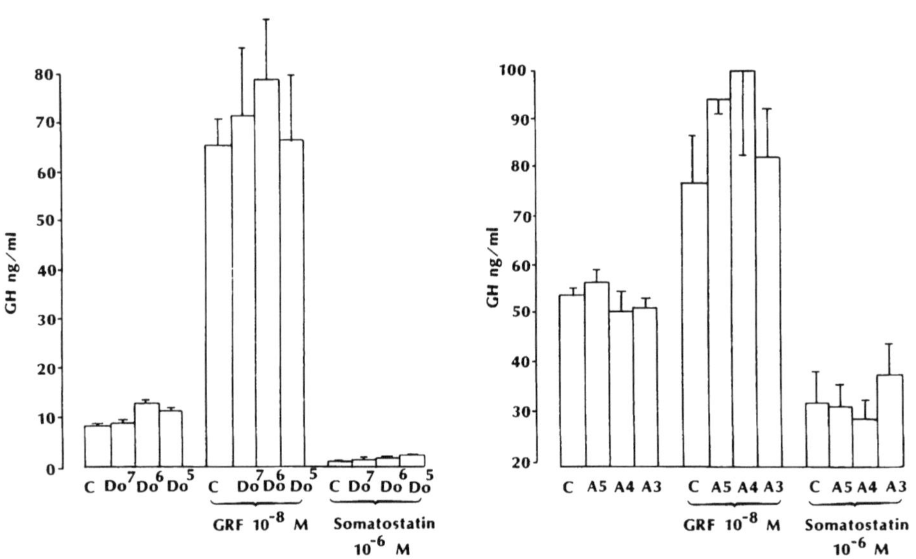

Figure 4. Effect of different concentrations of L-dopa (Do, 10^{-7}M-10^{-5}M) and arginine (A, 10^{-5}M-10^{-3}M) on basal GH secretion (C) and GH responses to GRF (10^{-8}M) and somatostatin (SRIF, 10^{-6}M) by rat anterior pituitary cells in monolayer culture. Values are mean ± SEM of 2-4 different experiments.

DISCUSSION

These data confirm previous reports that pretreatment with GRF completely abolishes the GH responses to the subsequent administration of GRF two hr later (2). In contrast, arginine, propranolol plus adrenaline, and clonidine each caused a clear rise in GH levels, irrespective of whether GRF or placebo was administered previously. In view of this, it is unlikely that the GH responses to arginine, propranolol plus adrenalin, and clonidine are mediated by endogenous GRF, at least at the AP level. However, it should be pointed out that our experimental design does not allow us to exclude completely any action of these substances on GRF release. It is possible that they may exert a dual action to increase GRF and decrease SRIF release from the hypothalamus. Furthermore, we found no effect of clonidine (unpublished observations) or arginine on GH release in vitro, which argues against a direct AP action of these agents in the stimulation of GH release. It is interesting that hypoglycemia also stimulates GH release two hr after GRF pretreatment (2). It has been suggested that hypoglycemia abruptly reduces somatostatinergic tone, allowing rebound release of GH (4-5). This could also explain the enhanced GH responses to arginine and hypoglycemia observed after GRF pretreatment. It is unlikely, however, that α-adrenergic agonists, arginine and hypoglycemia act through similar mechanisms since the continued administration of α-adrenergic agents together with arginine or hypoglycemia causes an additive effect on GH release (6). Furthermore, cholinergic blockade, the administration of exogenous free fatty acids, and TRH infusion completely abolish the GH response to arginine, but have little effect on the GH response to hypoglycemia (1).

Our findings of normal GH responses to α-adrenergic agonism after GRF pretreatment, at a time when GH responses to a supramaximal dose of GRF are absent, was somewhat unexpected. Until now most available data, albeit obtained in rats, indicated that α-2 adrenergic activation increases plasma GH levels by stimulating GRF release from the hypothalamus: clonidine-induced GH release in normal rats is abolished by prior administration of anti-rGRF antibodies (7), and recent data have indicated that α-2 adrenergic agonism causes GRF release from perfused rat hypothalami (8). Although our data show that the GH response to α-adrenergic agonism is not GRF-mediated, at least at the pituitary level, we cannot exclude the possibility that GRF may be involved in this response at the hypothalamic level since GRF can also stimulate SRIF release from the hypothalamus in vitro (9).

A possible explanation for these somewhat conflicting data can be proposed if there exist separate populations of intrahypothalamic GRF-producing neurons: one group that releases GRF into hypophyseal portal blood, which is under stimulatory α-2 adrenergic control, and a separate group of GRF-producing neurons that stimulate SRIF release into hypophyseal portal blood, which is under inhibitory α-2 adrenergic control. Hence, in addition to the direct stimulatory role of α-2 adrenergic pathways on GRF release from the hypothalamus, α-2 adrenergic activation would also lead to reduced SRIF and increased GH release. Therefore, administration of anti-GRF antibodies could abolish the GH response to α-2 adrenergic agonists by blocking both the central effects of GRF on SRIF release as well as the direct pituitary effects of GRF. Furthermore, although peripheral administration of GRF abolishes the pituitary GH response to subsequent GRF administration, it may not influence postsynaptic GRF effects at a central level, and hence one would observe maintained GH responses to adrenergic agonism as reported in this study.

In contrast to arginine and α-2 adrenergic agonism, GH responses to L-dopa were completely abolished by GRF pretreatment, indicating that GH responses to L-dopa are, at least to some extent, GRF-mediated. Furthermore, our findings of normal GH responses to α-adrenergic agonists with absent GH responses to L-dopa argue against the suggestion that L-dopa-induced GH release is mediated via activation of α-adrenergic pathways and support the hypothesis that this agent acts by increasing dopaminergic tone.

Although our findings, together with a recent report by others (10) of increased plasma GRF levels after L-dopa administration, support the view that dopamine may increase plasma GH levels by stimulating GRF release from the hypothalamus, it is possible that other mechanisms such as the inhibition of SRIF release may also be involved. In support of this, it has been demonstrated recently that GH responses to a maximal dose of GRF are enhanced after dopamine or bromocriptine administration (11-12). In any event, it is unlikely that endogenous dopaminergic tone plays a major role in the physiological control of GH secretion since administration of dopamine antagonist drugs does not affect basal GH levels or GH responses to GRF in normal adult subjects (4).

In summary, our data demonstrate that GH responses to L-dopa are GRF-mediated, at least in part, whereas GH responses to arginine, propranolol plus adrenalin, and clonidine are not. It is likely that hypoglycemia, arginine and α-2-adrenergic agonism release GH through different mechanisms, which could explain the clinical observation that a single test may fail to elicit a GH response in a patient who subsequently shows a perfectly normal response to other stimuli. A better understanding of the mechanisms by which these tests elicit GH secretion should improve our knowledge of the pathophysiological alterations in the neuroregulation of GH secretion in different disease states.

ACKNOWLEDGMENTS

This work was supported by the "Comision Asesore de Investigacion Cientifica y Tecnica" (CAI CYT 0681/84) and the Medical Research Council. M. D. Page has a Wellcome Trust Fellowship. We are grateful to Sanofi UK for supplying GRF 1-44, and to Dr. A. F. Parlow and the NIADDK for rGH assay reagents.

REFERENCES

1. Dieguez C, Page MD, Scanlon MF (1987) Growth hormone neuroregulation and its alterations in disease states. Clin Endocrinol (in press)
2. Shibasaki T, Hotta M, Masuda A, Imaki T, Obara N, Demura H, Ling N, Shizume K (1985) Plasma GH responses to GHRH and insulin-induced hypoglycemia in man. J Clin Endocrin Metab 60: 1265-1267
3. Dieguez C, Foord SM, Peters JR, Hall R, Scanlon MF (1985) The effects of thyroid hormone deprivation in vivo and in vitro on growth hormone (GH) responses to human pancreatic GH-releasing factor (GRF 1-40) by dispersed rat anterior pituitary cells. Endocrinology 116: 1066-1070
4. Jordan V, Dieguez C, Lafaffian I, Rodriguez-Arnao MD, Gomez-Pan A, Hall R, Scanlon MF (1986) Influence of dopaminergic, adrenergic and cholinergic blockade and TRH administration on GH responses to GRF 1-29. Clin Endocrinol 24: 291-298
5. Vance ML, Kaiser DL, Rivier J, Vale W, Thorner MO (1986) Dual effects of growth hormone (GH)-releasing hormone infusion in normal men: somatotroph desensitisation and increase in releasable GH. J Clin Endocrin Metab 62: 591-594
6. Parra A, Schultz RB, Foley TB, Blizzard RM (1970) Influence of epinephrine-propranolol infusions on growth hormone release in normal and hypopituitary subjects. J Clin Endocrin Metab 30: 134-137
7. Miki N, Ono M, Shizume K (1984) Evidence that opiaergic and α-adrenergic stimulate rat growth hormone release via growth hormone-releasing factor (GRF). Endocrinology 114: 1950-1952
8. Kabayama Y, Kato Y, Murakami Y, Tanaka H, Imura H (1986) Stimulation by α-adrenergic mechanisms of the secretion of growth hormone releasing factor (GRF) from perifused rat hypothalamus. Endocrinology 119: 432-434
9. Aguila MC, McCann SM (1985) Stimulation of somatostatin release in vitro by synthetic growth hormone-releasing factor by a non-dopaminergic mechanism. Endocrinology 117: 762-765
10. Donnadieu M, Evain-Brion D, Tonon MC, Vaudry H, Job JC (1985) Variations of plasma growth hormone (GH)-releasing factor levels during GH stimulation tests in children. J Clin Endocrin Metab 60: 1132-1134
11. Vance ML, Kaiser DL, Frohman LA, Rivier J, Vale WW, Thorner MO (1987) Role of dopamine in the regulation of growth hormone secretion: dopamine and bromocriptine augment growth hormone (GH)-releasing hormone-stimulated GH secretion in normal men. J Clin Endocrin Metab 64: 1136-1141
12. Delitala G, Palermo M, Ross R, Coy D, Besser M, Grossman A (1987) Dopaminergic and cholinergic influences on the growth hormone response to growth hormone-releasing hormone in man. Neuroendocrinology 45: 243-247

DIRECT GROWTH INHIBITORY ACTION OF SANDOSTATIN ON RAT PITUITARY 7315b TUMOR CELLS IN VITRO

J. W. Koper, P. M. van Koetsveld,
L. J. Hofland and S. W. J. Lamberts

Department of Medicine, Erasmus University,
Rotterdam, the Netherlands

INTRODUCTION

With the introduction of somatostatin analogs that are more stable in the circulation than the hormone itself, an additional tool has been found in the treatment of (neuro-)endocrine tumors, i.e. the use of Sandostatin (SMS 201-995) in the treatment of acromegaly (1). The full mechanisms by which somatostatin and its analogs exert their tumor growth inhibition remains to be elucidated. Two remaining questions can be formulated: 1) In some of the acromegalic patients treated with Sandostatin, an actual shrinkage of the tumor is observed in addition to a reduction of the circulating levels of growth hormone and somatomedin-C (IGF-I) (2); does this represent a direct antiproliferative action of the hormone (-analog) or a secondary effect which results from the action of the hormone on growth hormone secretion? 2) In some patients with metastatic endocrine pancreatic or gastro-intestinal tumors, the tumors escape from Sandostatin treatment with regard to hormone secretion (3); can this phenomenon be prevented by different regimes of Sandostatin administration?

The 7315b transplantable rat pituitary tumor was previously shown to be responsive to Sandostatin in vivo (4), both with regard to its secretion of prolactin and its growth. It was also shown to escape from the inhibitory action of Sandostatin after a relatively short period of treatment (4). In the present study, we have investigated the effects of Sandostatin on the prolactin secretion by and growth of 7315b tumor cells in vitro, with the aim of seeing whether this system provides a suitable model to study somatostatin action on tumor cells.

MATERIALS AND METHODS

Tumors

We originally obtained the 7315a tumor from Dr. R. M. MacLeod (Charlottesville, VA); the 7315b tumor used in this study originates from the 7315a tumor, but has lost its capacity to secrete ACTH as well as its receptors for glucocorticoids (5). It is unchanged with respect to prolactin secretion and estrogen- and progesterone-receptor content (5).

Short Term Cell Culture

Rat pituitary 7315b tumors were routinely maintained in female Buffalo rats (TNO, Rijswijk, The Netherlands). For experiments in vitro, the tumor was removed, cleaned and mechanically dissociated. Viable cells were isolated by Ficoll density-gradient centrifugation, washed and suspended in Minimal Essential Eagle Medium (MEM) (Grand Island Biological Co., Breda, The Netherlands) containing 10% fetal bovine serum (FBS) (Flow, Irvine, Scotland) (for a more complete description see ref. 5). The cells were seeded on poly-L-lysine-coated (5 μg in 0.5 ml of distilled water per well) 24-well tissue culture plates (Costar Europe, Badhoevedorp, The Netherlands) at 1 x 10^5 cells per well in 1 ml of medium.

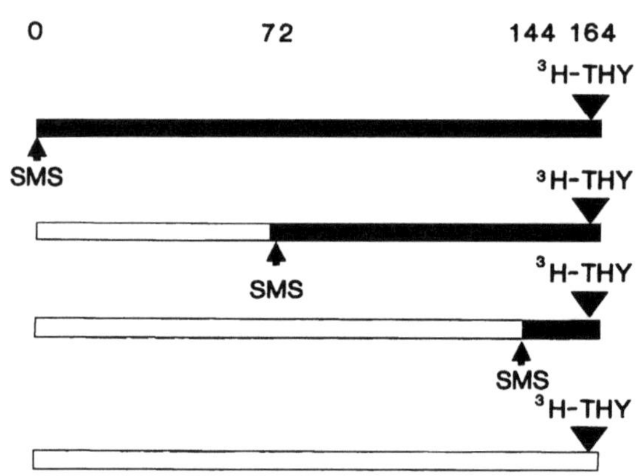

Figure 1. Incubation Conditions. 7315b cells were seeded at 100,000 cells per well in 24-well tissue culture plates at t = 0 hr. The four groups were: 1) treatment with SMS from t = 0 hr; 2) treatment with SMS from 72 hr; 3) treatment with SMS from 144 hr; 4) untreated control cultures. Culture medium was changed at 72 hr and at 144 hr.

Incubation Conditions

Figure 1 illustrates the subsequent incubation scheme: cultures were preincubated with Sandostatin (SMS 201-995, Sandoz, Basel, Switzerland) during different periods and the effect of this treatment with respect to prolactin secretion and the incorporation of [^{3}H]thymidine was measured during the seventh day of culture. For the measurement of prolactin secretion, 24 hr media were collected. Prolactin concentrations in the culture media were measured as described previously (5).

For the measurement of [^{3}H]thymidine-incorporation, 1 μCi of tritiated thymidine (91 Ci/mmol, Amersham, Houten, The Netherlands) was added during the last 4 hr of culture. At the end of the incubation period, the medium was removed and the cells were washed twice with ice cold 0.9% NaCl. They were then exposed to 10% trichloroacetic acid for 30 min and washed an additional time with 0.9% NaCl. The cells were transferred to scintillation counting vials after solubilization in 1 M NaOH and radioactivity was measured after neutralization and the addition of scintillation fluid.

Long Term Cell Culture

For more prolonged studies, cells were seeded at 1 x 10^5 cells per well in Transwell tissue culture inserts (Costar Europe, Badhoevedorp, The Netherlands). These inserts have a microporous membrane instead of a plastic bottom. The cells rest on this membrane and the entire insert, including the cells, can be transferred easily to a well containing fresh culture medium. Thus loss of the loosely attached cells is prevented without the necessity of centrifugation at each change of medium. Using this system, 7315b cells can be maintained for at least 6 weeks in culture. The medium was changed two times per week. At the end of the experiment, prolactin was measured in all collected media.

RESULTS

As can be seen in Figure 1, we have measured the status of the cultured cells with respect to the secretion of prolactin and the incorporation of [^{3}H]thymidine after a variable period of pretreatment with Sandostatin (SMS 201-995). Figures 2 and 3 show the results of a representative experiment. Figure 2 shows that the secretion of prolactin by the cells during the last 24 hr of culture was reduced by SMS. This reduction was dose-dependent, but not time-dependent: reaching a similar extent irrespective of whether the cells were treated from day 1 or only during the last 24 hr. Similarly, Figure 3 shows the SMS-induced changes in the incorporation of tritiated thymidine. The effects of

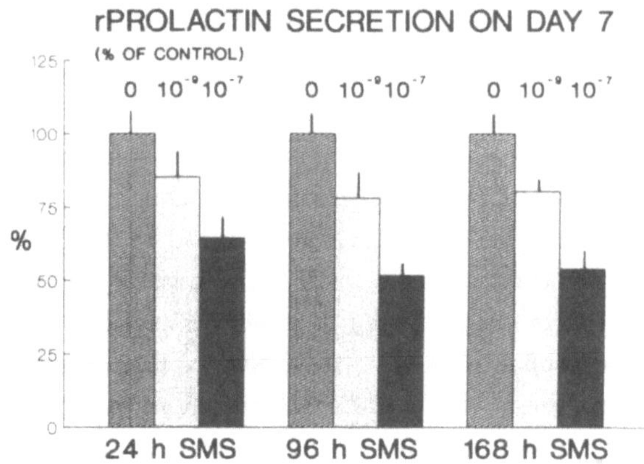

Figure 2. Prolactin secretion by 7315b cells in culture after treatment with 1 nM or 100 nM Sandostatin (SMS 201-995). Incubations were carried out as described in Figure 1. Prolactin secretion during the last 24 hr (t = 144-178 hr) of culture is expressed as the percentage of secretion by untreated control cultures.

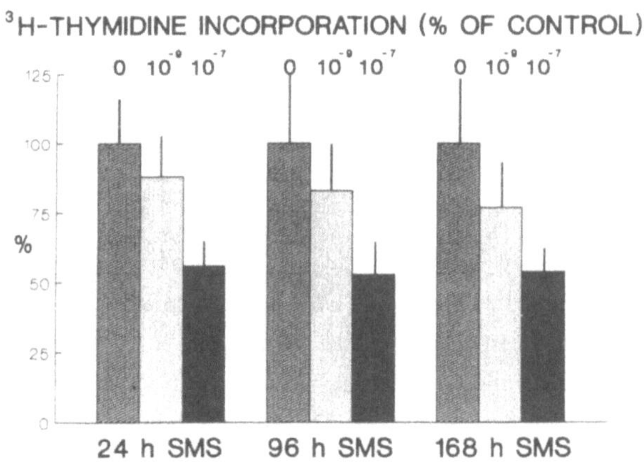

Figure 3. Incorporation of thymidine by 7315b cells in culture after treatment with 1 nM or 100 nM Sandostatin (SMS 201-995). Incubations were carried out as described in Figure 1. Tritiated thymidine incorporation during the last 4 hr of culture (t = 174-178) is expressed as the percentage of incorporation by untreated control cultures.

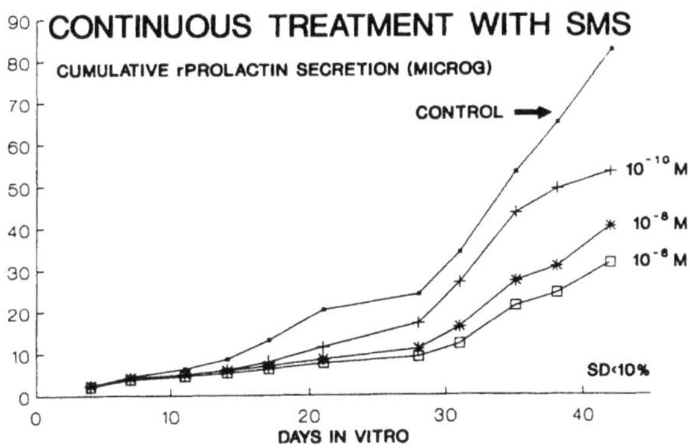

Figure 4. Cumulative prolactin secretion by long-term cultures of 7315b cells. Four groups of cultures were maintained during 6 weeks: controls, 0.1, 10 and 1000 nM Sandostatin (SMS 201-995). The culture media were changed two times per week and assayed for prolactin. Average, cumulative prolactin secretion is shown for each of the groups.

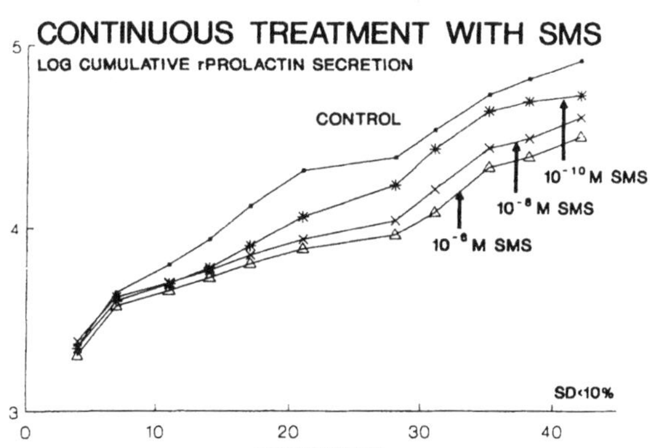

Figure 5. Cumulative prolactin secretion by long-term cultures of 7315b cells. The same data as in Figure 4, but with the data logarithmically plotted.

SMS-treatment on thymidine incorporation were almost identical to those seen on prolactin secretion. During the relatively short experimental period slight, but statistically nonsignificant, reductions were observed in the number of cells per culture and in the protein content of cultures. However, we have indications that in more prolonged incubations, these parameters are also affected significantly by SMS-treatment.

Figure 4 shows the cumulative prolactin secretion by 7315b cells maintained in long-term culture. As can be seen, Sandostatin caused a dose-dependent inhibition of total prolactin secretion. When these data are expressed logarithmically, however, as has been done in Figure 5, it can be seen that the rate of prolactin release is the same in treated and untreated cultures after day 21. This indicates that the tumor cells escape from the control of SMS.

DISCUSSION

We have previously shown that the 7315a transplantable rat pituitary tumor can be inhibited temporarily in its growth in vivo, by the administration of Sandostatin to the rat (4). However, the tumor escapes rapidly from SMS-control. In view of the observation that the levels of IGF-I and growth hormone in the tumor-bearing rats also cease to respond to the administration of SMS, it seemed possible that the reduction of tumor growth was a secondary effect, mediated by the temporary reduction in the levels of circulating IGF-I, growth hormone or other trophic factors. In this study, we show that Sandostatin has a direct inhibitory effect on the growth of 7315b tumor cells in vitro. Nevertheless, it is possible that this effect is not due to a direct inhibition of cell growth, but to the inhibition of the secretion of autocrine or paracrine growth promoting substances.

Interestingly, loss of sensitivity to Sandostatin occurs in these cultured tumor cells almost as rapidly as it does in the tumor when maintained in the rat. We are at present investigating if this escape can be correlated to a decrease in the number of affinity of somatostatin receptors on the cells. The observation that desensitization also occurs in vitro offers the possibility of using cultured 7315b cells as a model system to study the mechanism of desensitization and to investigate treatment modalities that offer a reduced risk of its occurrence.

REFERENCES

1. Lamberts SWJ, del Pozo E (1986) Long-term treatment of acromegaly with the somatostatin analog SMS 201-995. N Engl J Med 314: 1391-1392
2. Lamberts SWJ, Uitterlinden P, del Pozo E (1987) SMS 201-995 induces a continuous decline in circulating growth hormone and somatomedin-C levels during therapy of acromegalic patients for over two years. J Clin Endocrinol Metab 65: 703-710
3. Lamberts SWJ (1986) Non pituitary actions of somatostatin. A review on the therapeutic role of SMS 201-995 (Sandostatin). Acta Endocrinol 112: 41-56
4. Lamberts SWJ, Reubi JC, Uitterlinden P, Zuiderwijk J, vd Werff P, van Hal P (1986) Studies on the mechanism of action of the inhibitory effect of the somatostatin analog SMS 201-995 on the growth of the PRL/ACTH pituitary tumor 7315a. Endocrinology 118: 2188-2194
5. Lamberts SWJ, van Koetsveld P, Verleun T (1987) Prolactin release-inhibitory effects of progesterone, megestrol acetate and mifepristone (RU 38486) by cultured rat pituitary tumor cells. Cancer Res 47: 3667-3671

HUMAN PITUITARY TUMORS IN CELL CULTURE: MODULATION OF PROLACTIN AND GROWTH HORMONE MESSENGER RNA LEVELS IN VITRO

J. R. E. Davis, E. M. Wilson
and M. C. Sheppard

Department of Medicine, Queen Elizabeth Hospital
Edgbaston, Birmingham, U.K.

INTRODUCTION

Neither the pathogenesis nor the treatment of pituitary tumors in man is yet well understood. The mechanisms controlling both hormone release and gene transcription in the rat pituitary are becoming increasingly better understood; however, knowledge concerning hormonal control of the human pituitary gland is still largely limited to studies of hormone release from cultured tissue (1,2), and there are very few studies of the regulation of human pituitary gene transcription. In this paper, we report preliminary data from studies of two unusual and aggressive human tumors grown in short-term monolayer cell culture. We have applied the techniques of DNA/RNA hybridization analysis to study the control of cytoplasmic messenger RNA levels in order to give some index of pretranslational stages of hormone synthesis.

METHODS

Pituitary adenoma tissue was obtained for cell culture from routine transsphenoidal or transfrontal hypophysectomy operations. The tissue was chopped into small fragments and enzymatically dispersed using collagenase and hyaluronidase as previously described for rat pituitary tissue (3). Dispersed adenoma cells were plated into plastic culture dishes at a density of approximately 2 x 10^5 cells per well in 2 ml of final culture fluid (Dulbecco's modified Eagle's medium, supplemented with 10% horse serum and 2.5% fetal calf serum with antibiotics). Cells were incubated at 37°C in a humidified atmosphere of 5% CO_2 - 95% air, and allowed to grow in monolayer culture for at least five days before experiments commenced. Cultured cells were then washed by repeated medium exchanges, and treated with test substances for 24 hr. After the treatment periods, supernatant culture fluid was removed for assay of released growth hormone or prolactin and cell cytoplasmic extracts were prepared for messenger RNA estimations.

The prolactin and growth hormone (GH) that were released into the culture fluid were measured by radioimmunoassay, using as standards MRC 66/217 for GH and 83/562 for prolactin. Mid-range intra- and interassay coefficients of variation were less than 10%.

The relative concentrations of messenger RNA were measured by cytoplasmic dot hybridization assays, using the technique of White and Bancroft (4). Cells were harvested from culture dishes and lysed with nonionic detergent; nuclei were pelleted by centrifugation, and cytoplasmic preparations were denatured in formaldehyde at 60°C. Cytoplasmic samples were applied in duplicate to a hybridization transfer membrane, and mRNAs were immobilized by baking, before hybridization for 24 hr with [^{32}P]-labeled complementary DNA probes specific for human prolactin (5) or human growth hormone (6). Relative levels of prolactin or growth hormone mRNA were quantitated by autoradiography and scanning densitometry. For Northern blot analysis, mRNA species were extracted from the cytoplasm of cultured cells, and then separated by agarose gel electrophoresis before hybridization.

RESULTS

The first patient had an unusual and aggressive prolactinoma. He had presented in 1982 with a large tumor, which had only partly responded to surgery followed by radiotherapy. The patient had also received bromocriptine, but showed only a partial reponse, with serum prolactin levels falling from 92000 to 30000 mU/l, then subsequently rising again despite continued bromocriptine therapy. The tumor continued to extend and in 1986 a second hypophysectomy operation was performed, at which time pituitary tissue was obtained for in vitro studies.

The prolactinoma was established in primary monolayer culture as described above, and after five days cells were washed and then treated

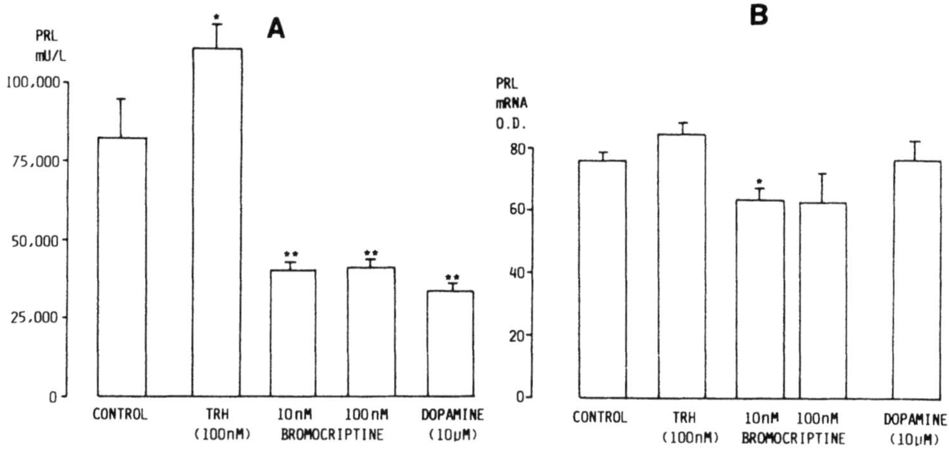

Figure 1. Prolactin release (A) and mRNA levels (B) from prolactinoma cells in culture. Cells were treated for 24 hr as shown, and mRNA data are shown in arbitrary units of optical density. Data are means ± SEM, n = 4 per group; *$P < 0.05$, **$P < 0.002$ compared to control.

for 24 hr with 100 nM TRH, 10 nM and 100 nM bromocriptine, or 10 μM dopamine. Figure 1 shows prolactin release compared with relative prolactin mRNA levels from these cells. TRH caused a moderate stimulation in prolactin release; the two doses of bromocriptine halved this release, and a high dose of dopamine had a similar effect. In contrast, prolactin mRNA levels showed much less marked changes. TRH caused only minimal and nonsignificant stimulation, and the two doses of bromocriptine caused only marginal reductions. Dopamine was without effect. These responses contrast with those seen using normal rat pituitary tissue, where clear stimulation or inhibition of prolactin mRNA levels with these agents can be readily demonstrated using similar experimental techniques.

The second patient had acromegaly, and like the previous patient, had an aggressive pituitary tumor. He initially presented in 1974 with visual loss and had a transfrontal hypophysectomy followed by radiotherapy; he failed to respond, however, and required a further hypophysectomy operation in 1979. A subsequent trial of bromocriptine (30 mg daily) had no effect on serum growth hormone levels, which remained between 1000 and 4000 mU/l, and following the development of further visual failure in 1986, he underwent a third craniotomy from which tumor tissue was obtained for culture.

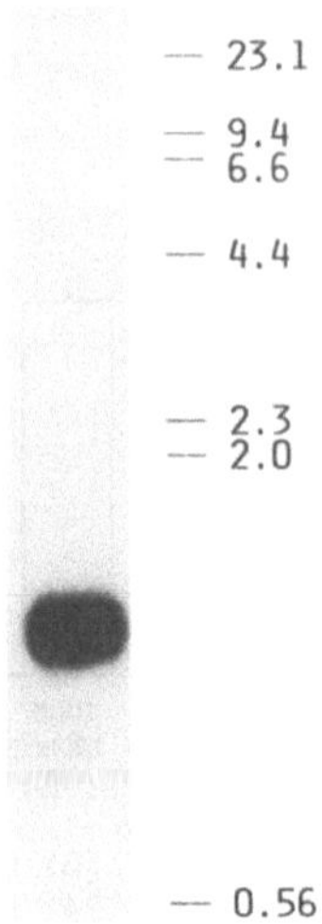

Figure 2. Northern blot analysis of human growth hormone mRNA, showing a single species of approximately 1 kb in length. Molecular size markers are shown in kb on the right.

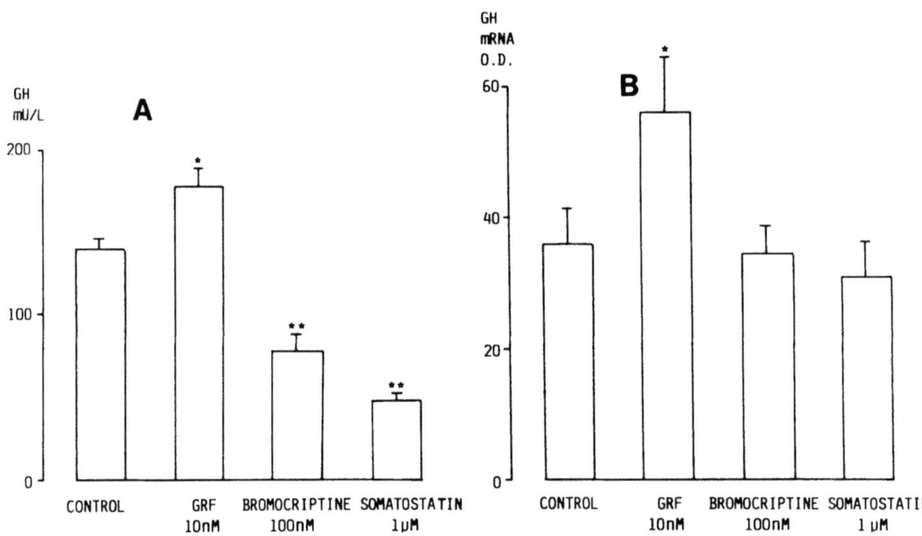

Figure 3. Growth hormone release (A) and mRNA levels (B) from cultured somatotrophinoma cells. Cells were treated for 24 hr as shown, and mRNA data are shown in units of optical density. Data are means ± SEM, n = 4 per group; *P < 0.02, **P < 0.005 compared to control.

Somatotrophinoma cells were established in monolayer culture for several days before experiments commenced. Figure 2 shows a Northern blot analysis of cytoplasm from cultured cells taken from this tumor and demonstrates that the cDNA probe to human growth hormone hybridizes to a single mRNA species of about 1 kilobase in length, the correct size for human growth hormone mRNA and in agreement with the one previous report of GH mRNA regulation in human tumor tissue (7). Figure 3 shows data from dot hybridization experiments: Growth hormone-releasing factor (GRF, 10 nM), stimulated both GH release and mRNA accumulation. Bromocriptine (10 nM) reduced GH release to 60% of control, but had no effect on mRNA levels. In a similar manner, 1 μM somatostatin reduced the release to 39% of control values, but had only a marginal and nonsignificant effect on mRNA levels.

In order to examine the possible mechanisms for the stimulatory effect of GRF in this tumor, we used three agents to mimic the action of

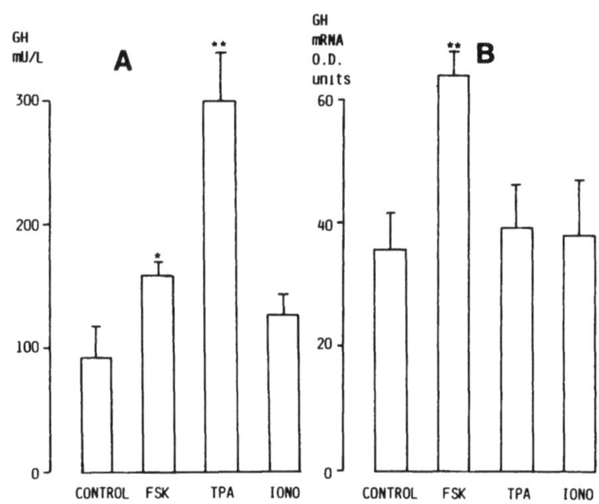

Figure 4. Growth hormone release (A) and mRNA levels (B) from cultured somatotrophinoma cells, treated as shown for 24 hr. Data are means ± SEM, n = 4 per group, *P < 0.05, **P < 0.005 compared to control.

the three major intracellular pathways that may mediate hormone action: forskolin (FSK) stimulates adenylate cyclase and raises cellular cyclic AMP content; the phorbol ester TPA was used to stimulate protein kinase C; and ionomycin (IONO) was used to cause calcium influx and thus raise intracellular free calcium concentrations. Figure 4 shows GH release and mRNA levels from this experiment: Both stimulation of cyclic AMP content and activation of protein kinase C caused GH release, while ionomycin had little effect. In contrast, mRNA levels were increased only by activation of adenylate cyclase by forskolin, suggesting that the cyclic AMP system may be involved in stimulation of both hormone release and gene transcription, whereas the protein kinase C system (acting alone) affects only hormone release.

DISCUSSION

These preliminary results indicate that it is possible to study the mechanisms controlling mRNA accumulation in human pituitary tumor tissue by using simple DNA/RNA hybridization techniques. Previously, such techniques have been applied mainly to rat pituitary tissue, where the factors controlling gene transcription are much more clearly understood.

In the present studies, we have examined two unusually aggressive tumors. In both patients' tumors, mRNA levels appeared to be relatively less sensitive than hormone release to inhibition, and this may relate to the aggressive nature of these tumors. Prolactin mRNA accumulation in rat pituitary cells is normally stimulated by TRH (8) and inhibited by bromocriptine and dopamine (9); we have readily been able to confirm these findings using the techniques outlined above (unpublished data). The lack of response of the human lactotrophs to both stimulation and inhibition suggests that the prolactinoma tissue may have lost normal regulatory mechanisms. In the second study, growth hormone mRNA accumulation was stimulated appropriately by GRF, but showed no response to inhibition by bromocriptine or somatostatin. We previously showed in rat tissue that GH mRNA levels are unaffected by bromocriptine, and inhibited in a dose-dependent manner by somatostatin (10), although others have found no effect of short-term somatostatin treatment on GH gene transcription rate (11). The lack of response of this somatotrophinoma to inhibition by these agents therefore requires confirmation by further experiments to discover whether this pattern is typical in less aggressive tumors. The finding that forskolin stimulated GH mRNA accumulation as well as GH release suggests that cyclic AMP is involved in the control of GH gene transcription, as has been shown in rat pituitary tissue (11). The lack of response of mRNA to stimulation of protein kinase C or calcium influx also confirms data from rat studies (11), although it remains to be seen whether combined stimulation of both intracellular pathways may be effective.

In conclusion, we have presented preliminary data on the control of gene expression in two aggressive human pituitary tumors. The data suggest that gene transcription in these tumors may be relatively insensitive to normal regulatory mechanisms, and this may be related to their clinical behavior. It is also clear that pretranslational stages of hormone synthesis may be regulated independently of hormone release by different intracellular mediators. Such findings may give important insights into the pathogenesis of pituitary tumors in man.

ACKNOWLEDGMENTS

Drs. J. A. Martial and D. D. Moore generously provided the cDNA probes for human prolactin and GH, respectively. We are grateful to Drs. D. A. Heath, D. R. London, J. J. McMillan and P. Rhys-Evans who had clinical care of these patients, and to Dr. S. S. Lynch for radioimmunoassays.

REFERENCES

1. Adams EF, Bhuttacharji SC, Halliwell CLJ, Loizou M, Birch G, Mashiter K (1984) Effect of pancreatic growth hormone releasing factors on GH secretion by human somatotrophic pituitary tumours in cell culture. Clin Endocrinol 21: 709-718
2. Lamberts SWJ, Verleun T, Hofland L, Del Pozo E (1987) A comparison between the effects of SMS 201-995, bromocriptine and a combination of both drugs on hormone release by the cultured pituitary tumour cells of acromegalic patients. Clin Endocrinol 27: 11-23
3. Davis JRE, Sheppard MC (1986) Modulation of thyrotrophin release from an intracellular pool by pre-exposure to thyrotrophin-releasing hormone and dibutyryl cyclic AMP. J Endocrinol 108: 211-217
4. White BA, Bancroft FC (1982) Cytoplasmic dot hybridization. Simple analysis of relative mRNA levels in multiple small cell or tissue samples. J Biol Chem 257: 8569-8572
5. Truong AT, Duez C, Belayew A, Renard A, Pictet R, Bell GI, Martial JA (1984) Isolation and characterization of the human prolactin gene. EMBO J 3: 429-437
6. Martial JA, Halliwell RA, Baxter JD, Goodman HM (1979) Human growth hormone: complementary DNA cloning and expression in bacteria. Science 205: 602-607
7. Yamashita S, Weiss M, Melmed S (1986) Insulin-like growth factor I regulates growth hormone secretion and messenger ribonucleic acid levels in human pituitary tumor cells. J Clin Endocrinol Metab 63: 730-735
8. White BA, Bancroft FC (1983) Epidermal growth factor and thyrotropin-releasing hormone interact synergistically with calcium to regulate prolactin mRNA levels. J Biol Chem 258: 4618-4622

9. Maurer RA (1980) Dopaminergic inhibition of prolactin synthesis and prolactin messenger RNA accumulation in cultured pituitary cells. J Biol Chem 244: 8092-8097
10. Wood DF, Docherty K, Ramsden DB, Sheppard MC (1987) A comparison of the effects of bromocriptine and somatostatin on growth hormone gene expression in the rat anterior pituitary in vitro. Mol Cell Endocrinol 52: 257-261
11. Barinaga M, Bilezikjian LM, Vale WW, Rosenfeld MG, Evans RM (1985) Independent effects of growth hormone releasing factor on growth hormone release and gene transcription. Nature 314: 279-281

INFLUENCE OF CHRONIC DEPOLARIZATION ON SYNTHESIS AND THE RELATIVE AMOUNTS OF DIFFERENT FORMS OF SOMATOSTATIN IN CULTURED FETAL CEREBROCORTICAL CELLS

L. Cacicedo, M. T. de los Frailes, M. J. Lorenzo,
G. Fernandez and F. Sanchez-Franco

Servicio de Endocrinología, Hospital Ramón y Cajal
Madrid, Spain

INTRODUCTION

The neuropeptide somatostatin is widely distributed throughout the nervous system. In addition to its well known role as a hypophysiotropic hormone, somatostatin also appears to serve as a neurotransmitter or neuromodulator (1,2). Although the effects of conventional neural depolarizing agents on the release of somatostatin (SRIF) have been extensively studied (3), little is known about the influence of these agents on SRIF synthesis and processing. We and others previously reported that in using cultured fetal cerebral cortical cells, immunoreactive (ir)-SRIF release occurred in response to acute depolarization induced by high potassium concentrations or veratridine (VTD) (3,4). In the present work, we sought to investigate to what extent SRIF synthesis is regulated by factors that influence its secretion, by studying the effect of chronic exposure to depolarizing agents on ir-SRIF synthesized by cultured cortical cells. As an index of synthesis, we followed the accumulation of ir-SRIF (cell content + media). We examined the response of the cultured neurons to high concentrations of K^+ and found that K^+ increased total ir-SRIF and that this effect is calcium-dependent.

MATERIAL AND METHODS

Seventeen day old fetuses were removed from Wistar rats that had been raised in our laboratory. The cerebral cortices were taken and collected in Hank's Balanced Solution. The cortical cells were dispersed by mechanical techniques and cultured in Minimum Essential Medium supplemented with 10% fetal calf serum, 10% horse serum as described previously (4). Final cell suspensions were plated at 5 x 10^6 cells/35 mm plate. Media were changed twice weekly. At the beginning of each experiment, old media were removed and fresh medium, containing test substances but no sera, was added. Exposure to depolarizing agents was continuous for 24 hr. At the termination of the experiments, cells and media were acid extracted, affinity purified, gel filtered and prepared for immunoassay. Immunoreactive-SRIF present in the media and the cell extracts was purified by affinity chromatography. S-14 antiserum was coupled to Affi-Gel 10 using 1 M ethanolamine. The specifically bound material was eluted with 1 M acetic acid and fractionated on Bio-Gel P-10 using 2 N acetic acid 0.1% BSA buffer.

Mean and standard errors of samples were calculated. The student's "t" test was used to evaluate the differences between the control and the experimental groups.

RESULTS

Cerebral cortical cells kept in culture for 7 days were exposed to high potassium concentrations (30 and 56 mM). In some cases, the calcium channel blocking agent verapamil or the sodium channel blocking agent tetrodotoxin (TTX) was added simultaneously. Cell exposure to potassium resulted in a significant dose-dependent increase of total and released somatostatin, which was reversed by the addition of VPM (20 μM), but not by TTX (1 μM). ir-SRIF in cell extracts was not modified by any treatment (Fig. 1).

Radioimmunoassay of SRIF released into the media from control cultures purified by Bio-Gel P-10 revealed the presence of five forms of ir-SRIF, 14% corresponding to somatostatin-28 (S-28) and 64% to somatostatin-14 (S-14). A comparison with ir-SRIF from high potassium treated cultures demonstrated that much more of the total ir-SRIF in the potassium-treated media is found in the high mol wt forms. The proportion of S-14 decreased to 9%, increasing S-28 from 14% to 40% and the void volume forms from 11% to 20% (Fig. 2). Characterization of ir-SRIF in extracts of control and potassium-exposed cultured cerebral cortical cells by gel filtration showed, as depicted in Figure 3, the existence of five peaks. The percentages of both S-14 and S-28 decreased

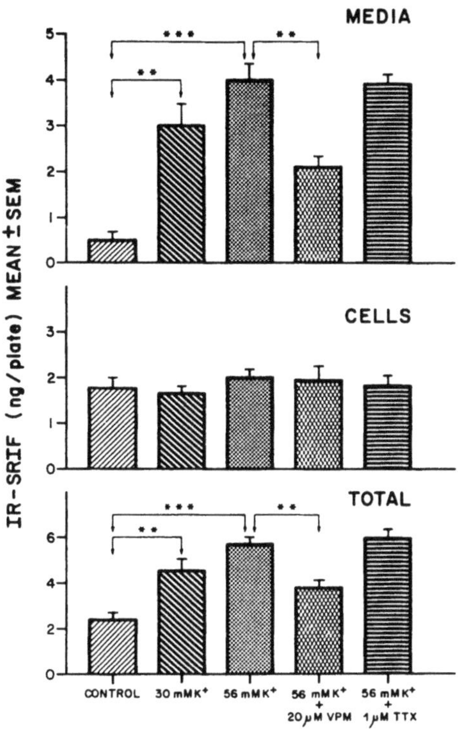

Figure 1. ir-SRIF production (cell content + media) by cerebrocortical cultures after a 24-hr exposure to 30 and 56 mM K^+. Responses in the presence of a calcium channel blocker (VPM) and a sodium channel blocker (TTX) were also tested.

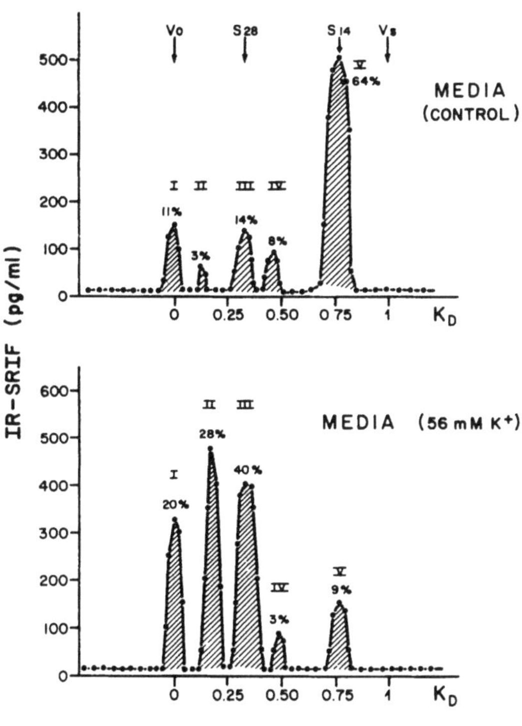

Figure 2. Elution profiles on Bio-Gel P-10 of ir-SRIF present in pooled media after 24-hr incubation in control media and in media containing 56 mM K^+. The Bio-Gel P-10 column (70 x 1 cm) was eluted with 2.0 N acetic acid 0.1% BSA.

with prolonged exposure to high potassium. Forty-six percent of ir-SRIF from potassium treated cells eluted in the void volume, while only 14% was present in the unstimulated cultures.

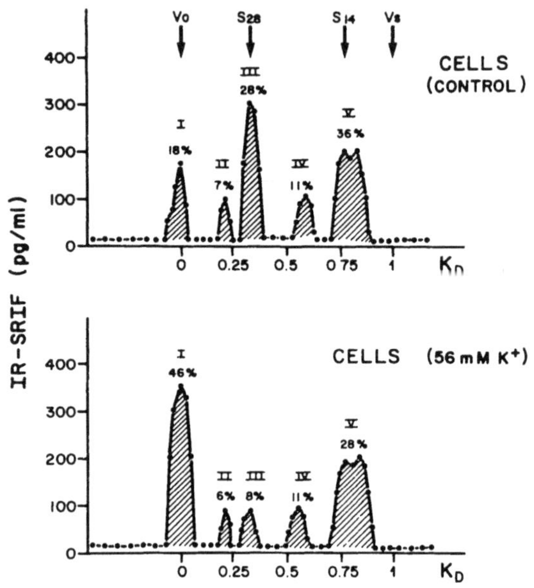

Figure 3. Characterization of ir-SRIF in extracts of control and 24-hr K^+ exposed cultured cerebrocortical cells by gel filtration. The Bio-Gel P-10 column was eluted with 2.0 N acetic acid 0.1% BSA.

DISCUSSION

The data presented in this study demonstrate that chronic stimulation with depolarizing concentrations of K^+ leads to a dose-related enhancement of ir-SRIF secretion and synthesis, this effect being calcium-dependent. It has previously been shown that acute membrane depolarization, induced by potassium and veratridine, stimulated ir-SRIF release in various neuronal systems (3,4,6,7), but no data has been reported concerning the effect of chronic depolarization on SRIF synthesis. It has been postulated that intracellular depletion itself may stimulate synthesis, suggesting that synthesis might be coupled to secretion of the neuropeptide. Increase in vasoactive intestinal peptide mRNA accumulation has been observed in cultured neuroblastoma cells after dibutyral cAMP and phorbol ester treatment (8,9). Stimulation of somatostatin mRNA accumulation has also been reported in cultured cerebral cortical cells in response to forskolin, but not to high K^+, during a 4 hr incubation period (10). These data lead to the conclusion that somatostatin secretion could be uncoupled from mRNA accumulation. However, our data showed that when exposure to K^+ is extended for 24 hr, the release of SRIF is not accompanied by intracellular depletion, which accounts for an increase in total ir-SRIF. This strongly suggests that prolonged release of ir-SRIF may stimulate its synthesis.

The finding of multiple forms of ir-SRIF by gel filtration analysis in cell cultures confirms many previous studies that suggested a precursor role for such a species (11). The presence of S-28 in the media indicates that this form is secreted by the cells. The chromatographic analysis of ir-SRIF from media and cell extracts revealed that prolonged K^+ treatment modified the elution profile on Bio-Gel P-10, causing an increase in high mol wt forms. These findings would support the hypothesis that an increase in the rate of synthesis caused by the releasing stimulus would lead to the appearance of immature forms, some of them such as S-28 possibly having an extracellular modulatory role. It may also represent changes in post-translational processing or changes in peptidase and protease activity by the modification of intracellular calcium concentrations produced by chronic depolarization.

SUMMARY

Although the factors controlling the secretion of the neuropeptide somatostatin (SRIF) have been extensively studied, little is known about the mechanisms that control somatostatin biosynthesis. Somatostatin release is regulated by numerous agents that cause depolarization of the cellular membrane, such as high concentrations of K^+ and vertridine (VTD). We investigated whether prolonged exposure to high K^+ concentrations regulates SRIF synthesis. We found that 30 and 56 mM K^+ concentrations were effective in causing an increase in total (cells + media) ir-SRIF. This effect could be prevented by the addition of the calcium channel locker verapamil (VPM). The nature of this ir-SRIF was characterized by gel filtration. Chronic exposure to high potassium reduced the percentage of S-14 in the media and increased the percentages of S-28 and high mol-wt forms; percentages of S-14 and S-28 decreased in cell extracts, whereas the void volume forms increased after prolonged high potassium exposure. These data indicate that chronic exposure to high potassium increases ir-SRIF synthesis and causes a major synthesis of its high molecular weight forms.

ACKNOWLEDGMENTS

The technical assistance of M. Paz Muñoz is gratefully acknowledged. We also thank Anna Steele for the preparation of this manuscript.

This work was supported by Grant PB 85-0026 from the Comision Asesora de Investigación Científica y Técnica and by Grant CCB 8409008 from the US-Spain Joint Committee for Scientific and Technological Cooperation.

REFERENCES

1. Delfs J, Dichter M (1983) Effects of somatostatin on mammalian cortical neurons in culture: physiological actions and unusual dose-response characteristics. J Neurosci 3: 1176-1188
2. Dichter MA, Delfs JR (1981) Somatostatin and cortical neurons. In: Marin JB, Reichlin S, Bick KL(eds) Neurosecretion and Brain Peptides, New York: Raven Press: pp 145-157
3. Robbins RJ, Sutton RE, Reichlin S (1982) Sodium and calcium dependent somatostatin release from dissociated cerebral cortical cells in culture. Endocrinology 110: 496-499
4. Cacicedo L, de los Frailes MT, Lorenzo MJ, Sanchez Franco F (1985) Prolonged exposure to membrane depolarization agents stimulates somatostatin synthesis in cerebral cortical cells in culture. J Endocr Invest 8: Suppl 3,158
5. Sanchez-Franco F, de los Frailes MT, Cacicedo L (1986) Inhibitory effect of thyroid hormone on protein synthesis by fetal rat neurons in primary culture. In: Madeiros Neto G, Gaitan E (eds) Frontiers in Thyroidology. New York: Plenum Medical: pp 647-650
6. Gamse R, Vaccaro DE, Gamse G, DiPace M, Fox TO, Leeman S (1980) Release of immunoreactive somatostatin from hypothalamic cells in culture: inhibition by v-aminobutyric acid. Proc Natl Acad Sci 77: 5552-5556
7. Bennett G, Edwardson H, Marcano De Cotte D, Berelowitz M, Pimstone B, Kromheim S (1979) Release of somatostatin from rat brain synaptosomes. J Neurochem 32: 1127-1130
8. Hayakawa Y, Obata K, Itoh N, Yanaihara N, Okamoto H (1984) Cyclic-AMP regulation of pro-vasoactive intestinal polypeptide/PHM-27 synthesis in human neuroblastoma cells. J Biol Chem 259: 9207-9211
9. Ohsawa K, Hayakawa Y, Nishizawa M, Yamagami T, Yamamoto H, Yanaihara N, Okamoto H (1985) Syneristic stimulation of VIP/PHM-27 gene expression by cyclic AMP and phorbol esters in human neuroblastoma cells. Biochem Biophys Res Comm P132: 885-891
10. Montminy MR, Low MJ, Tapia-Arancibia L, Reichlin S, Maendel G, Goodman R (1986) Cyclic-AMP regulates somatostatin mRNA accumulation in primary diencephalic cultures and in transfected fibroblast cells. J Neurosciences 6: 1171-1176
11. Spiess J, Vale W (1980) Multiple forms of somatostatin-like activity in the rat hypothalamus. Biochemistry 19: 2861

D. THE CONTROL AND FUNCTION OF THE OPIATE PATHWAYS

INTRODUCTION

It is now known that there are three principal families of opiate peptides that are present in high concentration in the hypothalamus. These derive from three separate large precursor proteins: preproenkephalin, preprodynorphin and preproopiomelanocortin. Post-translational enzymatic cleavage of these proteins gives rise to the enkephalins, dynorphin and β-endorphin. Large numbers of enkephalin-producing neurons are found in many hypothamic nuclei, and these are the most widely distributed of the opioid peptides throughout both the hypothalamus and the extrahypothalamic C.N.S. It is now know that within the hypothalamus, particularly in the paraventricular nucleus, enkephalinergic neurons also produce corticotropin releasing factor (CRF). Direct pathways from these hypothalamic neurons containing enkephalin follow several different routes, for example, from the paraventricular nucleus to the brain stem and spinal cord. The highest concentration of dynorphin in the nervous system is found in the hypothalamus, and probably the most important cells containing it are within the arcuate nucleus. The distribution of dynorphin resembles that of vasopressin, and it is probable that most vasopressin-containing neurons also produce dynorphin. In the arcuate nucleus, it is likely that dynorphin co-exists with dopamine. The hypothalamus is also the principal site of endorphin production.

With such widespread distribution, it is not surpising that opiate peptides are involved in the control of many aspects of hypothalamic function. These include control not only of the posterior and anterior pituitary, but also other facets of hypothalamic function, including feeding behavior, pain sensation, sexual behavior and responses to stress. It is probable that coordination of stress responses is mediated by central CRF neurons connecting the limbic system, hypothalamus and brain stem, and that a network of opioid peptidergic neurons attentuates or buffers this system.

J. A. H. Wass

A SURVEY OF THE FUNCTIONAL ANATOMY OF OPIATE PATHWAYS RELATING TO THE HYPOTHALAMUS

M. V. Sofroniew

Anatomy School, Cambridge University
Cambridge, England, U.K.

In the fifteen years since the identification of the first opiate peptides in the central nervous system, much has been learned about their chemical nature and synthesis as well as about their distribution in specific neurons and pathways. Knowledge about the functions of opiate pathways and the specific actions of the opiate peptides at the celluar level is also expanding rapidly. This survey will concern itself with principal opiate peptide pathways relating to the hypothalamus as they have been thus far elucidated and some of the current concepts about the functions of these pathways.

At present three principal families of opiate peptides have been identified in the hypothalamus, deriving from three separate large precursor proteins: preprodynorphin, preproenkephalin and preproopiomelanocortin (1-3; Fig. 1). Post-translational enzymatic cleavage of these precursors gives rise to the various active forms of the opiate peptides: dynorphin (DYN), the enkephalins (ENK) or β-endorphin (β-End or END). Neurons and terminals containing these peptides have been identified throughout the hypothalamus; these will be described in the following sections.

DYNORPHIN

The hypothalamus contains one of the highest concentrations of DYN in the nervous system. Most of this appears to be in cell bodies. The most prominent hypothalamic cell types containing DYN are magnocellular neurohypophyseal neurons and neurons within the arcuate nucleus (Fig. 2). In

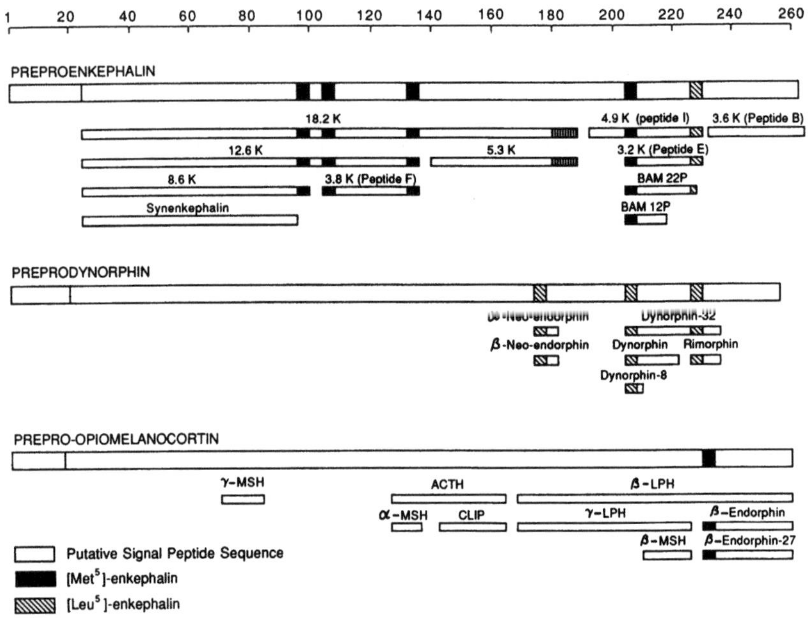

Figure 1. Schematic representation of the three opioid peptide prohormones and their principal pathways of proteolytic processing. Modified from Nakanishi et al. (2) and Petrusz et al. (6), with permission.

the magnocellular neurohypophyseal system, the distribution of cells producing vasopressin or oxytocin has been well documented and these neurons are found not only in the supraoptic and paraventricular nucleus, but also in numerous scattered groups throughout the hypothalamus (4). The distribution of DYN parallels that of vasopressin neurons, and studies using immunohistochemical techniques at the single cell level have shown that most, if not all, vasopressin neurons produce DYN in addition to vasopressin (5). Thus these neurons are capable of expressing at least two different peptide precursors, that for DYN and that for vasopressin.

In the arcuate nucleus, some DYN cells have been shown to contain tyrosine hydroxylase, and in these cells DYN is likely to coexist with dopamine (7).

Pathways from hypthalamic DYN neurons (Fig. 3) have been established as expected from the magnocellular neurons to the neurohypophysis, where the same

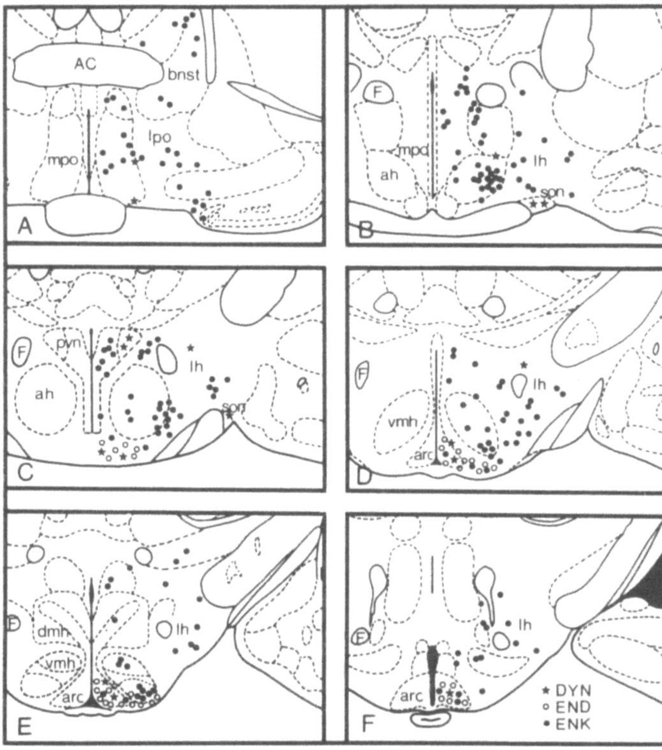

Figure 2. Schematic drawings depicting the location of neuronal cell bodies producing different types of opioid peptides at various rostral to caudal levels of the rat hypothalamus.

secretory granules contain and release both vasopressin and DYN (8,9), and where DYN may participate in regulation of hormone release from the posterior pituitary (10). DYN neurons in the arcuate nucleus are thought to project to portal vessels of the median eminence, where opiate peptides have been implicated in regulation of anterior pituitary secretion.

ENKEPHALIN

The hypothalmus contains large numbers of ENK-producing neurons distributed over many hypothalamic nuclei (Fig. 2) as well as extensive networks of fibers and terminals as shown by immunocytochemical as well as in situ hybridization studies (6,11). Such studies have also shown that ENK is the most

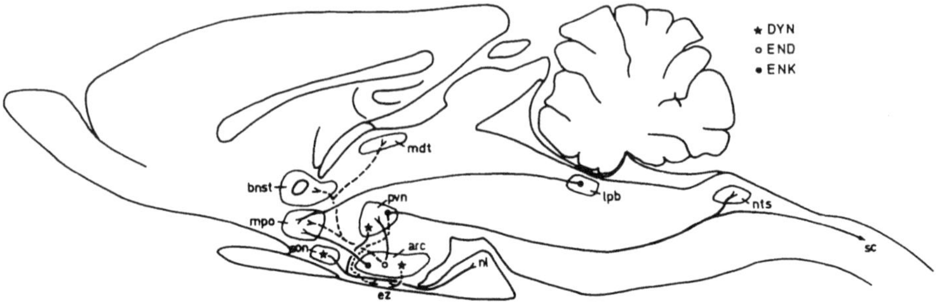

Figure 3. Schematic drawing of the sagittal rat brain depicting various pathways containing opiate peptides relating to the hypothalamus.

widely distributed of the opiate peptides within the nervous system, with large numbers of cells and extensive terminal networks in virtually all regions. Within the hypothalamus, certain ENK neurons in the paraventricular nucleus have also been shown to produce corticotropin-releasing factor (CRF) and possibly a PHI-like peptide (12). These neurons, which probably project to the portal capillaries in the external zone of the median eminence, may be involved in the coordinated simultaneous regulation of the release of several hormones from the anterior pituitary (12). While immunohistochemical reports exist suggesting that hypothalamic magnocellular neurons may also produce ENK (13), in situ hybridization results show that under normal conditions, these cells do not contain mRNA encoding for ENK (11), and that the immunohistochemical results may well be due to cross-reactivity.

Direct pathways from hypothalmic ENK neurons (Fig. 3) have also been shown from the paraventricular nucleus to the brain stem and spinal cord (14) and from the arcuate nucleus to the medial preoptic area (15). Such projections may be, respectively, involved in autonomic regulation or control of behavior. An incoming ENK pathway from the brain stem to the hypothalamus has also been shown from the lateral parabrachial nucleus to the medial preoptic area (16).

β-ENDORPHIN

The hypothalamus contains the principal group of neurons producing β-End, and these are in the arcuate nucleus (Fig. 2). These neurons have been shown by in situ hybridization and immunohistochemistry to contain the mRNA encoding for the preopiomelanocortin precursor (17), as well as all of the peptides deriving from that precursor (18-21). Some additional neurons have also been found in the brain stem (21,22). Within the hypothalamus, no evidence has yet been found to indicate that the β-End neurons also produce another transmitter candidate (23). The β-End neurons are clearly separate from dopamine neurons (24).

β-End fibers and terminals are found throughout the central nervous system from the olfactory bulb to the spinal cord (22,25; Fig. 4), much of which are thought to derive from β-End neurons in the hypothalamus. From the arcuate nucleus, an extensive and specific β-End projection has been experimentally demonstrated to the hypothalamic paraventricular nucleus (26), and principal β-End projections are thought to reach the medial preoptic area, bed nucleus of the stria terminalis, and dorsomedial thalamus (Fig. 3).

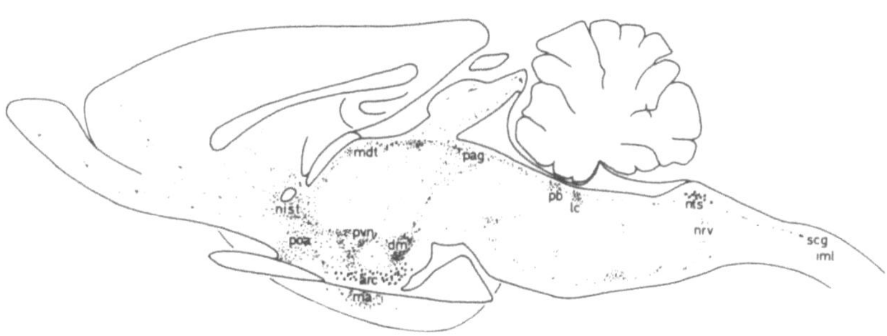

Figure 4. Schematic drawing of the sagittal rat brain depicting the distribution of neurons and fibers containing β-endorphin. Modified from Sofroniew (25), with permission.

FUNCTIONAL CONSIDERATIONS AND SUMMARY

The hypothalamus contains an extensive and specifically organized distribution of all three major types of opiate neurons (Fig. 2). While these peptides may co-exist with other types of neuroactive substances, the different types of opiate peptides do not at present appear to co-exist in any population of the same neurons. The arcuate nucleus is particularly rich in these neurons, containing extensive but separate populations of neurons that contain each type of opiate peptide. The hypothalamus is also rich in opiate-containing terminals, primarily ENK or β-End, diffusely distributed over virtually all regions with very dense plexuses of terminals in specific areas. Specific projections of opiate peptide-containing neurons are also now being identified experimentally, and extensive networks of internal hypothalamic pathways, as well as pathways to other portions of the brain are emerging (Fig. 3). Given the widespread anatomical connections, it is not surprising that opiate peptides should be implicated in virtually all major aspects of hypothalamic function. They appear to participate in regulation of secretion from both the posterior pituitary and anterior pituitary (10) via separate anatomic connections, and they are likely to be involved in autonomic regulation as well as in a number of different behaviors including feeding (27,28), the sensation of pain (29), and sexual behavior (30,31). At present it is unclear which specific pathways may be involved in these functions, but techniques are now becoming available that should allow discrimination of the precise functional anatomy of the various hypothalamic opiate peptides.

ACKNOWLEDGMENTS

The author thanks K. Baker for technical assistance, P. Campbell for editorial assistance, and C. Cook for assistance with typing; and the MRC, Wellcome Trust and TRCT for grant support. Figure 1 was partially modified from Nakanishi et al. (2) by permission from *Nature*, Vol. 278, No. 5703, pp. 423-427, Copyright (c) 1979 Macmillan Journals Limited; and from Petrusz et al. (6).

REFERENCES

1. Kakidani H, Furutani Y, Takahashi H, Noda M, Morimoto Y, Hirose T, Asai M, Inayama S, Nakanishi S, Numa S (1982) Cloning and sequence analysis of cDNA for porcine β-neo-endorphin/dynorphin precursor. Nature 298: 245-249
2. Nakanishi S, Inoue A, Kita T, Nakamura M, Chang ACY, Cohen SN, Numa S (1979) Nucleotide sequence of cloned cDNA for bovine corticotropin-β-lipotropin precursor. Nature 278: 423-427
3. Noda M, Furutani Y, Takahashi H, Toyosato M, Hirose T, Inayama S, Nakanishi S, Numa S (1982) Cloning and sequence analysis of cDNA for bovine adrenal preproenkephalin. Nature 295: 202-206
4. Sofroniew MV (1985) Vasopressin, oxytocin and their related neurophysins In: Björkland A, Hökfelt T (eds) Handbook of Chemical Neuroanatomy 4: GABA and Neuropeptides in the CNS, Part 1, Amsterdam: Elsevier: pp 93-165
5. Watson SJ, Akil H, Fischli W, Goldstein A, Zimmerman E, Nilaver G, Van Wimersma Greidanus TB (1982) Dynorphin and vasopressin: common localization in magnocellular neurons. Science 216: 85-87
6. Petrusz D, Merchenthaler I, Maderdrut JL (1985) Distribution of enkephalin-containing neurons in the central nervous system. In: Björklund A, Hökfelt T (eds) Handbook of Chemical Neuroanatomy 4: GABA and Neuropeptides in the Central Nervous System, Part 1. Amsterdam: Elsevier: pp 273-334
7. Everitt BJ, Meister B, Hokfelt T, Melander T, Terenius L, Rokaeus A, Theodorsson-Norheim E, Dockray G, Edwardson J, Cuello C, Elde R, Goldstein M, Hemmings H, Ouimet C, Walaas I, Greengard P, Vale W, Weber E, Wu J, Chang K (1986) The hypothalamic arcuate nucleus-median eminence compex: Immunohistochemistry of transmitters, peptides and DARPP-32 with special reference to coexistence in dopamine neurons. Brain Research Reviews 11: 97-155

8. Whitnall MH, Gainer H, Cox BH, Molineaux CJ (1983) Dynorphin-A-(1-8) is contained within vasopressin neurosecretory vesicles in rat pituitary. Science 222: 1137-1138
9. Millan MJ, Millan MH, Herz A (1983) Contribution of the supra-optic nucleus to brain and pituitary pools of immunoreactive vasopressin and particular opioid peptides, and the interreltionships between these, in the rat. Neuroendocrinology 36: 310-319
10. Bicknell RJ (1985) Endogenous opioid peptides and hypothalamic neuroendocrine neurones. J Endocrinol 107: 437-446
11. Harlan RE, Shivers BD, Romano GJ, Howells RD, Pfaff DW (1987) Localization of preproenkephalin mRNA in the rat brain and spinal cord by in situ hybridization. J Comp Biol 258:159-184
12. Hökfelt T, Fahrenkrug J, Tatemoto K, Mutt V, Werner S, Hulting A-L, Terenius L, Chang KJ (1983) The PHI (PHI-27)/corticotropin-releasing factor/enkephalin immunoreactive hypothalamic neuron: Possible morphological basis for integrated control of prolactin, corticotrophin, and growth hormone secretion. Proc Natl Acad Sci USA 80: 895-898
13. Martin R, Voigt KH (1981) Enkephalins co-exist with oxytocin and vasopressin in nerve terminals of rat neurohypohysis. Nature 289: 502-504
14. Sawchenko PE, Swanson LW (1982) Immunohistochemical identification of neurons in the paraventricular nucleus of the hypothalamus that project to the medulla or to the spinal cord in the rat. J Comp Neurobiol 205: 260-272
15. Yamamoto M, Tohyama M, Shiosaka S, Shiotani Y, Inagaki S, Kito S (1985) Leucine-enkephalin-like immunoreactive fibers in the medial preoptic area of the rat: their distribution and origin. Exp Brain Res 60: 535-541
16. Lind RW, Swanson LW (1984) Evidence for corticotropin releasing factor and leu-enkephalin in the neural projection from the lateral parabrachial nucleus to the median preoptic nucleus: a retrograde transport, immunohistochemical double labeling study in the rat. Brain Research 321: 217-224
17. Gee CE, Chen C-L, Roberts JL (1983) Identification of proopiomelanocortin neurones in rat hypothalamus by in situ cDNA-mRNA hybridization. Nature 306: 374-375
18. Watson SJ, Richard CW, Barchas JD (1978) Adrenocorticotropin in rat brain: immunocytochemical localization in cells and axons. Science 200:1180-1182
19. Sofroniew MV (1979) Immunoreactive β-endorphin and ACTH in the same neurons of the hypothalamic arcuate nucleus in the rat. Am J Anat 154: 283-289
20. McGinty JF, Bloom FE (1983) Double immunostaining reveals distinctions among opioid neurons in the medial basal hypothalamus. Brain Research 278: 145-153
21. Khachaturian H, Lewis ME, Tsou K, Watson SJ (1985) β-endorphin, α-MAH, ACTH, and related peptides. In: Bjorklund A, Hokfelt T (eds) Handbook of Chemical Neuroanatomy 4: GABA and Neuropeptides in the Central Nervous System, Part 1. Amsterdam: Elsevier: pp 216-272
22. Romagnano MA, Joseph SA (1983) Immunocytochemical localization of $ACTH_{1-39}$ in the brainstem of the rat. Brain Research 276: 1-16
23. Sofroniew MV, Eckenstein F, Schrell U, and Cuello AC (1984) Evidence for colocalization of neuroactive substances in hypothalamic neurons. In: Chan-Palay V and Palay SL (eds) Coexistence of Neuroactive Substances in Hypothalamic Neurons. Chichester: Wiley: pp 73-90
24. Ibata Y, Watanabe K, Kinoshita H, Kubo S, Sano Y, Sakura N, Yanaihara C, Yanaihara N (1980) Dopamine and β-endorphin are contained in different neurons of the arcuate nucleus of hypothalamus as revealed by combined fluorescence histochemistry and immunohistochemistry. Neurosci Letters 17: 185-189
25. Sofroniew MV (1982) Vascular and neural projections of hypothalamic neurones producing neurohypophysial or ACTH-related peptides. In: Baertschi AJ and Dreifuss (eds) Neuroendocrinology of Vasopressin, Corticoliberin and Opiomelanocortins. New York: Academic Press: pp 73-86
26. Sawchenko LW, Swanson LW, Joseph SA (1982) The distribution and cells of origin of ACTH-stained varicosities in the paraventricular and supraoptic nuclei. Brain Research 232: 365-374
27. Margules DL, Moisset B, Lewis MJ, Shibuya H, Pert CB (1978) β-Endorphin is associated with overeating in genetically obese mice (ob/ob) and rats (fa/fa). Science 202: 988-999

28. Millan MJ, Millan MH, Reid LD, Herz A (1986) The role of the mediobasal arcuate hypothalamus in relation to opioid systems in the control of ingestive behavior in the rat. Brain Research 381: 29-42
29. Akil H, Richardson DE, Barchas JD, Li CH (1978) Appearance of β-endorphin-like immunoreactivity in human ventricular cerebrospinal fluid upon analgesic electrical stimulation. Proc Natl Acad Sci USA 75: 10: 5170-5172
30. Martensz ND, Vellucci SV, Keverne EB, Herbert J (1986) β-Endorphin levels in cerebrospinal fluid of male talapoin monkeys in social groups related to dominance status and the luteinizing hormone response to naloxone. Neuroxcience 8: 651-658
31. Hughes AM, Everitt BJ, Herbert J (1987) Selective effects of β-endorphin infused into the hypothalamus, preoptic area and bed nucleus of the stria terminalis on the sexual and ingestive behavior of male rats. Neuroscience (in press)

POST-TRANSLATIONAL PROCESSING OF POMC IN BRAIN AND PITUITARY

X. Bertagna and Y. de Keyzer

Centre de Recherches sur les Maladies Endocriniennes
and INSERM U-129, Hopital Cochin
Paris, France

Recently, considerable progress has been made in understanding the molecular aspects of ACTH biosynthesis: a high molecular weight (HMW) ACTH-precursor molecule was demonstrated in the ACTH-producing AtT-20/D16-V mouse tumor cell line (1,2). Recombinant DNA methods lead to the characterization of this precursor, called proopiomelanocortin or POMC, in many species including man. The coordinate proteolysis, or processing, of POMC generates various peptide fragments: the 16K fragment (16K), ACTH, β-lipotrophin (β-LPH), and a small but definite amount of β-endorphin (β-End) and γ-LPH are formed in the anterior lobe of the pituitary. In the intermediate lobe, POMC processing is different and smaller peptides are released such as β-melanocyte stimulating hormone (β-MSH), corticotropin-like-intermediate-lobe peptide (CLIP), and α-MSH. The concept that POMC products may vary in different tissues also applies to man: several studies have shown that POMC-derived peptides are different in ACTH-producing pituitary and nonpituitary tumors (4-8). Because a single functional gene has been detected for rat, beef and human POMC, it has generally been assumed that variable POMC products result from different processing of the same precursor molecule in different tissues. However, variable expression of a single gene through an alternate mode of RNA splicing has recently been recognized as a new cause of peptide heterogeneity within the products of growth hormone and calcitonin genes, for example. In the case of POMC itself, various forms of mRNAs have been detected in rat and beef normal tissues, some of them resulting from an alternate mode of intron removal during RNA splicing (9).

The human POMC gene sequence also predicts a 30 amino acid joining peptide (JP) separating 16K and ACTH within their common precursor. We used an antiserum directed against the amidated COOH-terminal end of mouse JP to develop a radioimmunoassay for the predicted human JP molecule. Immunoreactive JP was detected in tissue extracts from human normal pituitary, ACTH-secreting pituitary and nonpituitary tumors, and in plasma from patients with ACTH hypersecretory syndromes. Its molar concentration was the same order of magnitude as - and correlated with - that of the other POMC peptides. Gel exclusion chromatography in 1% formic acid and 6 M guanidine-HCl revealed a predominant immunoreactive material with an apparent molecular weight of ca. 6,000. After reduction with dithiothreitol, this material was recovered in an elution volume identical to that of purified human JP and corresponding to a molecular weight of ca. 3,000. These data show that POMC processing generates a COOH-terminally amidated JP predominantly secreted as a homodimer, probably through disulfide bonding between the single Cys^9 residue of two molecules.

We looked at the transcriptional events in ACTH-producing tissues by specifically examining POMC-mRNA. We first used a DNA probe designed to cover most of the protein coding region of the POMC gene limited to the third exon and to avoid repetitive Alu sequences present in intron B.

Simple dot blot hybridization showed the presence of POMC-mRNA in all examined tissues, and this content was correlated with that of POMC peptides showing wide variations in nonpituitary tumors. These data strongly suggest that the rate of gene transcription is primarily responsible for the amount of peptide produced, although theoretically at least, decreased mRNA turnover could be an alternate possibility.

In normal pituitaries, a single POMC-mRNA species of ca. 1200 b was observed by Northern blot analysis. The POMC-mRNA of corticotropic adenomas appears indistinguishable from that of the normal pituitary. These data thus provide evidence that in human corticotropic adenomas, the mechanisms of POMC-mRNA generation are not qualitatively altered.

Four out of five nonpituitary tumors contained a single POMC-mRNA species identical to that of the normal human pituitary. Since these tumors had an abnormal POMC processing, it seems reasonable to assume that the second phenomenon is not related to an abnormality of gene transcription or RNA processing. It seems more likely that the exaggerated POMC processing that occurs in some nonpituitary tumors is caused by an abnormal enzymatic "environment," the action of which is not limited as neatly as it is in the normal (and tumoral) pituitary where only five out of nine potential "processing sites"- i.e., basic amino acids pairs - are cleaved. These data thus emphasize the role of the processing enzymes as the determinant factor of POMC processing specificity.

In several nonpituitary tumors, two POMC-mRNAs were found: in addition to the 1200 b mRNA species, a second larger mRNA of ca. 1450 b was detected in equivalent amounts.

This large POMC-mRNA species also hybridized with an extra-exonic probe, indicating that it was extended on the 5′ flanking region of the gene. Primer extension and S_1 mapping studies were subsequently performed, which established that the probe corresponded to a family of mRNAs with transcription initiation sites located at -369, -217, and -108 b from the 5′ end of exon 1 (10).

The mechanism of POMC gene expression was subsequently studied in normal human nonpituitary tissues, mainly the testis and the thymus. Northern blot analysis showed the presence of a predominant 800 b mRNA species, which eventually could be fully characterized using primer extension and S_1 mapping techniques. These short transcripts form a population of at least six different RNAs with heterogeneous 5′ extremities starting 41-162 b downstream from the 5′ end of exon 3 (11). This result raised the question of the mechanism responsible for this tissue-specific size polymorphism.

Since there is only one POMC gene per human haploid genome, this phenomenon could only result either from a variable splicing event or an alternate mode of transcription initiation. It is remarkable that the 5′ flanking region of the short POMC transcripts corresponds to a GC-rich region of the gene and does contain two typical "GC boxes" and several related sequences. It is suggested that the small POMC transcripts that are found in many, but not all, nonpituitary tissues result from an alternate mode of POMC gene transcription using a "GC box" type promoter sequence.

A major question raised by these results concerns the putative translation products of the truncated POMC transcripts. None of them can code for a complete POMC molecule because its normal initiation codon is located in exon 2. Analysis of the POMC reading frame of the small transcripts reveals two AUG codons, which according to Kozak's rule could initiate their translation. The first one, present only in the single longest transcript of 793 b, would initiate the translation of a 183 amino acid protein that lacks the NH_2-terminal part of γ-MSH and the 16K fragment. The second one would generate a 45 amino acid peptide reduced to the COOH-terminal parts of γ-LPH and to β-End. In both cases, the first amino acids are mostly hydrophilic and cannot constitute a valid signal peptide, thereby preventing the peptide translocation into the membrane compartment of the secretory pathway. Under these conditions, it is unlikely that these truncated POMC peptides can undergo an adequate post-translational processing.

These studies show that the human POMC gene can be expressed differently in various tissues through the use of alternate promoters. Further studies should be performed to elucidate the possible mechanisms leading to POMC (and ACTH) over-production by various tumors of pituitary or nonpituitary origins.

REFERENCES

1. Eipper BA, Mains RE (1980) Structure and biosynthesis of pro-adrenocorticotropin/endorphin and related peptides. Endocrinol Rev 1: 1-27
2. Nakanishi S, Inoue A, Kita T, Nakamura M, Chang ACY, Cohen SN, Numa S (1979) Nucleotide sequence of cloned cDNA for bovine corticotropin-β-lipotropin precursor. Nature 278: 423-427
3. Whitfeld P, Seeburg P, Shine J (1982) The human proopiomelanocortin gene: organization, sequence and interspersion with repetitive DNA. DNA 1: 133-143
4. Orth DN, Nicholson WE, Mitchell WM, Island DP, Liddle GW (1983) Biologic and immunologic characterization and physical separation of ACTH and ACTH fragments in the ectopic ACTH syndrome. J Clin Invest 52: 1756-1769
5. Pullan PT, Clement-Jones V, Corder R, Lowry PJ, Besser GM, Rees LH (1980) ACTH, LPH and related peptides in the ectopic ACTH syndrome. Clin Endocrinol 13: 437-445
6. Wilson RE, Orth DN, Nicholson WE, Mount CD, Bertagna XY (1981) Human γ-lipotropin radioimmunoassay: identification of immunoreactive γ-lipotropin in human plasma and tissue. J Clin Endocrinol Metab 53: 1-9
7. Ueda MT, Takeuchi T, Abe K, Miyakawa S, Ohnami S, Yanaihara N (1980) β-melanocyte-stimulating hormone immunoreactivity in human pituitaries and ectopic adrenocorticotropin-producing tumors. J Clin Endocrinol Metab 50: 550-556
8. Bertagna XY, Seurin D, Pique L, Luton JP, Bricaire H, Girard F (1983) Peptides related to the NH_2-terminal end of proopiocortin in man. J Clin Endocrinol Metab 56: 489-495
9. Oates E, Herbert E (1984) 5′ sequence of porcine and rat proopiomelanocortin mRNA. J Biol Chem 259: 7421-7425
10. de Keyzer Y, Lacaze T, Girard F, Luton JP, Kahn A, Bertagna X (1987) Variable mechanisms of proopiomelanocortin gene expression in human tissues. The Endocrine Society, Indianapolis, June 10-12, 1987, Abstract 431
11. Lacaze-Masmonteil T, de Keyzer Y, Luton JP, Kahn A, Bertagna X (1987) Characterization of proopiomelanocortin transcripts in human non-pituitary tissues. Proc Natl Acad Sci (USA): 7261-7265

DOPAMINERGIC AGENTS HAVE DIFFERENTIAL EFFECTS ON BOTH CONTENT AND PROCESSING OF ALPHA-N-ACETYLATED ENDORPHIN IN THE OVINE PITUITARY INTERMEDIATE LOBE

A. I. Smith, C. A. Wallace, I. J. Clarke and J. W. Funder

Medical Research Centre, Prince Henry's Hospital, Melbourne, Australia

INTRODUCTION

The intermediate lobe (IL) of the pituitary is directly innervated by neurons with cell bodies lying within the brain. A dopaminergic tract has been identified in rats (1,2), and recent studies also indicate that γ-amino butyric acid (GABA)-containing neurons in the brain also project to the pars intermedia (3,4). Dopamine and its agonists have been shown to inhibit the release of α-MSH and β-endorphin (β-End) from the IL (5,6), whereas dopamine antagonists such as haloperidol stimulate the release of α-MSH and β-End from the rat IL (6,7). More recent studies have demonstrated dopaminergic regulation of secretion of POMC-derived peptides from the intermediate lobe in both sheep (8) and dog (9).

There are conflicting data as to the regulatory role of GABA in the IL. Two studies showed GABA-mediated inhibition of α-MSH release from the IL (10,11); in a third study no effects of GABA on IL secretion could be demonstrated (12).

More recently it has been demonstrated that β-End processing, as well as its biosynthesis and release, may be at least in part under dopaminergic control in the rat IL. Ham et al. demonstrated in rats chronically treated with the dopamine antagonist haloperidol that IL β-End content is increased, this rise being accounted for by increases in the levels of the acetylated forms of β-End_{1-26}, β-End_{1-27} and β-End_{1-31} (13). Studies by Millington et al. have shown that similar treatment causes selective increases in α-N-acetylated (Nac)End_{1-27} and lowers $NacEnd_{1-26}$ (14). In monolayer cultures of rat melanotrophs, β-End appears less extensively processed, with the principal species being $NacEnd_{1-31}$ rather than the C-terminally shortened forms (15). When these cells are incubated with dopamine or dopamine agonists, the processing pattern is altered to resemble more closely that which is characteristic of the IL in situ, i.e., $NacEnd_{1-26}$ and $NacEnd_{1-27}$ become the predominant forms (15).

In addition to changes in the tissue levels of these peptides, it has been demonstrated that chronic treatment with the dopamine antagonist haloperidol not only increases levels of mRNA coding for POMC in the IL, but also causes an essentially parallel rise in α-N-acetyltransferase activity (16). In a complementary study in the frog (Rana Ridibunda) IL, Jenks et al. demonstrated that dopamine inhibits the acetylation of α-MSH and, by extrapolation, that of β-End (17). It therefore appears that dopamine not only regulates biosynthesis and secretion of POMC and POMC-derived peptides, but may also play a role in regulating the precise pattern of their processing.

We have previously reported that hypothalamo-pituitary disconnection (HPD) in the sheep results in a two-fold increase in IL immunoreactive-(ir)-NacEnd and a markedly altered pattern of NacEnd processing, in that $NacEnd_{1-27}$ becomes the dominant molecular species (18). To determine if these effects reflect the loss of descending dopaminergic neuronal input to the IL, two groups of normal sheep were treated with either the dopamine agonist bromocriptine or the dopamine antagonist haloperidol; a third group of HPD sheep were also treated with bromocriptine. Acid extracts of IL were diluted and both

TABLE 1

IL CONTENT OF ir-NacEnd AND ir- α-MSH FOLLOWING HPD OR TREATMENT WITH DOPAMINERGIC AGENTS

	ng/mg tissue ir-NacEnd	ir- α-MSH
Control	110 ± 14	62 ± 19
HPD (90 day)	263 ± 17	136 ± 12
Haloperidol (1 mg, bd, sc, 10d)	160 ± 13	86 ± 14
Bromocriptine (2.5 mg, bd, sc, 10d)	75 ± 9	45 ± 16

Mean ± SEM
n = 4, mature merino ewes/group.

ir-NacEnd and α-MSH content determined by RIA with a NacEnd-specific (R92) and α-MSH-specific antiserum; aliquots were submitted to RP-HPLC and the ir-NacEnd profile determined by assaying collected fractions.

RESULTS

Tissue Content of ir-NacEnd and ir-α-MSH

The procedure of HPD is followed by more than doubling the IL content of ir-NacEnd and ir-α-MSH. Treatment of intact sheep with the dopamine antagonist haloperidol caused a parallel, ~ 40% increase in IL content of both ir-NacEnd and ir-α-MSH (Table 1); the increases observed following treatment with haloperidol, however, were not as great for either ir-NacEnd or ir-α-MSH as those observed following HPD. After chronic treatment with the dopamine agonist bromocriptine, the IL content of both ir-NacEnd and ir-α-MSH decreased in parallel (Table 1) to about two-thirds of the control values.

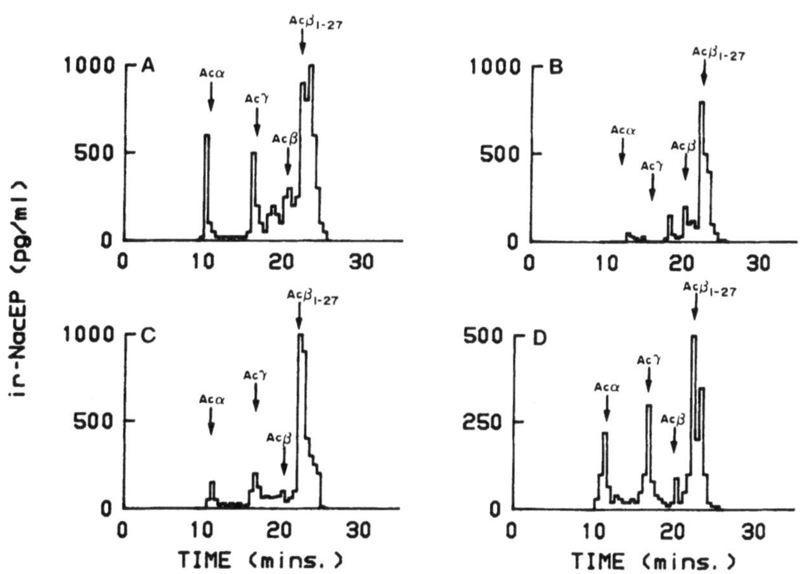

Figure 1. Typical RP-HPLC elution profiles of α-N-acetylated End (NacEnd) immunoreactivity from control (1A), HPD (90 day) (1B), haloperidol-treated (1 mg, bd, sc, 10 d) (1C), and bromocriptine treated (2.5 mg, bd, sc, 10 d) (1D) sheep IL. Sheep IL was extracted in 2 M acetic acid (1 ml), and an aliquot applied to a radially compressed Nova Pak column as described previously (19). The retention times of synthetic NacEnd standards are arrowed.

RP-HPLC of Tissue Extracts

In the control IL, the RP-HPLC ir-NacEnd profile showed the major species as NacαEnd (30%), NacγEnd (30%) and NacβEnd$_{1-27}$ (35%), with NacβEnd (5%) as a minor component (Fig. 1A). In the HPD animal (Fig. 1B), the dominant molecular species is NacEnd$_{1-27}$, representing 65% of the total ir-NacEnd as described previously (18). Chronic treatment with haloperidol produced an ir-NacEnd profile in the IL very similar to that seen in HPD animals (Fig. 1C), with NacEnd$_{1-27}$ (80%) becoming the principal molecular component. Although bromocriptine lowers the total IL content, on RP-HPLC the ir-NacEnd profile was unchanged (Fig. 1D), with NacαEnd, NacγEnd and NacEnd$_{1-27}$ the major species, and NacEnd$_{1-31}$ a relatively minor component. The data derived from the HPD animals treated chronically with bromocriptine are equivocal. In two of the HPD sheep (Fig 2B and 2C), bromocriptine at least in part appears to

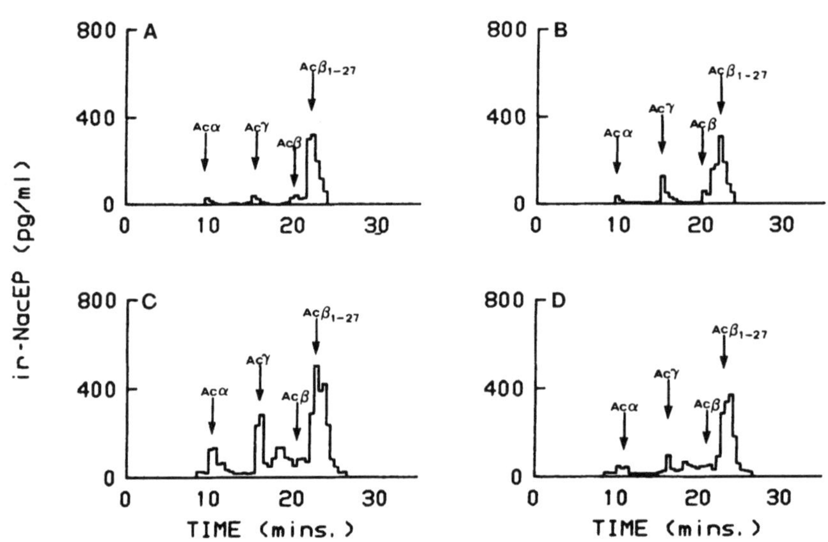

Figure 2. RP-HPLC elution profile of NacEnd immunoreactivity from 4 HPD (90 day) sheep IL treated with bromocriptine (2.5 mg, bd, sc, 10 d). Tissues were extracted and chromatographed as in Figure 1. The retention times of synthetic standards are arrowed.

reverse the HPD-induced processing changes, with the ir-NacEnd profile showing significant amounts of NacαEnd and NacγEnd. The ir-NacEnd HPLC profiles of IL from the two other HPD sheep in this group, however, appear to be unaffected by bromocriptine administration, with $NacEnd_{1-27}$ remaining the only major immunoreactive species (Figs. 2A and 2D).

DISCUSSION

This study has not only demonstrated an association between dopamine and POMC biosynthesis in the sheep IL, but also that dopamine may play a regulatory role in POMC post-translational processing. The increased IL content of both ir-NacEnd and ir-α-MSH, following the chronic administration of haloperidol, presumably reflects a blocking of the tonic dopaminergic inhibition both of POMC synthesis and of product release, as has been previously described by others (6,7,20). The reduction in IL content of both ir-α-MSH and NacEnd following bromocriptine treatment is further evidence for a dopamine involvement in sheep IL regulation. Recent studies in which cAMP turnover and α-MSH secretion from IL cells were measured in response to both D_1- and D_2-specific dopaminergic agents suggest that these inhibitory effects are D_2-receptor mediated events (21,22).

The dramatic changes in NacEnd processing observed in the IL following chronic administration of haloperidol are very similar to those observed in the IL of sheep following HPD - further evidence that dopamine is involved in regulating post-translational processing of POMC in the sheep. These data are also consistent with the observations of Ham et al., who reported an increase in the longer acetylated species, $NacEnd_{1-26}$, $NacEnd_{1-27}$ and $NacEnd_{1-31}$, in the rat IL, following the chronic administration of haloperidol (13).

To confirm that the changes observed in the HPD sheep result from a loss of dopaminergic input, a group of these animals was treated with the dopamine agonist bromocriptine in an effort to restore both processing and content. The results were equivocal, with two animals in the group appearing to respond (Fig. 2B,C) and two animals unaffected (Fig. 2A,D). This disparity may be due to a "down regulation" of dopamine receptors, following a 90 day withdrawal from the natural ligand; an additional, more likely explanation is that there are other neural factors involved in IL control. Further evidence for the latter is that the effects of haloperidol on IL content of both ir-α-MSH and ir-NacEnd were not as dramatic as those observed in the HPD animals. Another candidate likely to be involved in IL neural control is GABA; GABAergic fibers have been shown to innervate the rat IL (3,4), and GABA has been shown to inhibit α-MSH release (9).

As well as pharmacological agents, certain stressors can also have profound effects on POMC processing in the IL. In a study designed to examine the effect of acute and chronic stress on both content and released forms of intermediate lobe β-End, Akil et al. demonstrated a marked difference in the immunoreactive molecular profile of both stored and released β-End between the two groups (23). In animals subjected to acute footshock stress, neither IL content or molecular forms of β-End were significantly different from controls. In the chronically stressed group, however, intermediate lobe content was increased nearly two-fold, with a marked change in the pattern of processing.

The conversion of $NacEnd_{1-27}$ to Nacα-End and Nacγ-End requires cleavage between residues Leu_{17} and Phe_{18} by an enzyme termed the gamma endorphin generating endopeptidase (24), an enzyme shown to be sensitive to glucocorticoids in the anterior pituitary (25). Whether the accumulation of $NacEnd_{1-27}$ following HPD or haloperidol treatment reflects the specific inhibition of this enzyme, as other POMC processing enzymes have been shown to be sensitive to haloperidol in the IL (16), or merely represents a change in intracellular steady state, remains to be established.

It is interesting to note that during both surgical and pharmacological manipulations, $NacEnd_{1-31}$ remains unchanged in the ir-NacEnd RP-HPLC profiles. $NacEnd_{1-27}$ is flanked by paired basic residues, the sites of trypsin-like cleavage; at positions 28 and 29 of $NacEnd_{1-31}$ the basic residues are Lys-Lys, which appear to be preferentially cleaved in the IL. This suggests that the IL $NacEnd_{1-27}$ is a primary POMC product, and is not produced by sequential processing of $NacEnd_{1-31}$.

In conclusion, this study has demonstrated that in the sheep, as in the rat, dopamine is a crucial factor in regulating POMC biosynthesis and product release, and may also play a role in regulating post-translational processing. In this way physiological regulators such as dopamine may maintain a balance between the multitude of bioactivities potentially contained within a precursor sequence, in response to stress or other physiological stimuli.

ACKNOWLEDGMENTS

The authors wish to thank Sue Smith for secretarial assistance and the National Health and Medical Research Council of Australia for financial support.

REFERENCES

1. Bjorklund A, Moore RV, Nobin A, Stenevi U (1973) The organization of tubero-hypophyseal and reticulo-infundibular catecholamine neuron systems in the rat brain. Brain Res 51: 171-191

2. Saavedra JM (1985) Central and peripheral catecholamine innervation of the rat intermediate and posterior pituitary lobes. Neuroendocrinology 40: 281-284
3. Oertel WH, Mugnaini E, Tappaz ML, Weise VK, Dahl AL, Schmechel DE, Kopin IJ (1982) Central GABAergic innervation of neurointermediate pituitary lobe: biochemical and immunocytochemical study in the rat. Proc Natl Acad Sci USA 79: 675-679
4. Vincent SR, Hokfelt T, Wu JY (1982) GABA neuron systems in hypothalamus and the pituitary gland. Neuroendocrinology 34: 117-125
5. Tilders FJH, Mulder AH, Smelik PG (1975) On the presence of a MSH-release inhibiting system in the rat neuro-intermediate lobe. Neuroendocrinology 18: 125-130
6. Penny RJ, Thody AJ (1978) An improved radioimmunoassay for α-melanocyte-stimulating hormone (α-MSH) in the rat: serum and pituitary α-MSH levels after drugs which modify catecholaminergic neurotransmission. Neuroendocrinology 25: 193-203
7. Hollt V, Bergmann M (1982) Effects of acute and chronic haloperidol treatment on the concentrations of immunoreactive β-endorphin in plasma, pituitary and brain of rats. Neuropharmacology 21: 147-154
8. Newman CB, Wardlaw SL, Stark RI, Daniel SS, Frantz AG (1987) Dopaminergic regulation of α-melanocyte-stimulating hormone and N-acetyl- β-endorphin secretion in the fetal lamb. Endocrinology 120: 962-966
9. Kemppainen RJ (1986) In vivo evidence for dopaminergic regulation of the canine pituitary intermediate lobe. Acta Endocrinol 113: 471-478
10. Verburg-Van Kemenade BML, Tappaz M, Paut L, Jenks BG (1986) GABAergic regulation of melanocyte-stimulating hormone secretion from the pars intermedia of xenopus laevis: immunocytochemical physiological evidence. Endocrinology 118: 260-267
11. Tomiko SA, Taraskevich PS, Douglas WW (1983) GABA acts directly on cells of pituitary pars intermedia to alter hormone output. Nature 301: 706-707
12. Petraglia F, Cella SG, Radice L, Genazzani AR, Muller EE (1986) γ-Aminobutyric acid inhibits β-endorphin secretion from the anterior pituitary but not the neurointermediate lobe in the rat. Endocrinology 118: 360-366
13. Ham J, Smyth DG (1985) β-endorphin processing in pituitary and brain is sensitive to haloperidol stimulation. Neuropeptides 5: 497-500
14. Millington WR, Chappell MC, O'Donohue TL, Mueller GP (1984) Chronic haloperidol selectively alters the molecular forms of β-endorphin in the intermediate lobe of the rat pituitary. Proc 7th Int Congr Endocr, Quebec City, Abstract 1385, p 953.
15. Ham J, Smyth DG (1984) Regulation of bioactive β-endorphin processing in rat pars intermedia. FEBS Letters 175: 407-411
16. Millington WR, O'Donohue TL, Chappell MC, Roberts JL, Mueller GP (1986) Coordinate regulation of peptide acetyltransferase activity and proopiomelanocortin gene expression in the intermediate lobe of the rat pituitary. Endocrinology 118: 2024-2033
17. Jenks BG, Verburg van Kemenade BML, Tonon MC, Vaudry H (1985) Regulation of biosynthesis and release of pars intermedia peptides in rana ridibunda: dopamine affects both acetylation and release of α-MSH. Peptides 6: 913-921
18. Smith AI, Wallace C, Autelitano DJ, Cheng MC, Clarke IJ, Funder JW (1986) α-N-acetyl β-endorphin is differentially processed in the normal and hypothalamo-pituitary-disconnected (HPD) ewe. Neurosci Letters 65: 229-233
19. Smith AI, Wallace CA, Cheng MC, Autelitano DJ (1987) The combined use of size-exclusion and reversed-phase high-performance liquid chromatography for characterization of β-endorphin processing pathways. J Chromatog 416: 255-264
20. Chen CLC, Dionne FT, Roberts JL (1983) Regulation of the pro-opiomelanocortin mRNA levels in rat pituitary by dopaminergic compounds. Proc Natl Acad Sci USA 80: 2211-2215
21. Munemura M, Eskay RL, Kebabian JW, Long R (1980) Release of α-melanocyte-stimulating hormone from dispersed cells of the intermediate lobe of the rat pituitary gland: involvement of catecholamines and adenosine 3',5'-monophosphate. Endocrinology 106: 1795-1803

22. Beaulieu M, Felder R, Kebabian JW (1986) D-2 dopaminergic agonists and adenosine 3',5'-monophosphate directly regulate the synthesis of alpha-melanocyte-stimulating hormone-like peptides by cultured rat melanotrophs. Endocrinology 118: 1032-1039
23. Akil H, Shiomi M, Matthews J (1985) Induction of the intermediate pituitary by stress: synthesis and release of a nonopioid form of β-endorphin. Science 227: 424-426
24. Lebouille JLM, Burbach JPH, De Kloet ER (1984) Quantitation of the endopeptidase activity generating γ-endorphin from β-endorphin in rat brain synaptic membranes by a radiometric assay. Anal Biochem 141: 1-9
25. Lebouille JLM, Burbach JPH, De Kloet ER (1985) Adrenalectomy changes γ-endorphin generating endopeptidase activity, β-endorphin and γ-endorphin levels in rat brain and pituitary. Proc 26th Dutch Fed Meeting, Amsterdam, Abstract 203

DOPAMINERGIC MODULATION OF PITUITARY ALPHA-N-ACETYL-TRANSFERASE ACTIVITY AND ADRENARCHE

R. Pérez-Fernández*, F. Facchinetti, M. J. Lechuga*, A. R. Genazzani and J. Devesa*

Department of Physiology, School of Medicine,
University of Santiago de Compostela, Spain*
Department of Obstetrics and Gynecology,
School of Medicine, University of Modena, Italy

INTRODUCTION

In many mammals including man and dogs (1), adrenarche is a key step in sexual development. However, the mechanisms triggering this process still remain unclear.

We previously reported (2) that chronic dopaminergic receptor blockade in pre-adrenarchal dogs induced a significant enlargement of the adrenal zona reticularis, which was accompanied by an increased δ-5 androgens response to ACTH challenge - two main features of the adrenarche (3,4). Hormones subject to tonic inhibitory control of dopamine, such as prolactin or α-MSH (5,6), could theoretically be responsible for this process. Moreover, it has been shown that α-MSH selectively stimulates the growth of the adrenal zona glomerulosa (7) and the secretion of adrenal androgens from human adrenal cells in culture (8). On this basis, and given that in the dog, the intermediate lobe of the pituitary is in general negatively regulated by dopamine (9), we decided to examine the dopaminergic blockade in this animal, in order to analyze its effects on some pituitary proopiomelanocortin-related peptides.

MATERIALS AND METHODS

Protocol

Twelve pre-adrenarchal dogs (8 weeks) from three different litters, were treated with domperidone (DOM, n=6, 3 mg/kg p.o. 3 times/day, plus 0.6 mg/kg i.m. at 2100 hr) or placebo (C, n=6) for 15 days. Twelve hr after the last DOM treatment, the animals were sacrificed and immediately thereafter the whole pituitaries were removed, weighed and boiled for 10 min in 0.05 M acetic acid to block protease activities. The organs were homogenized and centrifuged (12000 x g); the pellet was discarded and the supernatant dried and reconstituted with 0.4 ml acetonitrile/0.01 M HCl (18:82). Two-tenths ml of each homogenate was injected into an HPLC (Waters, UV detector) and eluted (reverse phase C-18 column; μBondapack, 3.9 x 300 mm, 10 micron size) in a linear gradient from 10% to 40% acetonitrile in 0.01 M HCl (22 min, flow rate: 1.5 ml/min). The retention times (min) of the main peptides tested were: β-LPH (15.8 ± 0.09), hβ-endorphin (β-EP) (18.5 ± 0.1), acetyl-β-EP (20.7 ± 0.2), α-MSH (10.4 ± 0.3), des-acetyl-α-MSH (8.1 ± 0.2), ACTH (15.2 ± 0.2). For each sample, 48 fractions were collected (each 20 min, starting from the 7th min of elution), dried and redissolved in 1 ml 0.12 M phosphate buffer, pH 7.4, 0.1% BSA, before their analysis by specific RIA. Elution recoveries were: β-EP (88.1 ± 10.4%); ACTH (89.99 ± 12.4%); α-MSH (92.1 ± 11.4%); des-acetyl-α-MSH (87.4 ± 8.1%).

Radioimmunoassays

β-Endorphin-like immunoreactivity (β-End ir) was tested in each fraction using the anti-β-End serum B-4 (Dr. V. Wiegant, University of Utrecht, N.L.), which cross-reacts with many β-End related peptides (acetyl-β-End: 100%, hβ-LPH: 110%). Thus, according to this β-End ir spectrum and in view of the β- End retention time, fractions were tested again using a specific anti-hβ-End serum (Prof. C.H. Li, San Francisco, CA) that does not recognize the peptides listed above, with the exception of β-LPH (16.4%). Simultaneously, ACTH ir (in all fractions) and α-MSH ir (in the first 17 fractions) were tested using antiserum and standard provided by NIH (ACTH) and Dr. Wiegant (α-MSH). The anti-ACTH serum (West) cross-reacts with 1-39 ACTH (100%), 1-24 ACTH (6.7%) and 11-24 ACTH (10.5%), but does not recognize α-MSH, des-acetyl-α-MSH and CLIP. Anti-α-MSH serum (M2) cross-reacts with α-MSH (100%) and des-acetyl-α-MSH (24.8%), but does so less than 0.1% with other ACTH fragments.

RIAs for β-End, ACTH and α-MSH were carried out as previously described (10-12).

Statistical Analysis

Wilcoxon's test and ANOVA were used to establish the statistical significance of the data. Data are expressed as fmol/mg pituitary, X ± SEM.

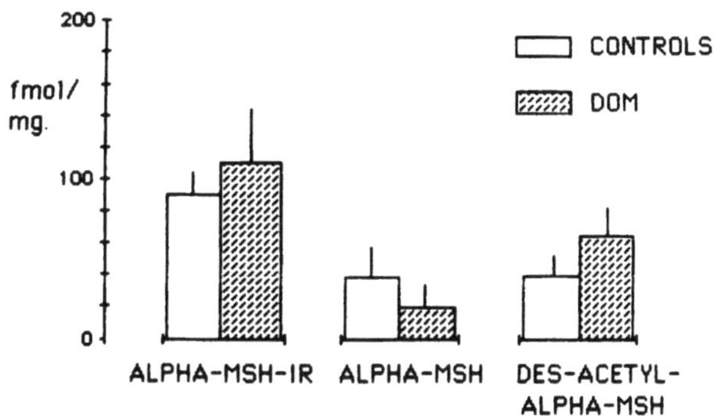

Figure 1. Pituitary concentration of total α-MSH ir, and acetylated and des-acetylated forms, in controls and DOM-treated dogs (X ± SE).

RESULTS

Neither body nor pituitary weight were modified after dopaminergic blockade in DOM-treated dogs, as compared to controls.

The HPLC elution pattern of α-MSH showed a first peak appearing at the retention time of des-acetyl-α-MSH and a second one corresponding to the acetylated form of the peptide. Although the total α-MSH ir was similar in DOM-treated and control dogs (109.2 ± 32.6 vs. 90 ± 25.7 fmol/mg, respectively), the fact that the pituitary concentration of des-acetyl-α-MSH was nearly doubled in DOM-treated animals, while the acetylated form was halved (Fig. 1), led to a significant ($P < 0.01$) reduction of the acetyl- to des-acetyl-α-MSH ratio in DOM-treated animals, as compared to controls (Fig. 2).

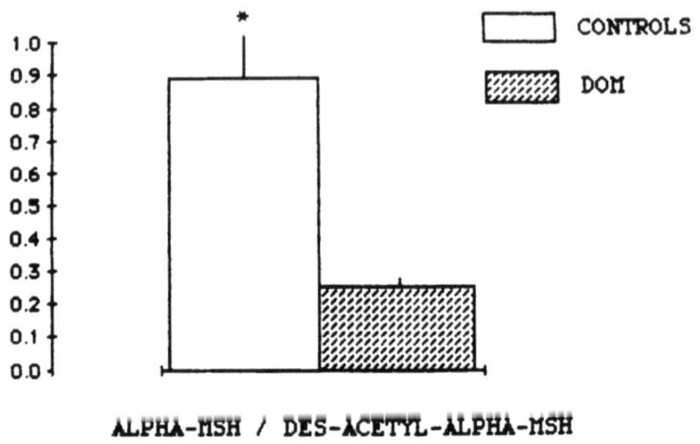

Figure 2. Pituitary ratio between acetyl and des-acetyl-α-MSH in controls and DOM-treated dogs (X ± E).

Pituitary concentrations of ACTH (31.4 ± 9.8 vs. 34.8 ± 14.2 fmol/mg) and β-End (11.1 ± 2.7 vs. 14.1 ± 4.1 fmol/mg) were similar in DOM-treated and control dogs, respectively (Fig. 3). Acetyl-β-End ir (measured in fractions 38-42) was significantly ($P < 0.05$) lower in DOM-treated than in control dogs (38.4 ± 5.4 vs 86.6 ± 19.2 fmol/mg, respectively) (Fig. 3).

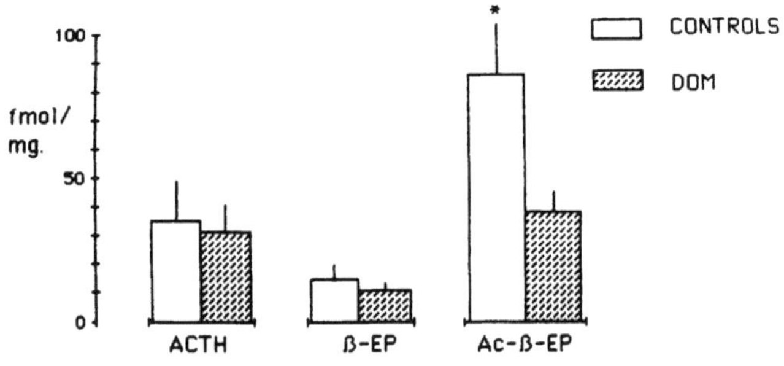

Figure 3. Pituitary content of ACTH, β-endorphin (total ir) and acetyl-β-endorphin in controls and DOM-treated dogs (X ± SE).

DISCUSSION

These data demonstrate that peripheral dopaminergic receptor blockade produces marked changes in the pituitary content of POMC-related peptides in pre-adrenarchal dogs. Specifically, while the total α-MSH ir was not affected by the treatment, DOM-treated dogs showed lower α-MSH and higher des-acetyl-α-MSH values than in controls. This led to a decreased ratio between the acetylated and des-acetylated forms of the peptide in the pituitary of those animals, as compared to that in controls. Pituitary acetyl-β-End was also diminished in DOM-treated dogs. These results, together with the fact that both ACTH and β-End were not affected by the treatment, indicate that dopaminergic blockade acts to decrease the acetylated forms of POMC-related peptides at the pituitary level.

Acetylation greatly interferes with the biological activity of POMC-peptides (13-15); thus the effect of dopamine on post-translational POMC processing can be of great physiological significance. The shift from α-MSH to des-acetyl-α-MSH that we observed in DOM-treated animals is reminiscent of the situation shown in the human fetal pituitary, where des-acetyl-α-MSH is the predominant form (16).

During fetal life, the adrenal cortex is mainly constituted by a "specialized" zona reticularis, producing δ-5 androgens instead of cortisol (17), which involutes after birth and slowly reappears years later and during adrenarche (3). Because we previously produced morphological and functional development of this androgen-secreting zone in pre-adrenarchal dogs through DOM-treatment (2), it is tempting to speculate that the changes we now observed on the pituitary forms of POMC-peptides, could play an important role in the evolution of adrenal zona reticularis.

In conclusion, our study suggests that there is a dopaminergic modulation of the pituitary α-N-acetyl-transferase activity. The blockade of peripheral dopaminergic receptor interferes with the acetylation processes of POMC-related peptides, producing a predominance of des-acetylated forms at the pituitary level. This may be a key process in the adrenal development leading to adrenarche in dogs.

SUMMARY

We report the effects of chronic peripheral dopaminergic receptor blockade using domperidone (DOM) on the pre-adrenarchal dog pituitary content of POMC-related peptides. No significant differences were observed between DOM-treated and control dogs with respect to pituitary concentrations of ACTH and β-endorphin. The total α-MSH immunoreactivity (α-MSH ir) was also similar in both groups, but the percentage of α-MSH of total α-MSH ir in DOM-treated dogs was significantly lower than in controls, which oppositely showed a higher percentage of des-acetyl- α-MSH. Thus the ratio between the acetylated and the des-acetylated forms of the hormone was considerable in DOM-treated dogs. Acetyl-β-endorphin ir was also lower in treated than in control animals. These data indicate that dopamine plays an important role in the control of acetylation processes of POMC-related peptides in the pituitary. Changes in pituitary dopaminergic activity may be a key process in the adrenal development leading to adrenarche in dogs.

ACKNOWLEDGMENTS

We thank Prof. C. H. Li, University of California (USA), for his supply of anti-β-endorphin serum. α-MSH and des-acetyl- α-MSH standards as well as anti-α-MSH and anti-β-endorphin-like sera were kindly supplied by Dr. V. Wiegant, Institute of Pharmacology, University of Utrecht (NL). We also acknowledge the receipt of hACTH standard and anti-hACTH serum from the National Pituitary Agency (Bethesda, MD, USA). Partially supported by Xunta de Galicia, Grant No. 609-026-1066.

REFERENCES

1. Schiebinger RJ, Albertson BD, Barnes KM, Cutler GB, Loriaux DL (1981) Developmental changes in rabbit and dog adrenal function: a possible homologue of adrenarche in the dog. Am J Physiol 240: 694-699
2. Pérez-Fernández R, Facchinetti F, Beiras A, Lima L, Gaudiero GJ, Genazzani AR, Devesa J (1987) Morphological and functional stimulation of adrenal zona reticularis by dopaminergic blockade in dog. J Steroid Biochem 28: 465-470
3. Dhom G (1973) The perpubertal and pubertal growth of the adrenal (adrenarche). Beitr Path Bd 150: 357-366
4. Ducharme JR, Forest MG, De Peretti E, Sempre M, Collu R, Bertrand J (1976) Plasma adrenal and gonadal sex steroids in human pubertal development. J Clin Endocrinol Metab 42: 468-476
5. Ben Jonathan N (1985) Dopamine: a prolactin inhibiting hormone. Endocrine Rev 6: 564-591
6. Tilders FJH, Smelik PG (1977) Direct neural control of MSH secretion in mammals: The involvement of dopaminergic tubero-hypophyseal neurons. Front Horm Res 4: 80-93
7. Robba C, Rebuffat P, Mazzocchi G, Nussdorfer GG (1986) Long-term trophic action of α-melanocyte-stimulating hormone on the zona glomerulosa of the rat adrenal cortex. Acta Endocrinol 112: 404-408
8. Baird A, Kan KN, Solomon S (1983) Role of pro-opiomelanocortin derived peptides in the regulation of steroid production by human fetal adrenal cells in culture. J Endocrinol 97: 357-363
9. Kemppainen RJ, Sartin JL (1986) In vivo evidence for dopaminergic regulation of the canine pituitary intermediate lobe. Acta Endocrinol 113: 471-478

10. Genazzani AR, Nappi G, Facchinetti F, Mazzarella GL, Parrini D, Sinforiani E, Petraglia F, Savoldi F (1982) Central deficiency of β-endorphin in alcohol addicts. J Clin Endocrinol Metab 55: 583-586
11. Facchinetti F, Petraglia F, Nappi G, Parrini D, Martignoni E, Antoni G, Genazzani AR (1983) Different pattern of central and peripheral endorphins through the life. Peptides 4: 744-749
12. Facchinetti F, Perez-Fernandez R, Toma MO, Gaudiero GJ, Lechuga MJ, Devesa J, Genazzani AR (1987) Dopamine acts on acetylation of pro-opiomelanocortin products in dog pituitary. Acta Endocrinol, (in press)
13. Smyth DG, Massey DE, Zacharian S, Finnie MDA (1979) Endorphins are stored in biologically active and inactive forms: isolation of α-N-acetyl peptides. Nature 279: 252-254
14. O'Donohue TL, Handellman GE, Mieler RL, Jacobowitz DM (1982) N-acetylation regulates the behavioral activity of α-melanotropin in a multineurotransmitter neuron. Science 215: 125-127
15. Rudman D, Hollins BM, Kutner MH, Moffit SD, Lynn MJ (1983) Three types of α-melanocyte-stimulating hormone: bioactivities and half-lives. Am J Physiol 245: E47-E51
16. Tilders FJH, Parker CR, Barnea A, Porter JC (1981) The major immunoreactive α-melanocyte-stimulating hormone (α-MSH)-like substance found in human fetal pituitary tissue is not α-MSH but may be des-acetyl α-MSH (adrenocorticotropin 1-13 NH2). J Clin Endocrinol Metab 52: 319-323
17. Fujieda K, Faiman C, Reyes FI, Thliveris J, Winter JSD (1981) The control of steroidogenesis by human fetal adrenal cells in tissue culture. II. Comparison of morphology and steroid production in cells of the fetal and definitive zones. J Clin Endocrinol Metab 53: 401-405

LHRH PRIMING OF CRF-STIMULATED ACTH OUTPUT IN VITRO: A POSSIBLE PARACRINE EFFECT?

T. T. Chuang, A. Di Sciullo**, B. Gillham* and M. T. Jones

Departments of Gynaecology and *Biochemistry
U.M.D.S. (St. Thomas's Campus), London, U.K.
**Medical Faculty, University of Rome "La Sapienza," Rome, Italy

INTRODUCTION

The existence of paracrine interactions in the regulation of hormonal outputs from the anterior pituitary gland has long been suspected, and has more recently been extensively studied (for review, see ref. 1). While these studies have revealed some "immediate" paracrine effects, there has been no report of paracrine "priming" effects - "priming" refers to the prior exposure of tissue to a substance, leading to an enhancement of the response to a subsequent stimulation (2). We now report the possible presence of an LHRH priming effect on CRF-stimulated ACTH release from male rat anterior pituitary fragments statically incubated in vitro.

MATERIALS AND METHODS

Animals

Male Wistar-derived CSE rats (200-250 g) were obtained from a colony bred in a pathogen-free animal house at United Medical and Dental Schools and subsequently kept for at least seven days in a light-cycle controlled, air-conditioned room before experimentation. Water and rat chow were available ad libitum.

Incubation of Tissues

Rats were removed from cages and decapitated by guillotine within 15 sec; this was followed by removal of the anterior pituitary glands. The glands were quartered and transferred into tissue-holders of the incubation wells made from 24 well-tissue culture dishes (Costar). Each well contained 1 ml of Kreb's bicarbonate medium with 0.1% BSA and 1 μg/ml of ascorbate. These modified culture dishes were placed in a 37°C water bath and the surrounding atmosphere gassed continuously with a mixture of 95% O_2-5% CO_2.

Two different protocols were used for priming the anterior pituitary quarters following two hours of preincubation; in each 20 min period, tissues were subjected to various treatments:

Protocol one -prime, rest, first CRF stimulation, rest, prime, rest, second CRF stimulation.

Protocol two -different groups of tissues were exposed to different numbers of primings, and stimulated only once at the end of experimentation.

Immediately following experimentation, tissues were removed and the incubation media placed on ice for a short period prior to assay by radioimmunoassay.

Radioimmunoassays

(i) LH: This was done using reagents obtained from the National Institute of Diabetes, Digestive and Kidney Diseases (NIDDK), U.S.A. The intraassay variations at LH concentrations of 78 pg-tube and 5000 pg/tube were 6.6% and 8.7%, respectively, while the interassay variations at the same LH concentrations were 8.0% and 12.3%, respectively. Sensitivity was 39 pg-tube.
(ii) ACTH: The antibody was a gift of Dr. G. Makara. The intraassay variations at ACTH concentrations of 460 pg/ml and 97.4 pg/ml were 7.6% and 5.8%, rspectively, while the interassay variations at the same concentrations were 9.9% and 8.2%, respectively. The sensitivity was 6 pg/tube.

Materials

LHRH - Sigma; r/h CRF - Peninsula; AVP - Penninsula; angiotensin II-Sigma; 8-Br cAMP - Sigma.

RESULTS

In preliminary experiments, protocol 1 was used, in which each priming with LHRH was followed by a challenge with one of two doses of CRF (100 pM or 10 nM). The CRF-induced release of ACTH following the first priming appeared to be higher than the corresponding response from unprimed tissue, but this failed to reach significance (Fig. 1). However, following the second priming with LHRH, the ACTH response to CRF significantly exceeded that of unprimed tissue for the lower (100 pM) but not the higher (10 nM) concentration of CRF (Fig. 1).

Two factors render experiments such as those in protocol 1 difficult to interpret: the tissue is exposed to CRF at different times in vitro and multiple exposures to CRF in static incubations are known to result in desensitization (3). Protocol 2 was devised to avoid these problems. Once again, significant elevation of CRF-stimulated ACTH release followed two-fold priming, while a single priming caused an effect that was not significant (Fig. 2). In a more extended protocol, the response to CRF following a triple round of priming did not differ significantly from that of unprimed control tissues.

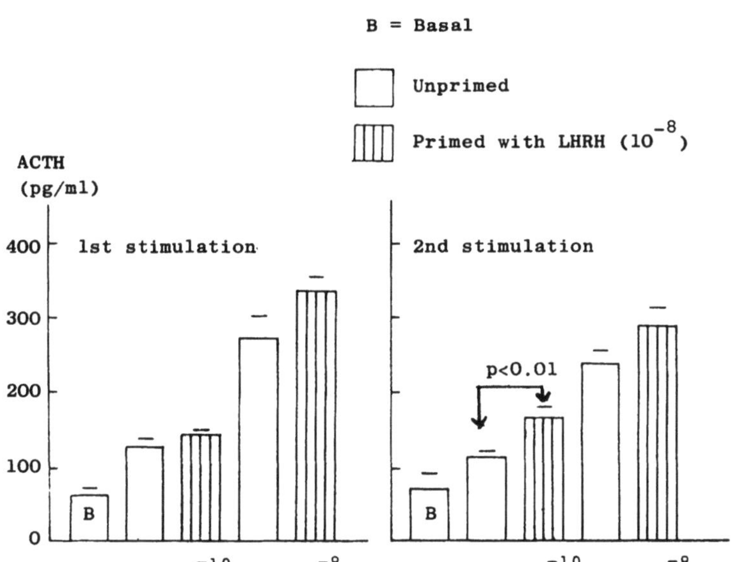

Figure 1. ACTH output (primed before each stimulation or unprimed) in response to CRF. n = 4-6.

Although the results indicate that LHRH is able to prime the response to CRF, the converse is not true: there is no reciprocal effect of CRF on LHRH-stimulated LH release (Fig. 3) when single priming of protocol 2 was used.

The priming effect of LHRH may be restricted to ACTH release provoked by CRF since ACTH output stimulated by two doses of AVP (1 and 10 nM), was unaffected by two-fold priming with LHRH (Fig. 4). Moreover, 8 Br-cAMP

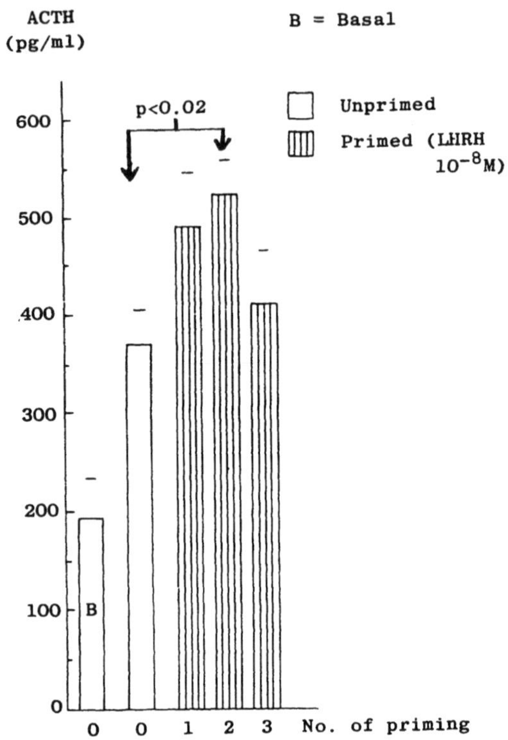

Figure 2. ACTH output in response to CRF 10^{-11}. n = 4 to 6.

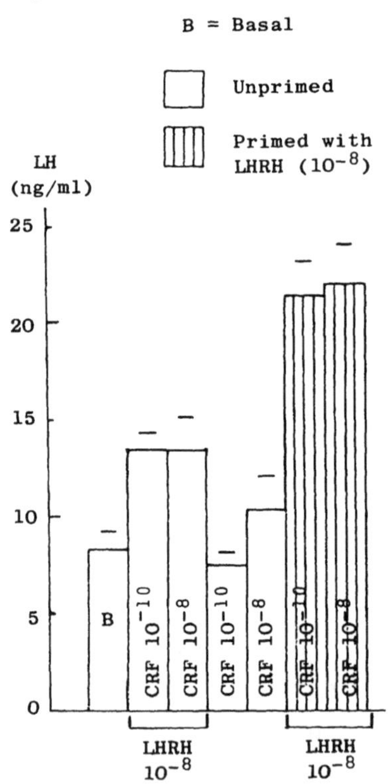

Figure 3. LH output in response to LHRH (with or without CRF). n = 5 or 6.

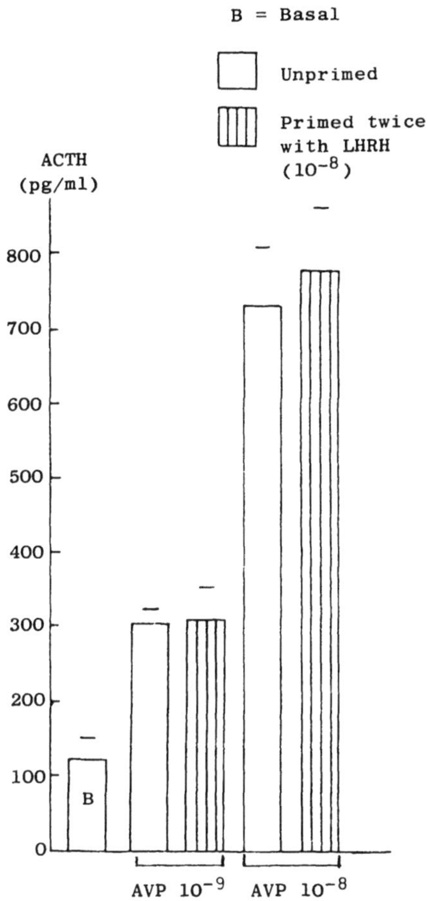

Figure 4. Effects of priming with LHRH on AVP-stimulated ACTH output.
n = 5 or 6.

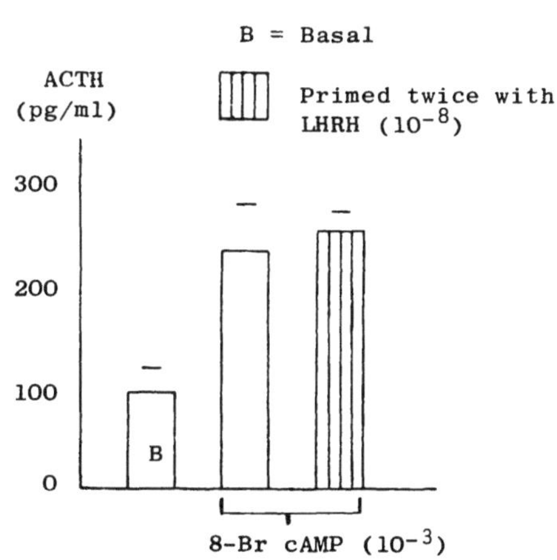

Figure 5. ACTH output in response to 8-Br cAMP with or without priming with LHRH.
n = 5 or 6.

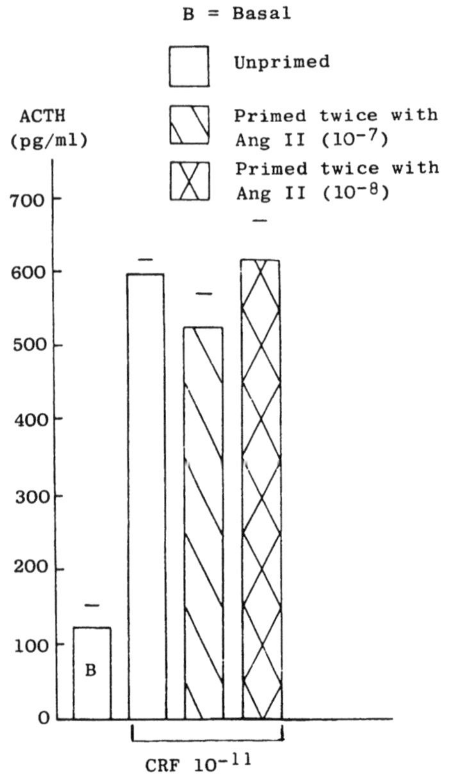

Figure 6. Effects of priming with angiotensin II on CRF-stimulated ACTH output. n = 5 or 6.

(1 mM)-stimulated ACTH release was also unaffected by two-fold priming with LHRH (Fig. 5).

In a search for possible paracrine agents that might mediate the priming, angiotensin II (two doses of 10 and 100 nM) was used instead of LHRH to prime the tissues. It was found to have no priming effect (Fig. 6).

DISCUSSION

The present results show that exposure of anterior pituitary gland tissue to LHRH in vitro results in a small, but reproducible priming effect on ACTH responses to challenges with CRF. It appears that two rounds of priming with LHRH are required for an optimal effect, although it is true to say that the experimental protocols used would not preclude the possibility that the timespan per se following a single priming is the critical variable.

Since there is no evidence for functionally-coupled LHRH receptors (or, indeed, for LHRH binding sites) resident in corticotroph plasma membranes, it seems clear that the priming effect must be mediated by some, as yet unidentified, paracrine agent released from gonadotrophs. A likely candidate for such an agent appeared to be angiotensin II for three reasons: (a) It is known to be synthesized in gonadotrophs (4) and is releasable upon stimulation by LHRH; (b) in physiological doses, it stimulates ACTH output from dispersed rat anterior pituitary cells (5,6); (c) high affinity binding sites for angiotensin II are present on corticotrophs (7). However, the current work failed to provide evidence in favor of this hypothesis. Another possibility considered here was that the "paracrine agent" was released adventitiously from gonadotrophs as a result of large changes in its intracellular concentrations. In this case, LHRH-induced second messengers, such as inositol polyphosphates (IP) or diacylglycerols, seem to be possibilities for mediating such paracrine priming. This view is strengthend by two observations: first, AVP is known to prime the response to CRF (3), and it has been demonstrated that AVP promotes IP formation in corticotrophs (8); second, LHRH causes rapid and profound changes in IP turnover in gonadotrophs.

This paracrine priming may not be of great physiological importance in normal male rats. However, it may serve to enhance ACTH output from the anterior pituitary gland in conditions where there is enhanced LHRH release from the hypothalamus - e.g. Buckingham et al. (9) observed the presence of elevated plasma ACTH and corticosterone levels in the female rats on the afternoon of proestrus day, accompanying the LH surge (9).

From a topographical point of view, in vivo gonadotrophic paracrine regulation of corticotrophs is made possible by the presence of large gonadotrophs in the vicinity of the portal vessels that are fenestrated (10,11). The secretory products of the gonadotrophs are, therefore, able to reach the corticotrophs via the vessels.

It has been suggested that AVP-stimulated ACTH release is mediated by the production of inositol trisphosphate (8,12,13), while CRF stimulation is mediated by the production of cAMP (14). The lack of any LHRH-priming effect on the AVP or 8-Br cAMP-stimulated ACTH release indicates that the CRF receptor complex may be the target of this priming effect.

Paracrine effects apart, the underlying factor for the LHRH priming effect may well be the ACTH/LH co-containing cells (15). Childs and her group have used immunocytochemical staining to show that, in the anterior pituitary glands of nine-day old male rats, about 80% of cells staining for ACTH also stained for LH (15). The proportion of these cells declined to 22% in adults. Neill et al. in studies using reversed-phase plaque assay combined with immunocytochemistry, showed that in neonatal male rats of six, nine and fifteen days, less than 5% of the cells that formed plaques or stained for ACTH, also formed plaques for LH (16). Despite the discrepancy between these two findings in terms of percentages, it is clear that a portion of the immunoreactive ACTH measured in our studies may have originated from these ACTH/LH cells. The possibility of LHRH's priming the CRF-stimulated ACTH output from these cells cannot be dismissed.

In conclusion, an LHRH-induced priming of CRF-stimulated ACTH release was observed. This could be mediated by a paracrine agent sensitizing the CRF-receptor complex. It could alternatively be the result of LHRH priming of ACTH/LH cells.

REFERENCES

1. Denef C (1986) Paracrine interactions in the anterior pituitary. Clinics in Endocrinol and Metab 15: 1-32
2. Jones MT, Gillham B (1988) Factors involved in the regulation of ACTH/ βLPH. Physiol Rev (in press).
3. Buckingham, JC (1985) Two distinct corticotrophin-releasing activities of vasopressin. Br J Pharmacol 84: 213-219
4. Deschepper, CF, Crumrine DA, Ganong WF (1986) Evidence that the gonadotrophs are the likely site of production of angiotensin II in the anterior pituitary of the rat. Endocrinology 119: 36-43
5. Gaillard RC, Grossman A, Gillies G, Rees LH, Besser GM (1981) Angiotensin II stimulates the release of ACTH from dispersed rat anterior pituitary cells. Clin Endocrinol 15: 573
6. Sobel D, Vagnucci A (1982) Angiotensin II mediated ACTH release in rat pituitary cell culture. Life Sci 30: 1281-1286
7. Paglin S, Stukenbrook H, Jamieson JD (1984) Interaction of angiotensin II with dispersed cells from the anterior pituitary of the male rat. Endocrinology 114: 2284
8. Todd K, Lightman SL (1987) Vasopressin activation of phosphotidylinositol metabolism in rat anterior pituitary in vitro and its modification by changes in the hypothalamus-pituitary-adrenal axis. Neuroendocrinology 45: 212-218
9. Buckingham JC, Dohler KD, Wilson CA (1978) Activity of the pituitary-adrenocortical system and thyroid gland during the oestrous cycle of the rat. J Endocrinol 78: 359-366
10. Gross DS (1985) Hormone production in the hyophysial pars tuberalis of intact and hypophysectomized rats. Endocrinology 112: 733-744
11. Gross DS, Turgeon JL, Waring DW (1984) The ovine pars tuberalis: a naturally occurring source of partially purified gonadotropes which secrete LH in vitro. Endocrinology 114: 2084-2091
12. Abou-Samra AB, Catt KJ, Aguilera G (1986) Involvement of protein kinase C in the regulation of adrenocorticotropin release from rat anterior pituitary cells. Endocrinology 118: 212-217
13. Thomas AP, Alexander J, Williamson JR (1984) Relationship between inositol polyphosphate production and the increase of cytosolic-free Ca^{2+} induced by vasopressin in isolated hepatocytes. J Biol Chem 259: 5574-5584
14. Aguilera G, Harwood JP, Wilson JX, Morell J, Brown JH, Catt KJ (1983) Mechanisms of action of CRF and other regulators of corticotropin release in rat pituitary cells. J Biol Chem 258: 8039-8045

15. Childs GU, Ellison DG, Ramaley JA (1982) Storage of anterior lobe adrenocorticotropin in corticotropes and a sub-population of gonadotropes during the stress-non-responsive period in the neonatal male rat. Endocrinology 110: 1676-1692
16. Neill JD, Smith PF, Luque EH, Munoz de Toro M, Nagy G, Mulchahey JJ (1987) Detection and measurement of hormone secretion from individual pituitary cells. Recent Prog Horm Res 43: 175-229

SECRETION OF CRF-41, AVP AND SRIF BY FETAL HYPOTHALAMIC NEURONS CULTURED IN DEFINED MEDIUM

G. Gillies and M. Clarke

Protein Hormone Unit, St. Bartholomew's Hospital
London, U.K.

INTRODUCTION

The precise mechanisms regulating the secretion of the 41 amino acid corticotrophin-releasing factor (CRF-41) and arginine vasopressin (AVP) into the hypothalamo-hypophysial portal circulation remain undetermined. This is attributable in part to the inherent complications of in vivo studies on parameters particularly sensitive to external manipulations and to the indirect methods of CRF measurements both in vivo and in vitro. These rely on biological activities, e.g. changes in levels of adrenocorticotrophic hormone (ACTH), corticosteroids or adrenal ascorbic acid, which tell us nothing of the individual components of the hypothalamic CRF complex (1,2). In order to study its control, therefore, we established a system of long-term primary culture of rat hypothalamic neurons coupled with the direct measurement of released CRF-41 and AVP by immunoassay. As an internal control, we simultaneously measured somatostatin (SRIF). Initially, the neurons were grown in a serum-containing medium (3). Detection of released and extracted CRF-41 and AVP depended first, on the concentration of very small amounts of peptides on silica resins; second, on the maintenance of the cultures beyond two weeks; and third, on highly sensitive assays for CRF-41 (4) and AVP. A large number of reports have concentrated on the release of SRIF from similar serum-grown cultures between days 10 and 14 in vitro, when it is present in abundance (5-7). However, serum is likely to represent a nonphysiological component of the central neuron environment. Therefore, we have investigated the responses of neurons grown in a defined medium based on those used by Bottenstein and Sato (8) and by Faivre-Bauman and colleagues (9).

METHODS

Dispersed hypothalamic cells from fetal rats at 18 days' gestation were prepared as described elsewhere (3). The final cell suspension was divided into three separate units, then resuspended, plated and maintained in a serum-containing medium (S) (3), a basic defined medium (basic DM), or a basic defined medium plus supplements (DM+) (See Table 1). Cells were grown in 35 mm dishes previously coated with poly-L-lysine and 10% serum. Medium was changed every 4 to 5 days or as required for release experiments. Release of peptides into an isotonic balanced salt solution (3) was assessed under basal conditions [release medium (RM) containing 5.6 mM K^+], depolarizing conditions (56 mM K^+ RM or RM plus veratridine) or after adding neurotransmitters to RM. Unless otherwise stated, test incubations lasted 1 hr.

TABLE 1

MEDIA COMPOSITIONS

	Serum (S)	Basic DM	Basic DM & Supplements (DM+)
Penicillin	100 IU/ml	100 IU/ml	100 IU/ml
Streptomycin	100 μg/ml	100 μg/ml	100 μg/ml
Fungizone	0.5 μg/ml	0.5 μg/ml	0.5 μg/ml
Dulbecco's Modified Medium	100 ml	50 ml	50 ml
Ham's F.12	-	50 ml	50 ml
Fetal Calf Serum	10 ml	-	-
Horse Serum	10 ml	-	-
Insulin	-	5 μg/ml	5 μg/ml
Transferin	-	100 μg/ml	100 μg/ml
Progesterone	-	20 nM	20 nM
Selenium	-	30 nM	30 nM
Putrescine	-	100 μM	100 μM
17 β estradiol	-	1 pM	1 pM
Arachidonic Acid	-	-	1 μg/ml
Docosahexaenoic Acid	-	-	0.5 μg/ml
Triiodothyronine	-	-	1 nM

TABLE 2

EFFECT OF CULTURE MEDIA COMPOSITION ON NEUROPEPTIDE RESPONSES

	Days in Culture	Peptide Release pg/hypothalamus Basal	Peptide Release pg/hypothalamus 56 mM K^+
(a) CRF-41			
S	14	0.54 ± 0.03	1.26 ± 0.06
	21	0.58 ± 0.03	3.19 ± 0.30
	31	0.81 ± 0.05	3.90 ± 0.32
Basic DM	14	0.68 ± 0.02	1.51 ± 0.17
	21	0.72 ± 0.03	2.18 ± 0.18
	31	0.68 ± 0.09	2.02 ± 0.05
DM^+	14	1.00 ± 0.03	4.67 ± 0.29
	21	1.06 ± 0.13	5.79 ± 0.14
	31	1.45 ± 0.11	10.50 ± 0.92
(b) AVP			
S	14	0.20 ± 0.01	0.98 ± 0.01
	21	0.50 ± 0.06	3.97 ± 0.10
	31	1.73 ± 0.17	11.30 ± 0.58
Basic DM	14	0.25 ± 0.03	0.66 ± 0.05
	21	0.22 ± 0.01	1.02 ± 0.07
	31	0.26 ± 0.02	1.88 ± 0.11
DM^+	14	0.40 ± 0.02	2.08 ± 0.06
	21	0.42 ± 0.01	6.23 ± 0.15
	31	1.42 ± 0.11	26.80 ± 0.58
(c) SRIF			
S	14	19.50 ± 1.54	57.10 ± 4.62
	21	26.90 ± 0.58	104.50 ± 9.28
	31	26.20 ± 1.86	102.40 ± 12.80
Basic DM	14	18.50 ± 0.03	24.50 ± 2.67
	21	11.40 ± 0.08	40.30 ± 2.39
	31	33.00 ± 2.28	97.20 ± 8.05
DM^+	14	18.50 ± 1.25	25.20 ± 2.20
	21	24.40 ± 1.52	206.30 ± 8.53
	31	28.50 ± 2.65	272.20 ± 7.90

RESULTS

In defined medium it was possible to maintain primary cultures of fetal rat hypothalamic neurons for 5 to 6 weeks. They develop morphologically, extending processes and making intercellular contacts. The overgrowth of non-neuronal cells is markedly reduced when compared with serum-grown cultures, as described by others for cultures maintained up to 2 weeks (9). Table 2 a, b and c demonstrates the release of CRF-41, AVP and SRIF, respectively, into RM and 56 mM K^+ RM from neurons grown in S, basic DM and DM+ after 14, 21 and 31 days in culture. Generally, basic DM does not support neuropeptide release as well as S, the AVP-producing neurons being most susceptible to the lack of serum and the SRIF neurons least susceptible. However, cells grown in DM+ respond to 56 mM K^+ RM not only with an increased peptide output, but also with an increased magnitude of response, as can be seen if stimulated release is expressed as a percentage of basal. Superimposed on this effect is a dramatic increase in stimulated peptide output between days 14 and 31 in culture. In S this effect is slight for SRIF (3), but in DM+ it ranges from a two-fold increase for CRF-41 to a ten-fold increase for AVP and SRIF. In subsequent experiments, cells were cultured in DM+ unless otherwise stated. The 56 mM

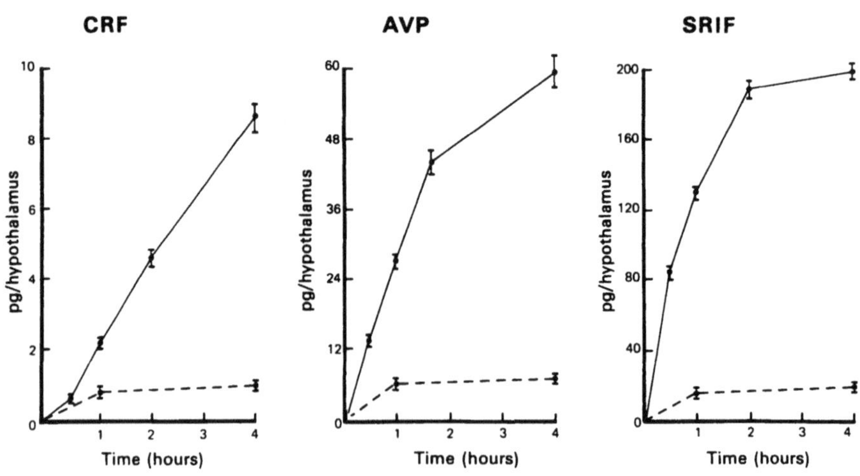

Figure 1. Time course of peptide release under basal (•----•) and 56 mM K^+ depolarizing (•——•) conditions from fetal rat hypothalamic neurons cultured in fully supplemented defined medium (mean ± SEM, n = 3). Reprinted from ref. 13, with permission.

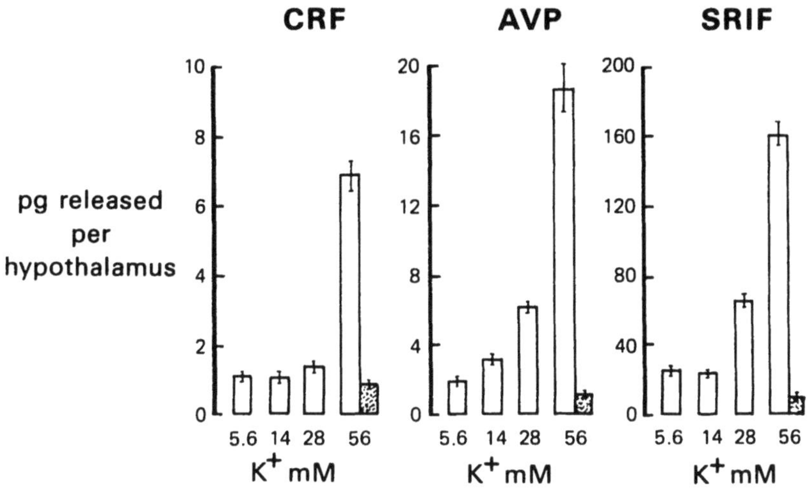

Figure 2. Dose dependency and extracellular Ca^{2+} requirement for K^+-stimulated neuropeptide release from fetal rat hypothalamic neurons cultured in fully supplemented defined medium. Stippled bar represents Ca^{2+}-free medium. (mean ± SEM, n = 3). Reprinted from ref. 13, with permission.

K^+-stimulated release of CRF-41, AVP and SRIF continues for at least 4 hr as shown in Figure 1. This suggests active synthesis and release of these peptides from neurons in culture. This release, and that stimulated by veratridine (10-100 μM, not shown) is concentration-dependent and Ca^{2+}-dependent, as shown in Figure 2. Furthermore, the inclusion of 100 nM corticosterone in DM+ from the time of plating significantly reduces the stimulated release of CRF-41 and AVP ($P < 0.01$), but not SRIF.

DISCUSSION

The present work has shown, for the first time, that serum is not essential for the long-term maintenance and development in vitro of hypothalamic neurons producing CRF-41, AVP and SRIF. The peptide output from cultured hypothalamic neurons and the magnitude of responses are considerably increased when cells are cultured in a defined medium rather than one containing serum (Table 2). As a result, no concentration step is necessary before detection of CRF-41 and AVP. The neurons appear to respond in a physiologically relevant manner, as demonstrated by their dependence on Ca^{2+} and the responsiveness of CRF-41 and AVP release to the continued presence of corticosterone in the culture medium. The cultures also respond to neurotransmitters (Clarke and Gillies, unpublished observations). For example, acetylcholine stimulates CRF-41 and AVP release, but not SRIF release; serotonin stimulates CRF-41 release. This is in agreement with results obtained from intact adult hypothalamic tissue, either in vivo or in vitro (10-12) and suggests that the relevant receptors are present on the cultured neurons after several weeks in vitro. Furthermore, the responsiveness to nonspecific depolarizing stimuli (3) and to neurotransmitters (Clarke and Gillies, unpublished observations) increases with time in culture, as does the cellular content of CRF-41, AVP and SRIF. This suggests a functional maturation of the endocrine hypothalamus in vitro. We conclude that the in vitro model described here is suitable to study further the development and control of hypothalamic peptidergic neurons in a manner that cannot be achieved in vivo or in acute in vitro preparations.

REFERENCES

1. Gillies GE, Linton EA, Lowry PJ (1982) Corticotropin releasing activity of the new CRF is potentiated several times by vasopressin. Nature 299: 355-357
2. Gillies GE, Puri A, Linton EA, Lowry PJ (1984) Comparative chromatography of hypothalamic corticotrophin-releasing factors. Neuroendocrinology 36: 17-24
3. Clarke MJO, Lowry PJ, Gillies GE (1987) Assessment of corticotropin-releasing factor, vasopressin and somatostatin secretion by fetal hypothalamic neurons in culture. Neuroendocrinology 46: 147-154
4. Linton EA, Lowry PJ (1986) Comparison of a specific two-site immunoradiometric assay with radioimmunoassay for rat/human CRF-41. Regulat Pept 14: 69-74
5. Richardson SB, Hollander CS, D'Eletto R, Greenleaf PW, Thaw C (1980) Acetylcholine inhibits the release of somatostatin from rat hypothalamus in vitro. Endocrinology 197: 122-128
6. Peterfreund RA, Vale W (1982) High molecular weight somatostatin secretion by cultured rat brain cells. Brain Research 239 : 463-477
7. Scanlon MF, Robbins RJ, Bolaffi JL, Jackson IMD, Reichlin S (1983) Characterisation of somatostatin and TRH release by rat hypothalamic and cerebral cortical neurons maintained on a capillary membrane. Neuroendocrinology 37: 269-274
8. Bottenstein JE, Sato GH (1979) Growth of a rat neuroblastoma cell line in serum-free supplemented medium. Proc Natl Acad Sci 76: 514-517
9. Faivre-Baumann A, Rosenbaum E, Puymirat J, Grouselle D, Tixier-Vidal A (1981) Differentiation of fetal mouse hypothalamic cells in serum-free medium. Dev Neurosci 4: 118-129
10. Hedge GA, Smelik PG (1968) Corticotropin-release: inhibition by intrahypothalamic implantation of atropine. Science 159: 191-192
11. Hillhouse EW, Burden J, Jones MT (1975) The effect of various putative neurotransmitters on the release of corticotropin releasing hormone from the hypothalamus of the rat in vitro. I. The effect of acetylcholine and noradrenaline. Neuroendocrinology 17: 1-11
12. Buckingham JC, Hodges JR (1979) Hypothalamic receptors influencing the secretion of corticotrophin releasing hormone in the rat. J Physiol 290: 421-431
13. Clarke MJO, Gillies GE (1988) Comparison of peptide release from fetal rat hypothalamic neurones cultured in defined media and serum-containing medium. J Endocrinol 116: 349-356

THE EFFECTS OF ENDOGENOUS CATECHOLAMINES ON THE HYPOTHALAMO-PITUITARY ADRENAL AXIS IN MAN

S. Al-Damluji, L. Perry, L. Rees and G. M. Besser

Department of Endocrinology, St. Bartholomew's Hospital
London, U.K.

INTRODUCTION

Our previous investigations of the effects of catecholamines on the secretion of ACTH in man demonstrated that intravenous infusion of the α_1-adrenoceptor agonist methoxamine stimulated the secretion of ACTH in a dose-dependent manner and the effect was blocked by a concomitant administration of the α_1-adrenoceptor antagonist thymoxamine. The site of action of these stimulant α_1-adrenoceptors appears to be in the central nervous system; the effects of methoxamine are not reproduced by equipotent doses of the more hydrophylic α_1-adrenoceptor agonist noradrenaline, which reaches the pituitary gland and the median eminence following a systemic injection, but does not cross the blood-brain barrier (1). Further studies demonstrated that the α_2- and β-adrenoceptor agonist properties of noradrenaline did not account for the differences from methoxamine (1). Intravenous infusions of adrenaline had no enhancing effect on the activity of synthetic ovine corticotrophin releasing factor in man (2), which is in contrast to the findings in cultured rat adenohypophysial cells in vitro (3,4). The physiological significance of the stimulant α_1-adrenoceptor mechanism on ACTH secretion was demonstrated in two situations: the cortisol secretory pattern during waking hours and the ACTH and cortisol responses to food ingestion were both enhanced by intravenous infusions of the α_1-adrenoceptor agonist methoxamine and reduced by the α_1-adrenoceptor antagonist thymoxamine, suggesting that under these two physiological circumstances, ACTH secretion is mediated by α_1-

adrenoceptors (5,6). In our current studies, we have been examining the effects of endogenous catecholamines on the hypothalamo-pituitary adrenal axis in man. The aims of this study were: 1) To investigate whether the ACTH response to food ingestion is caused by absorption of the catecholamine precursor and dietary amino acid tyrosine, and its subsequent conversion to noradrenaline. This has been done by administering a dose of tyrosine that is similar to the estimated tyrosine content of a standard meal that had previously been shown to stimulate ACTH secretion. 2) To examine the effects on the hypothalamo-pituitary adrenal axis of supraphysiological doses of tyrosine. 3) To examine the role of α_2-adrenoceptors in modulating the actions of tyrosine. The synthesis and release of noradrenaline is controlled by at least two negative feedback processes. First, the activity of the enzyme tyrosine hydroxylase is inhibited by the end products of the catecholamine synthesis pathway (7,8) Since tyrosine hydroxylase is the rate-limiting step in the synthesis of catecholamines, this forms a negative feedback process in the rate of conversion of tyrosine to noradrenaline. Second, noradrenaline released from the nerve terminals activates presynaptic α_2-adrenoceptors, which limit the further release of neuronal noradrenaline (9). We have examined the possible role of this latter mechanism in modulating the responsiveness to tyrosine, by examining the effects of the α_2-adrenoceptor antagonist idazoxan on the hormonal effects of tyrosine.

METHODS

Tyrosine Dose-response Study

Eight normal subjects (including 4 females) were studied on 4 occasions each, starting at 1000 hr. The subjects received, in double-blind, random order, tyrosine 50, 250 and 500 mg/kg and placebo orally. The 50 mg/kg dose of tyrosine is similar to the estimated tyrosine content of a standard meal which had previously been shown to stimulate ACTH and cortisol secretion in normal subjects (6).

Interaction of Tyrosine and Idazoxan

Seven normal male subjects were studied on 6 occasions, each receiving: 1) idazoxan 0.1 mg/kg iv over 12.5 min, followed by tyrosine 250 mg/kg orally; 2) idazoxan followed by tyrosine 500 mg/kg; 3) normal saline iv followed by tyrosine 250 mg/kg; 4) normal saline followed by tyrosine 500 mg/kg; 5) idazoxan followed by oral placebo; 6) normal saline followed by placebo.

Concentrations

Plasma tyrosine concentrations were measured by high performance liquid chromatography (HPLC) followed by fluorimetric detection. Tyrosine was extracted from plasma using 3% salicylsulphonic acid, pH 1.8. Post-column derivitization was achieved with o-phthaldialdehyde and 3-mercaptopropinic acid. Plasma ACTH concentrations were measured by a double antibody immunoradiometric assay (IRMA) using reagents supplied by Dr. A. White, Manchester University (10). Plasma cortisol concentrations were measured by radioimmunoassay (2). The intrassay and interassay coefficients of variation are less than 10% for all these assays. Satistical examination was conducted by performing analysis of variance of changes in the area under the curve. The studies were approved by the Ethical Committee of St. Bartholomew's Hospital.

RESULTS

Tyrosine Dose-response Study

Plasma tyrosine concentrations rose in a dose-dependent manner following the administration of tyrosine. Tyrosine had no effect on plasma ACTH or cortisol concentrations in any of the doses given in comparison to placebo.

TABLE 1

THE EFFECTS OF TYROSINE, IDAZOXAN, PLACEBO AND THE COMBINATION OF TYROSINE AND IDAZOXAN ON THE SECRETION OF ACTH AND NORADRENALINE IN SEVEN NORMAL SUBJECTS

DRUG	PLASMA ACTH ng/l min	PLASMA NORADRENALINE nmol/l min
Placebo	-338 ± 165	+11.3 ± 22
Tyr 250 mg/kg	-189 ± 236	+19.8 ± 16.5
Tyr 500 mg/kg	-213 ± 345	-10.1 ± 20.9
Idazoxan 0.1 mg/kg	-544 ± 240	+29.0 ± 32.4
Tyr 250 mg/kg + idazoxan	+278 ± 170	+75.3 ± 33.2
Tyr 500 mg/kg + idazoxan	+724 ± 413	+130.0 ± 49.7

The data are expressed as mean ± SEM changes in the area under the curve from the baseline.

Interaction of Tyrosine and Idazoxan

Idazoxan and tyrosine had no effect on plasma ACTH, cortisol or noradrenaline concentrations when they were administered on their own. In contrast, the combined administration of tyrosine and idazoxan caused dose-dependent increases of plasma ACTH and noradrenaline concentrations in comparison to placebo (Table 1). The increments in plasma ACTH and noradrenaline concentrations were statistically significant following the 500 mg/kg dose of tyrosine ($P < 0.02$ and $P < 0.01$, respectively, in comparison to placebo).

Idazoxan caused a small increase in systolic blood pressure, both when given on its own and in combination with tyrosine. No other side effects or unpleasant sensations were reported.

DISCUSSION

We found that tyrosine had no stimulant action on ACTH secretion in physiological or in higher doses in man. The ACTH response to feeding is therefore unlikely to be due simply to the ingestion of tyrosine and its subsequent conversion to catecholamines, and the mechanism of food-induced ACTH secretion remains unknown. Ishizuka et al. reported that tyrosine had a small stimulant effect on the secretion of cortisol, but they did not discuss the statistical significance of their data (11).

Idazoxan is a highly selective α_2-adrenoceptor antagonist; it has minimal activity at α_1-adrenoceptors and no significant activity at β-adrenergic, opiate, histaminergic, cholinergic or serotonergic receptors (12). Intravenous injections of idazoxan in the rat cause significant α_2-adrenoceptor antagonist effects in the central nervous system (13), but the drug has no effect on the turnover of dopamine, in contrast to the previously used α_2-adrenoceptor antagonist, yohimbine (14). When administered alone in this study, idazoxan caused a small increase in systolic blood pressure, but had no effect on plasma noradrenaline concentrations, presumably because the latter represents a spill-over effect of noradrenaline released from the peripheral sympathetic neurons. The effectiveness of an α_2-adrenoceptor antagonist is dependent upon the concentration of noradrenaline at the receptor; since noradrenaline release from sympathetic neurons is minimal under basal circumstances, this may explain the relatively small changes in sympathetic activity observed in this study.

We postulated that the lack of effect of tyrosine on ACTH secretion may be due to activation of one of the negative feedback processes that control the rate of conversion of tyrosine to catecholamines and their subsequent release. In this study, we have examined the role of the presynaptic α_2-adrenoceptors, which inhibit the release of noradrenaline from the central and peripheral noradrenergic neurons. Following pretreatment with idazoxan, the administration of tyrosine caused a significant increase in plasma noradrenaline concentrations, indicating activation of the peripheral sympathetic nervous system, and of plasma ACTH concentrations. On the basis of our previous findings, we postulate that the increase in activity of the hypothalamo-pituitary adrenal axis represents an increase in noradrenaline turnover in the central nervous system; we have previously found that activation of peripheral adrenoceptors with a hydrophilic adrenoceptor agonist that does not cross the blood-brain barrier causes no increase in pituitary adrenal activity (1). Our data therefore demonstrate that, in the presence of α_2-adrenoceptor blockade, the administration of tyrosine results in an increase in the rate of release of noradrenaline and secretion of ACTH.

In our previous studies with the α_1-adrenoceptor agonist methoxamine, we found that this drug caused sensations related to its α_1-adrenoceptor action, including piloerection and a sensation of urethral tightness in males (1). In the current studies using a combination of tyrosine and idazoxan, no subjective side effects or sensations were experienced, confirming that the subjective sensations associated with the higher doses of methoxamine were unlikely to be responsible for the activation of the hypothalamo-pituitary adrenal axis.

We conclude that it is possible to activate endogenous catecholamine systems using the combination of a catecholamine precursor, tyrosine, and an α_2-adrenoceptor antagonist. This results in an increase in the activity of the hypothalamo-pituitary adrenal axis. The mechanisms and the site of action of this effect are the subject of continuing investigations.

SUMMARY

The purpose of this study was to investigate the effects of activation of endogenous catecholamine systems on the hypothalamo-pituitary adrenal axis in man. Eight normal subjects were given, in double-blind, random order, the catecholamine precursor tyrosine in doses of 50, 250 and 500 mg/kg and a placebo orally. Tyrosine had no effect on the secretion of ACTH or cortisol in any of these doses. We postulated that the lack of effect of tyrosine was due to activation of one of the negative feedback mechanisms that control the rate of synthesis and release of the catecholamines, and therefore tested this hypothesis by examining the effects of the α_2-adrenoceptor antagonist idazoxan on the action of tyrosine. Seven normal males were given on 6 separate occasions, placebo, idazoxan 0.1 mg/kg intravenously and two doses of tyrosine, 250 and 500 mg/kg orally, following pretreatment with placebo and idazoxan.

Following pretreatment with idazoxan, tyrosine stimulated the secretion of ACTH, cortisol and noradrenaline in a dose-dependent manner, although neither tyrosine nor idazoxan on its own had any effect on the secretion of these hormones. The lack of effect of tyrosine when given on its own appears to be due, at least in part, to activation of α_2-adrenoceptors, which inhibit the release of noradrenaline.

ACKNOWLEDGMENTS

This study was supported by the Medical Research Council, the Joint Research Board of St. Bartholomew's Hospital, and the Peel Medical Research Trust.

REFERENCES

1. Al-Damluji S, Perry L, Tomlin S, Bouloux P, Grossman A, Rees LH, Besser GM (1987) Alpha-adrenergic stimulation of corticotropin secretion by a specific central mechanism in man. Neuroendocrinology 45: 68-76
2. Al-Damluji S, Cunnan D, Grossman A, Perry L, Ross G, Coy D, Rees LH, Besser GM (1987) Effect of adrenaline on basal and ovine corticotrophin-releasing factor-stimulated ACTH secretion in man. J Endocrinol 112: 145-150
3. Vale W, Vaughan J, Smith M, Yamamoto G, Rivier J, Rivier C (1983) Effects of synthetic ovine corticotropin-releasing factor, glucocorticoids, catecholamines, neurohypophysial peptides, and other substances on cultured corticotropic cells. Endocrinology 113: 1121-1131
4. Giguere V, Labrie F (1983) Additive effects of epinephrine and corticotropin-releasing factor (CRF) on adrenocorticotropin release in rat anterior pituitary cells. Biochem Biophys Res Commun 110: 456-462
5. Al-Damluji S, Cunnah D, Perry L, Grossman A, Besser GM (1987) The effect of α-adrenergic manipulation on the 24 hour pattern of cortisol secretion in man. Clin Endocrinol 26: 61-66
6. Al-Damluji S, Iveson T, Thomas JM, Pendlebury DJ, Rees LH, Besser GM (1987) Food-induced cortisol secretion is mediated by central α_1-adrenoceptor modulation of pituitary ACTH secretion. Clin Endocrinol 26: 629-636
7. Nagatsu T, Levitt M, Udenfriend S (1964) Tyrosine hydroxylase. The initial step in norepinephrine biosynthesis. J Biol Chem 239: 2910-2917
8. Udenfriend S, Zaltzman-Nirenberg P, Nagatsu T (1965) Inhibitors of purified beef adrenal tyrosine hydroxylase. Biochem Pharmacol 14: 837-845
9. Langer SZ (1977) Presynaptic receptors and their role in the regulation of transmitter release. Brit J Pharmacol 60: 481-497
10. White A, Smith H, Hoadley M, Dobson SH, Ratcliffe JG (1987) Clinical evaluation of a two-site immunoradiometric assay for adrenocorticotrophin in unextracted human plasma using monoclonal antibodies. Clin Endocrinol 26: 41-52
11. Ishizuka B, Quigley ME, Yen SSC (1983) Pituitary hormone release in response to food ingestion: evidence for neuroendocrine signals from gut to brain. J Clin Endocrinol Metab 57: 1111-1116
12. Doxey JC, Roach AG, Smith CFC (1983) Studies on RX781094: a selective, potent and specific antagonist of α_2-adrenoceptors. Brit J Pharmacol 78: 489-505
13. Freedman JE, Aghajanian GK (1984) Idazoxan (RX781094) selectively antagonises α_2-adrenoceptors on rat central neurons. Eur J Pharmacol 105: 265-272
14. Walter DS, Flockhart IR, Haynes MJ, Howlett DR, Lane AC, Burton R, Johnson J, Dettmar PW (1984) Effects of idazoxan on catecholamine systems in rat brain. Biochem Pharmacol 33: 2553-2557

ACUTE AND LONG-TERM SUPPRESSION OF STRESS-INDUCED ACTH SECRETION BY A RAT MONOCLONAL ANTIBODY TO CORTICOTROPIN RELEASING FACTOR

J. W. A. M. van Oers, F. J. H. Tilders,
and F. Berkenbosch

Department of Pharmacology, Medical Faculty
Free University
Amsterdam, The Netherlands

INTRODUCTION

Stress-induced secretion of ACTH from the anterior pituitary appears to be under the control of several humoral and neural factors, including corticotropin releasing factor (CRF), vasopressin, oxytocin and angiotensin II (1,2). Our present knowledge of the physiological role of CRF in the control of pituitary and other functions is largely based on passive immunization studies with antibodies directed against CRF. The passive immunization approach is based on the generally accepted, but poorly evaluated, idea of competition of receptors and antibodies for the same peptide. Dependent on the antibody's affinity, local concentration and recognition site complexes appear to be formed, preventing the peptide from stimulating its receptor and, thereby, blocking its biological effects. The use of heterologous polyclonal antibodies is hampered by batch-to-batch variation in affinity and specificity. To overcome these drawbacks and to avoid induction of immune responses in long-term studies, we produced a rat monoclonal antibody (MoAb) directed to rat/human-CRF (rCRF).

METHODS AND RESULTS

Production of a Rat Monoclonal Antibody Directed to rCRF

Female Wistar rats were actively immunized ip at 4 week intervals with 25 μg of rCRF conjugated to bovine thyroglobulin by using glutaraldehyde. Ten days after the fourth booster, plasma was screened for circulating antibodies by studying binding of radioiodinated rCRF (for iodination procedure see ref. 3). In addition, plasma corticosterone responses to ether-stress were evaluated. This was based on data from passive immunization studies showing that CRF-antibodies can prevent ether-induced activation of the pituitary-adrenal system (4,5,6). The rat giving the strongest suppression of the stress response was again boosted ip and three days later, its spleen cells were isolated and fused with mouse myeloma P_3 cells, using a protocol essentially described (7). After three weeks in culture, medium samples were screened for their capacity to bind radioiodinated rCRF; positive wells were subcloned by limiting dilution and expanded. Subsequently, wells were tested again for rCRF-binding antibodies. In addition, culture supernates were tested for biological activity by determining their capacity to inhibit rCRF-induced ACTH release in primary anterior pituitary cell cultures (for details on primary cell cultures see ref. 8, on ACTH RIA see ref. 9). Positive wells were subcloned twice by limiting dilution, and assayed for rCRF binding and biological

TABLE 1

EFFECTS OF DIFFERENT DOSES OF PFU 83 ON ETHER-INDUCED ACTH SECRETION

PFU 83		Plasma $ACTH_i$ (pg/ml)		
Dose nmoles/rat	plasma nM	non stress	ether stress	% suppression ACTH response
0	< 0.4	54 ± 11	378 ± 48	0 ± 14
0.1	10 ± 2	-	363 ± 43	4 ± 13
0.3	22 ± 1	-	283 ± 43	29 ± 13
1.0	143 ± 7	-	227 ± 25*	46 ± 8
3.0	302 ± 22	15 ± 3*	106 ± 26**	84 ± 8
10.0	843 ± 23	-	48 ± 12**	102 ± 4
30.0	2260 ± 316	-	30 ± 7**	108 ± 2

Measurements 90 minutes after iv administration of PFU 83.
* $p < 0.05$; ** $p < 0.001$ vs. vehicle injected group.

activity. The clone finally selected, PFU 83, was injected ip into pristane-primed nude mice, and ascites fluid was obtained containing specific antibody concentrations of 50-150 μM, as determined by protein determination and CRF binding after FPLC purification.

ACUTE EFFECTS OF PFU 83 IN RATS

To test its potency to inhibit ether-induced ACTH secretion, various doses of PFU 83 ascites or vehicle were given iv to groups (n = 6) of handled male Wistar rats. Ninety minutes after injection of PFU 83 or vehicle, selected rats were exposed to ether for 2 minutes and then decapitated. Unstressed rats were not exposed to ether prior to decapitation (for details see ref. 10). As shown in Table 1, PFU 83 induces a dose-related suppression of ether-induced ACTH secretion to levels below that of unstressed vehicle-injected controls. A dose of 3 nmoles PFU 83 reduced the plasma ACTH concentration in unstressed rats and strongly inhibited the ACTH response to ether. It should be noted that ether-induced secretion of α-MSH and prolactin were not affected by PFU 83 at any of the doses tested. Moreover, no interference by PFU 83 could be detected in our radioimmunoassays for ACTH, α-MSH and prolactin.

PFU 83 - rCRF BINDING KINETICS

Determination of plasma PFU 83 concentrations in rats (Table 1) showed that ether-induced secretion of ACTH is blocked at circulating concentrations of approximately 1 μM PFU 83. Therefore, we questioned whether this concentrations of PFU 83 could fully bind rCRF during the short time interval from its secretion from the median eminence to its action on the pituitary gland - less than one second (2). To study the kinetics of the rCRF-PFU 83 binding, 1 nM radioiodinated rCRF was incubated with various concentrations of PFU 83 at 37°C in 0.05 M phosphate-buffered serum. After 5, 10, 15, 20 and 30 sec of incubation, bound and free hormone were separated by addition of 4 ml polyethylene glycol (PEG 6000, 20% w/v in phosphate buffer, pH 7.4) and 30 min of centrifugation. Determination of t zero binding was obtained by addition of PEG to radioiodinated rCRF-containing serum prior to addition of PFU 83-containing serum.

As illustrated in Figure 1, incubation of rCRF with PFU 83 results in typical saturation curves with high initial binding rates. During this initial phase, the binding rate is largely determined by the association constant

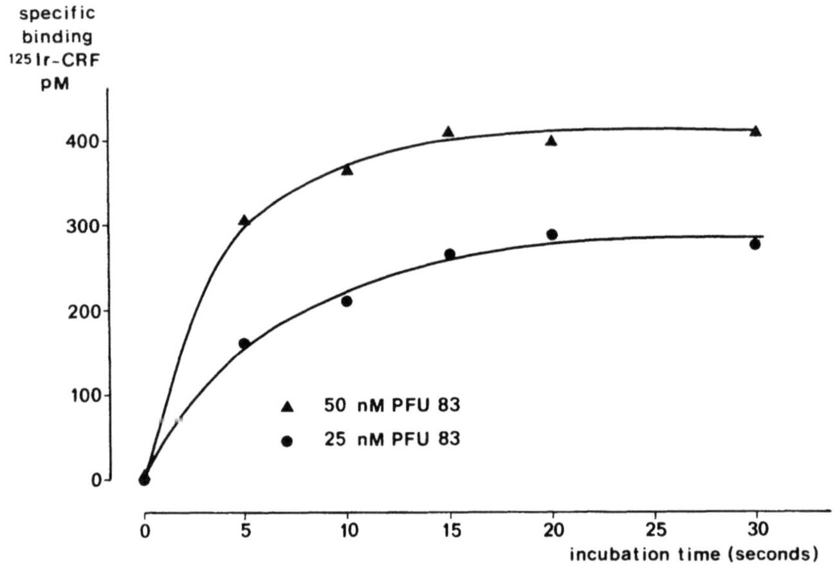

Figure 1. The effects of different concentrations of PFU 83 on the rCRF-PFU 83 binding kinetics, at 37°C in phosphate buffered serum containing 1 nM radio-iodinated rCRF.

K_{+1}, which can be calculated (11) from the classical second-order kinetic equation:

$$AR = K_{+1}[Ab]_0[Ag]_0$$

in which

AR is the association rate (binding of Ag, $M \cdot sec^{-1}$),
$[Ab]_0$ the initial antibody concentration, and
$[Ag]_0$ the initial antigen concentration.

As expected from this equation, the association rate as determined in our experiments was proportional to the antibody concentration. For example, 25 and 50 nM PFU 83 bind 162 and 308 pM rCRF, respectively, in 5 sec (Fig. 1). Interpolation of binding data over the first 5 sec allowed us to estimate the K_{+1} as $1.23 \cdot 10^6\ M^{-1} \cdot sec^{-1}$. Based on this value, it can be anticipated that PFU 83 at a concentration of 1 μM which is necessary to prevent ether-induced ACTH secretion (Table 1), may bind all rCRF in portal blood before reaching the pituitary gland.

TABLE 2

LONG-TERM EFFECTS OF PFU 83 ON ETHER-INDUCED ACTH SECRETION

PFU 83 dose nmoles/rat	time interval	PFU 83 plasma nM	% suppression ACTH response
10.0	90 min	714 ± 98	98 ± 2
10.0	9 days	239 ± 12	79 ± 3
0.5	90 min	43 ± 2	61 ± 6
0.5	9 days	15 ± 1	61 ± 9

Measurements 90 min or 9 days after iv administration of PFU 83. (For further legends, see Table 1.)

LONG-TERM EFFECTS OF PFU 83 (Table 2)

In order to evaluate the use of PFU 83 in chronic studies, we administered iv various doses of this antibody into chronically cannulated rats. Blood samples were collected 90 min later, and thereafter at 24-hr intervals. The decline of the circulating PFU 83 titer, as determined by binding of radioiodinated rCRF, is a biphasic phenomenon and is independent of the dose given. The first phase has a half-life of 1.4 days, the second phase 11.3 days. In view of the slow disappearance of CRF binding, it was interesting to study the duration of the biological effects of the antibody. Therefore, PFU 83 was given at a dose of 10 nmoles per rat and the animals were decapitated 90 min or 9 days later, with or without prior exposure to ether as described above. Nine days after administration of PFU 83, resting ACTH levels still tended to be lower than that in vehicle-injected controls and the ether-induced ACTH response was still suppressed by approximately 80%, which is in accordance with the residual PFU 83 titer found in the circulation. In a parallel experiment, a lower dose of PFU 83 was given (0.5 nmoles per rat), resulting in 60% suppression of ether-induced ACTH secretion after 90 minutes. Although the circulating antibody titer had dropped to 15 nM after 9 days, 60% suppression of ether-induced ACTH secretion was still found. This long-lasting suppression after a low dose of PFU 83 cannot be explained by the model described in the introduction, a case of competition of antibody and receptor for the same peptide, indicating that PFU 83 also blocks ACTH secretion by other mechanisms.

CONCLUSION

In conclusion, we have developed a rat monoclonal antibody (PFU 83) that binds rCRF and blocks CRF-induced ACTH secretion in vitro. In vivo studies with rats show that PFU 83 is a potent and specific tool that fully prevents ether-induced ACTH secretion. Inhibition of stress-induced ACTH secretion lasted for more than a week. This long duration of action is partly due to the slow clearance of PFU 83 from the circulation, and partly by other mechanisms.

ACKNOWLEDGMENTS

The authors wish to acknowledge R. Binnekade and P. A. J. Jonkergouw for their skillful technical assistance, Dr. A. J. R. de Jonge for his preliminary investigations, and H. Nordsiek for preparing the figure.

REFERENCES

1. Vale W, Rivier C, Brown MR, Spiess J, Koob G, Swanson L, Bilezikjian L, Bloom F, Rivier J (1983) Chemical and biological characterization of Corticotropin Releasing Factor. Recent Prog Horm Res 39: 245-270
2. Antoni FA (1986) Hypothalamic control of adrenocorticotropin secretion: advances since the discovery of 41-residue Corticotropin-Releasing Factor. Endocrine Reviews 7: 351-378
3. Linton EA, Lowry PJ (1982) Radioimmunoassay and chromatographic characterization of CRF-41-like immunoreactivity in hypothalami from several species. Neuropeptides 3: 45-52
4. Rivier C, Vale W (1983) Modulation of stress-induced ACTH release by corticotropin-releasing factor, catecholamines and vasopressin. Nature 305: 325-327
5. Linton EA, Tilders FJH, Hodgkinson S, Berkenbosch F, Vermes I, Lowry PJ (1985) Stress-induced secretion of Adrenocorticotropin in rats is inhibited by administration of antisera to ovine Corticotropin Releasing Factor and Vasopressin. Endocrinology 116: 966-970
6. Tilders FJH, Berkenbosch F, Vermes I, Linton EA, Smelik PG (1985) Role of epinephrine and vasopressin in the control of the pituitary-adrenal response to stress. Fed Proc 44: 155-160
7. Köhler G, Milstein C (1975) Continuous cultures of fused cells secreting antibody of predefined specificity. Nature 256: 495-497
8. Oosterom R, Verleun T, Lamberts SWJ (1983) Basal and dopamine-inhibited prolactin secretion by rat anterior pituitary cells: effects of culture conditions. Mol Cell Endocrinol 29: 197-212
9. Berkenbosch F, Van Oers JWAM, Del Rey A, Tilders FJH, Besedovsky H (1987) Interleukin-1 activates Corticotropin Releasing Factor producing neurons in rats. Science 238: 524-526
10. Berkenbosch F, Vermes I, Tilders FJH (1984) The Beta-adrenoceptor-blocking drug propranolol prevents secretion of immunoreactive beta-Endorphin and alpha-Melanocyte-Stimulating Hormone in response to certain stress stimuli. Endocrinology 115: 1051-1059
11. Mason DG, Williams AF (1980) The kinetics of antibody binding to membrane antigens in solution and at the cell surface. Biochem J 187: 1-20

DIAGNOSTIC DILEMMAS IN HYPERCORTISOLISM: INVESTIGATION AND MANAGEMENT

O.-A. Müller and K. von Werder

Medizinische Klinik Innenstadt, University of Munich
Munich, F.R.G.

Although Cushing's syndrome was described seventy years ago, diagnosis and management of this clinical entity remains one of the most difficult problems in clinical endocrinology (1,2). Cushing's syndrome may be caused by adrenal autonomous hypercortisolism, pituitary ACTH-dependent hypercortisolism, ectopic ACTH- or CRH-production and exogenous corticosteroids or ACTH. Hypercortisolism and its clinical expression may also be due to excessive alcohol intake - so-called alcoholic pseudo-Cushing - or due to endogenous depression (2).

ESTABLISHMENT OF THE DIAGNOSIS OF CUSHING'S SYNDROME

The first step of the diagnostic work-up of patients suspected of having endogenous Cushing's syndrome is the 1 or 2 mg dexamethasone over-night suppression test, which serves as a screening test for Cushing's syndrome (Table 1). Only patients in whom this test is compatible with Cushing's syndrome should be tested further with confirmatory tests to establish the diagnosis (Table 1). After establishment of the diagnosis, the next most difficult part is that of differential diagnosis of the various causes of Cushing's syndrome (Table 1). This must be done in order to provide the patients with the most appropriate treatment.

Diagnosis is confirmed by measuring free cortisol in the urine, examining diurnal variation of cortisol secretion, the cortisol and GH-responses to insulin hypoglycemia, and by using the dexamethasone

TABLE 1

THE DIFFERENT ADRENAL FUNCTION TESTS FOR THE SCREENING, ESTABLISHMENT OF DIAGNOSIS, AND DIFFERENTIAL DIAGNOSIS OF CUSHING'S SYNDROME

Screening	Establishment of Diagnosis	Differential Diagnosis
Over night Dex.-Suppression (1 or 2 mg)	Free urinary cortisol Diurnal variation Insulin hypoglycemia Dex.-Suppression (Liddle-Test)	ACTH-RIA CRH-Test High dose Dex.-suppression (8 mg overnight) ACTH-RIA after catheterisation X-ray Sonography

suppression test according to Liddle, which measures cortisol and cortisol metabolites in the urine (2).

The information in Table 2 summarizes the retrospective biochemical data of 70 patients in whom pituitary ACTH-dependent Cushing's disease was confirmed and treated by transsphenoidal surgery 3 to 13 years ago. It shows that with the exception of the screening test and the evaluation of the diurnal variation, all tests used for the establishment of the diagnosis may lead to false negative results. Thus, these tests are not absolutely certain and usually more than one test is needed to confirm the diagnosis of Cushing's syndrome (2).

DIFFERENTIAL DIAGNOSIS OF CUSHING'S SYNDROME

For differential diagnosis, the ACTH radioimmunoassay, particularly that after CRH-stimulation, is most important (3,4). However, since this test may also be misleading (4), other tests like the high dexamethasone suppression test and catheterization studies as well sophisticated x-ray evaluation may become necessary (5). Table 2 summarizes again the restrospective test results for the differential diagnosis in our 70 patients with Cushing's disease. Again, although the older lysine vasopressin or more recently the CRH-stimulation test and the high-dose dexamethasone suppression test are very effective in leading to a correct diagnosis, they do occasionally provide misleading results. As has been published by Hermus et al. (6) and Howlett et al. (7), the dexamethasone suppression test in combination with the CRH stimulation test is the most reliable test procedure.

TABLE 2

TEST RESULTS IN 70 PATIENTS WITH CONFIRMED PITUITARY ACTH-DEPENDENT CUSHING'S DISEASE

Establishment of Diagnosis

Test	Number of tests (N)	False negative (N/%)
Dexamethasone-Suppression-Test (2 mg over night)	70	0/0
Circadian Profile of Cortisol	54	0/0
Insulin-Hypoglycemia Test		
Cortisol	37	4/11
Growth Hormone	37	2/5
Liddle-Test (4x0.5 mg/day for 2 days)		
17-OH-CS	20	9/45
17-KS	17	6/35
Free Urinary Cortisol	10	3/30

Differential Diagnosis

Test	Number of tests (N)	Inappropriate results (N/%)
Dexamethasone-Suppression-Test (8mg)	59	6/10
Liddle-Test (4x 2 mg/day for 2 days)		
17-OH-CS	29	3/10
17-KS	21	7/33
Free Urinary Cortisol	18	0/0
ACTH-level (high normal or elevated)	59	0/0
Lysine-Vasopressin/ CRH-Test	42	2/5

In the upper part of the table, tests for the establishment of diagnosis and in the lower part, tests for the differential diagnosis are shown. The inappropriate results are expressed in numbers (n) and percent (%).

Figure 1 summarizes our experience with the CRH-test in establishing the differential diagnosis between adrenal hypercortisolism with

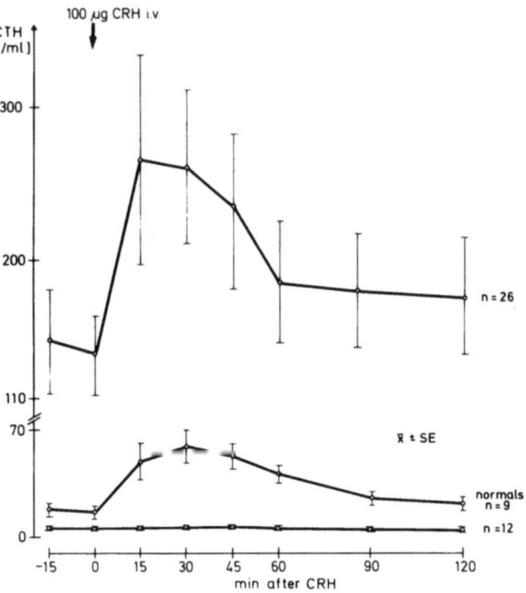

Figure 1. CRH-stimulation test in patients with Cushing's syndrome: mean values of the ACTH levels before and after stimulation with CRH are shown in normal controls, in patients with untreated pituitary ACTH-dependent Cushing's disease (n = 26), and in patients with autonomous cortisol secretion (n = 12).

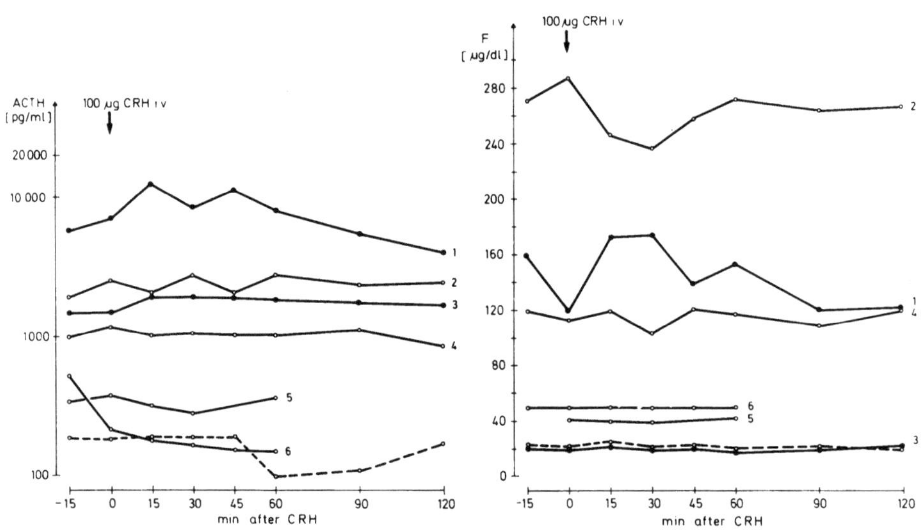

Figure 2. CRH stimulation test in patients with Cushing's syndrome: ACTH (left side) and cortisol (=F, right side) levels before and after CRH stimulation in seven patients with ectopic ACTH production. The dotted lines show the CRH stimulation test in one female patient with ectopic ACTH production due to a small lung carcinoid tumor.

suppressed ACTH-secretion and pituitary ACTH-dependent Cushing's disease with elevated basal ACTH levels that respond in an exaggerated fashion to CRH. It shows that the differential diagnosis of autonomous adrenal hypercortisolism is usually no problem.

However, occasionally patients with bilateral nodular hyperplasia (8) may be confused with patients who have ACTH-dependent Cushing's syndrome because of the inadequacy of many of the available ACTH assays that cannot differentiate between normal and suppressed ACTH levels (2,7). In cases of ectopic ACTH-syndrome, one usually finds that ACTH levels are excessively elevated and there is no CRH-induced ACTH increase (Fig. 2). However, two patients showed significant ACTH increases and one patient had only moderately elevated ACTH levels. This latter patient had a bronchial carcinoid tumor. The cortisol levels (Fig. 2, right side), which are often maximally stimulated, do not reflect changes of the ACTH levels after CRH.

The most difficult problem is the differential diagnosis of ACTH-dependent Cushing's syndrome. As yet, there is no test available to differentiate between an autonomous pituitary corticotroph adenoma and hypothalamic CRH hypersecretion leading to ACTH hypersecretion (2,4,7). In addition, although very rare, ectopic CRH-production may lead to pituitary ACTH secretion (9). Furthermore, a unique patient has been reported by Howlett et al. (10). This patient had a tumor that produced a bombesin-like peptide, which potentiated endogenous CRH-activity and led to pituitary ACTH-hypersecretion.

TABLE 3

TWO EXAMPLES FOR THE DIFFICULTY IN CORRECTLY DIAGNOSING ACTH-DEPENDENT CUSHING'S SYNDROME*

	Diagnosis	1. Therapy	Lag phase	2. Therapy	Outcome
1. Case	normal sella + CT ACTH moderately elevated t.s. surgery no tumour at exploration	bilateral adrenalectomy	6 years **slow increase of ACTH**	**pituitary tumour** detected and removed	decrease of ACTH
2. Case	normal sella + CT ACTH moderately elevated	bilateral adrenalectomy	6 years **ACTH continuously elevated**	**Carcinoid tumour** detected and removed	decrease of ACTH

*For details, see text.

The difficulty in correctly diagnosing ACTH-dependent Cushing's syndrome is exemplified by two cases (Table 3). In the first case, all test

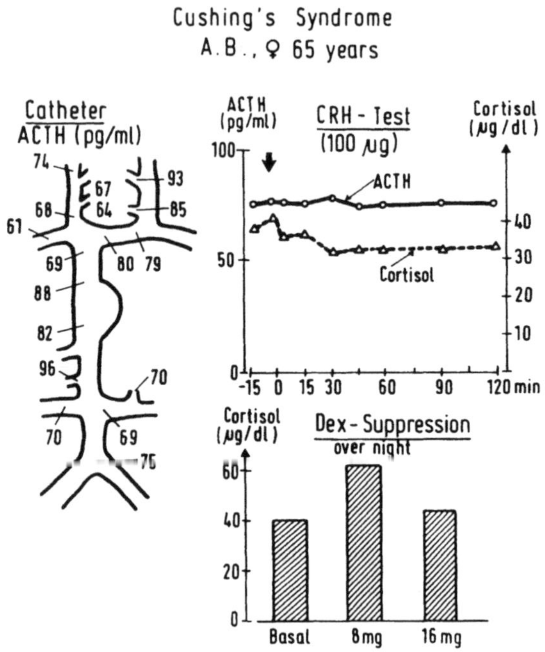

Figure 3. Test results in a female patient with ectopic ACTH production as the cause of her Cushing's syndrome.

results pointed to pituitary ACTH-dependent Cushing's disease, although no microadenoma was detected during sella exploration. Six years after bilateral adrenalectomy, ACTH levels further increased, a pituitary tumor was detected and removed. In the second case, tests did not suggest a pituitary source of ACTH, although no other ACTH source could be identified. Again, after six years, a small lung tumor was found and its removal led to a decrease in ACTH levels.

Figure 3 shows hormone data for a patient in whom the unequivocal diagnosis of ectopic ACTH syndrome was made. However, all available techniques (catheterization, computer tomography) did not reveal the source of ACTH production. The contribution of the CRH measurement in plasma to the diagnosis of the various causes of Cushing's syndrome is minimal (11).

THERAPY FOR CUSHING'S SYNDROME

Therapy for hypercortisolism is also controversial, even if the diagnosis is well established (1,2). Transsphenoidal surgery has become the most established form of treatment in patients with Cushing's disease, but

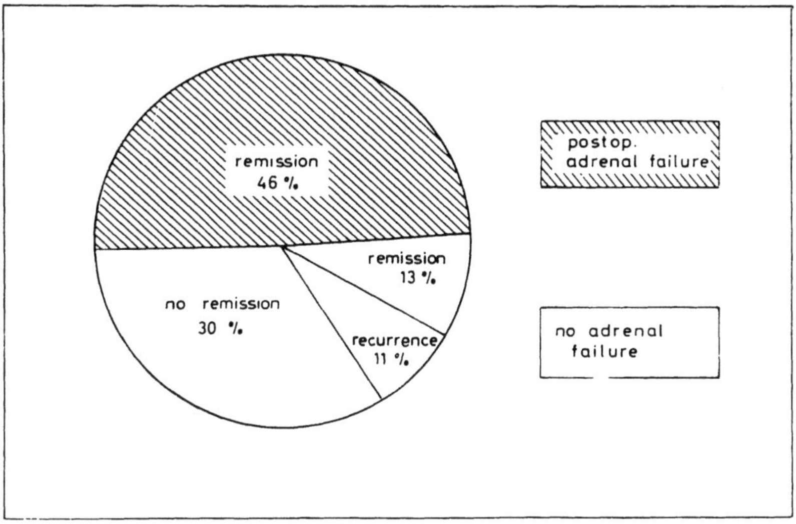

Figure 4. Follow-up of 70 patients with Cushing's disease 3-13 years after transsphenoidal surgery.

TABLE 4

THERAPEUTIC STEPS IN THE DIFFERENT FORMS OF CUSHING'S SYNDROME

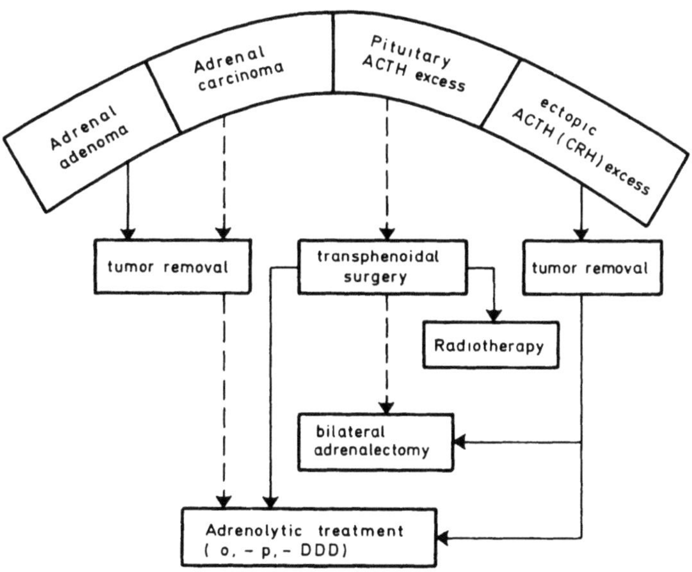

radiotherapy and bilateral adrenalectomy are still used, whereas the combination of radiotherapy and unilateral adrenalectomy are not performed anymore. Drug therapy is still in the experimental stage. The results reported by Fahlbusch et al. (12) show that about 25% of the patients are not cured by surgical intervention. The fact that some of these patients may in fact have an ectopic ACTH syndrome cannot be excluded completely. Even if there is remission of the disease documented by normal ACTH levels and regular dexamethasone suppressability, real recurrences are observed (13,14). It seems that postoperative secondary adrenal insufficiency is a good parameter for persisting remission (Fig. 4). Thus, 46% of all patients who had secondary adrenal failure immediately after the operation had persisting remission after transsphenoidal surgery, whereas 11% who later showed recurrence of active Cushing's disease had normal ACTH and cortisol levels postoperatively, but no adrenal insufficiency.

Secondary adrenal insufficiency after surgery is demonstrated early by the lack of an ACTH rise after CRH stimulation within one week after the operation (15).

Table 4 summarizes the therapeutic recommendation. In ACTH-dependent Cushing's disease, transsphenoidal surgery is the first choice, childhood radiotherapy is the second choice followed by bilateral adrenalectomy. Selected cases may be subjected to adrenolytic therapy with o′ p-DDD (16) or ketoconazole (17). The latter patients must be monitored carefully with respect to unwanted side effects.

CONCLUSIONS

It has become obvious that the title of this paper is justified and that we still face a dilemma with respect to diagnosis and management of Cushing's syndrome. Only if one is aware of the many exceptions that occur during the diagnostic process, can one be able to minimize the disappointment that may occur during the treatment of patients with this difficult clinical entity.

REFERENCES

1. Aron DC, Findling JW, Fitzgerald PA, Forsham PH, Wilson CB, Tyrrell JB (1982) Cushing's Syndrome: Problems in Management. Endocrine Reviews 3: 229-244
2. Krieger DT (1982) Cushing's Syndrome. New York: Springer

3. Müller OA, Stalla GK, von Werder K (1983) Corticotropin releasing factor: a new tool for the differential diagnosis of Cushing's syndrome. J Clin Endocrinol Metab 57: 227-229
4. Müller OA, Stalla GK, von Werder K (1987) Corticotropin releasing factor in humans. II. CRF stimulation in patients with diseases of the hypothalamo-pituitary-adrenal axis. Horm Res 25: 185-198
5. Howlett TA, Drury PL, Perry L, Doniach I, Rees LH, Besser GM (1986) Diagnosis and management of ACTH-dependent Cushing's syndrome: comparison of the features in ectopic and pituitary ACTH production. Clin Endocr 24: 699-713
6. Hermus ARMM, Pieters GFFM, Pesman GJ, Smals AGH, Benraad TJ, Kloppenborg PWC (1986) The corticotropin-releasing-hormone test versus the high-dose dexamethasone test in the differential diagnosis of Cushing's syndrome. The Lancet: 540-544
7. Howlett TA, Rees LH, Besser GM (1985) Cushing's syndrome. In: Besser GM, Rees LH (eds) Clinics in Endocrinology and Metabolism Vol 14 No. 4. The Pituitary Adrenocortical Axis. Toronto: WB Saunders: pp 911-945
8. van Berkhout FT, Crough RJM, Kater L, Schuurman HJ, Gmelig Meyling FJH, Kooyman CD, van der Gaag RD, Jolink D, Drexhage HA (1986) Familial Cushing's syndrome due to nodular adrenocortical dysplasia. A putative receptor-antibody disease? Clin Endocr 24: 299-311
9. Belsky JL, Cuello B, Swanson LW, Simmons DM, Jarrett RM, Braza F (1985) Cushing's syndrome due to ectopic production of corticotropin-releasing factor. J Clin Endocrinol Metab 60: 496-500
10. Howlett TA, Price J, Hale AC, Doniach I, Rees LH, Wass JAH, Besser GM (1985) Pituitary ACTH dependent Cushing's syndrome due to ectopic production of a bombesin-like peptide by a medullary carcinoma of the thyroid. Clin Endocr 22: 91-101
11. Stalla GK, Stalla J, Schopohl K, von Werder K, Müller OA (1986) Corticotropin-releasing factor in humans. I. CRF stimulation in normals and CRF radioimmunoassay. Horm Res 24: 229-245
12. Fahlbusch R, Buchfelder M, Müller OA (1986) Transsphenoidal surgery for Cushing's disease. J Roy Soc Med 79: 262-269
13. Müller OA, Fahlbusch R, von Werder K (1984) Real recurrences after transsphenoidal operation of ACTH-secreting pituitary adenomas? In: Lamberts SWJ, Tilders FJHT, van der Veen EA, Assies J (eds) Trends in diagnosis and treatment of pituitary adenomas. Amsterdam: Free University Press: pp 339-344
14. Lamberts SWJ, Klijn JGM, de Jong FH (1987) The definition of true recurrence of pituitary-dependent Cushing's syndrome after transsphenoidal operation. Clin Endocr 26: 707-712
15. Schrell U, Fahlbusch R, Buchfelder M, Riedl S, Stalla GK, Müller OA (1987) Corticotropin-releasing hormone stimulation test before and after transsphenoidal selective microadenomectomy in 30 patients with Cushing's disease. J Clin Endocrinol Metab 64: 1150-1159
16. Luton JP, Mahoudeau JA, Bouchard PH, Thiebolt PH, Hautecouverture M, Simon D, Laudat MH, Touitou Y, Bricaire H (1979) Treatment of Cushing's disease by o'p-DDD. N Engl J Med 300: 459-464
17. Sonino N, Boscaro M, Merola G, Mantero F (1985) Prolonged treatment of Cushing's disease by ketoconazole. J Clin Endocrinol Metab 61: 718-722

DUAL CONTROL OF ACTH SECRETION BY HYPOTHALAMIC AND PLACENTAL CORTICOTROPIN-RELEASING FACTOR (CRF) IN LATE PREGNANCY

B. Allolio*, H. M. Schulte**, J. Hoffmann*, M. Kusche+, W. Winkelmann*, E. A. Linton++ and P. J. Lowry++

*Medizinische Klinik II, Universitat zu Koln, **Medizinische Klinik, Christian-Albrecht-Universitat Kiel, +Universitats-Frauenklinik Koln, F.R.G., ++Department of Physiology and Biochemistry, University of Reading, U.K.

INTRODUCTION

In late pregnancy, high concentrations of circulating immunoreactive corticotropin-releasing hormone (CRH), which is secreted by the placenta, have been found (1-4). Chromatographic studies, as well as bioassays, suggest that most of the CRH-like material in the placenta is identical to hypothalamic CRH (5). These findings raise the question as to whether the placenta contributes to the regulation of the pituitary-adrenal-axis in the third trimester of pregnancy. The aim of the present study was to evaluate the role of the placenta in the control of adrenocortical function by measuring circulating hormones in healthy women before and shortly after delivery.

PATIENTS AND METHODS

After giving informed consent, 12 healthy, pregnant women aged 24-34 years were included in the study. All subjects were examined on two occasions: during the 37-40th week of gestation and 3-5 days after delivery. All women collected salivary cortisol samples at hourly intervals from 700 hr until 2300 hr. At 900 hr and 1900 hr, blood was drawn for plasma ACTH, plasma CRH, serum cortisol, and cortisol binding globulin (CBG). In addition, 24 hr urine was collected for determination of urinary free cortisol (n = 9).

Plasma ACTH was measured by a sensitive radioimmunoassay (RIA) after extraction with QUSO G_{32} (6). Plasma CRH was determined by a sensitive and highly specific two-site immunoradiometric assay (7). Serum cortisol and urinary cortisol were measured using commercially available reagents (New England Nuclear, FRG). Similarly, CBG was determined by RIA (IRE Medgenix, FRG). Salivary cortisol was measured by a direct RIA as described previously (8).

The area under the salivary cortisol profile was calculated using the trapezoid method. To assess the diurnal changes in salivary cortisol, the coefficient of variation was calculated for each profile.

Satistical evaluation was performed by the Student's t-test or by the nonparametric Wilcoxon test for paired data. Analysis of variance was used to compare salivary cortisol profiles before and after delivery. All values are expressed as mean± SEM.

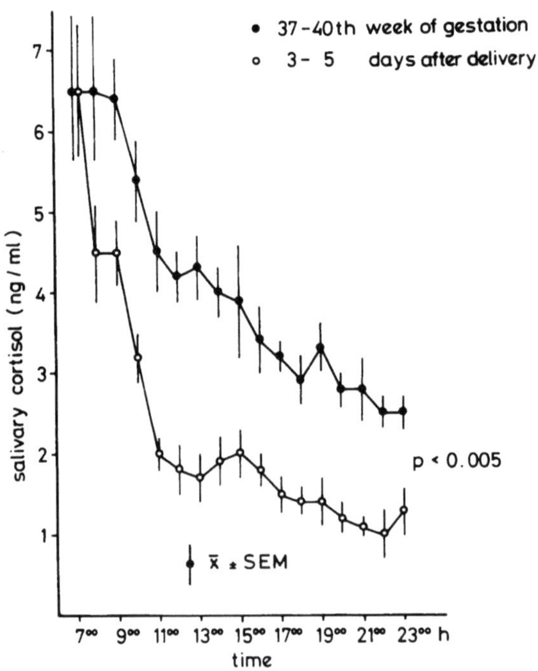

Figure 1. Salivary cortisol profiles (Mean ± SEM) before and after delivery in 12 healthy pregnant women.

TABLE 1

HORMONE CONCENTRATIONS IN 12 PREGNANT WOMEN BEFORE AND AFTER DELIVERY

		Before Delivery	After Delivery	
Salivary Free Cortisol (ng/ml)	900 hr	6.4 ± 0.5	4.5 ± 0.4	P < 0.05
	1900 hr	3.3 ± 0.3*	1.4 ± 0.3*	P < 0.005
Serum Cortisol (ug/dl)	900 hr	33.6 ± 2.2	23.6 ± 2.3	P < 0.05
	1900 hr	22.6 ± 1.2*	16.8 ± 1.5	P < 0.01
Plasma ACTH (pg/ml)	900 hr	10.7 ± 2.1	6.0 ± 1.3	P < 0.05
	1900 hr	8.7 ± 1.2*	4.3 ± 1.0	P < 0.005
Plasma CRH (pg/ml)	900 hr	1220 ± 265	17.8 ± 4.1	P < 0.001
	1900 hr	1296 ± 279	20.5 ± 5.5	P < 0.001
CBG (ug/ml)		91.3 ± 6.0	72.1 ± 6.5	P < 0.01
Urinary Free Cortisol (ug/24 hr)		92.4 ± 7.3	67.6 ± 8.0	P < 0.05

* P < 0.01; ** P < 0.05 compared to concentrations at 900 hr.

RESULTS

In all subjects, salivary free cortisol exhibited a clear diurnal rhythm both before and after delivery (Fig. 1). However, the area under the curve was significantly larger prior to delivery (66.1 ± 5.0 vs. 37.0 ± 2.6 ng/ml x hr, P < 0.01). Similarly, using analysis of variance, free cortisol profiles before and after delivery were significantly different (P< 0.005). Prior to delivery, the coefficient of variation for salivary cortisol was reduced, owing to an attenuated cortisol decrease in the evening hours (40.2 ± 3.4% vs. 77.6 ± 6.6% post partum, P < 0.01). Before delivery, the percentage increase in free cortisol was 140% at 1900 hr compared to only 40% at 900 hr (Table 1).

Both before and after delivery, plasma ACTH followed a diurnal rhythm with higher concentrations at 900 hr (Table 1). While plasma ACTH levels were within the normal range, prepartum values were significantly higher (P < 0.05). In contrast, no diurnal rhythm was found for plasma CRH concentrations (before delivery: 900 hr, 1220± 255 pg/ml; 1900 hr, 1296 ± 279 pg/ml; after delivery: 900 hr, 17.8± 4.1 pg/ml; 1900 hr, 20.5 ± 5.5 pg/ml). CBG was elevated before and after delivery with a significant decrease postpartum (Table 1). While 3-5 days after delivery salivary cortisol concentrations were in the normal range, total serum cortisol remained elevated due to the increased CBG concentrations.

Urinary free cortisol excretion was significantly higher in late pregnancy (92.4 ± 7.3 vs 67.6 ± 8.0 $\mu g/24$ hr, $P < 0.05$).

DISCUSSION

Salivary cortisol reflects the biologically active free fraction of total serum cortisol (8) and is, therefore, especially suited to investigate, in a noninvasive fasahion, the diurnal pattern of cortisol secretion in late pregnancy, when CBG concentrations are high. Our results confirm a substantial increase in free cortisol concentrations in the third trimester of pregnancy (9) and demonstrate a clear diurnal rhythm both before and after delivery. While plasma ACTH also showed a decrease in the evening, CRH remained unchanged during the day with concentrations about 100 fold higher before delivery. The rapid fall of CRH after delivery again points to the placenta as the source of the peptide (3,4). Recently, CRH messenger-RNA has been found in higher concentrations in the human placenta than in the human hypothalamus (10). However, because plasma CRH shows no diurnal variation, it is unlikely that placental CRH is the only regulating factor of the pituitary-adrenal-axis in late pregnancy. Instead, the diurnal changes in plasma ACTH and free cortisol suggest an important role for the hypothalamus in the control of adrenocortical function. This view is supported by our recent observation that exogenous CRH (1 $\mu g/kg$ iv) is unable to stimulate plasma ACTH or cortisol further in the third trimester of pregnancy, while different stressors (pain, labor) induce an increase in these hormones (11).

On the other hand, analysis of the pattern of free cortisol profiles showed an attenuation in the diurnal rhythm before delivery, with relatively high levels in the evening hours, when hypothalamic corticotropin-releasing factor (CRF) drive is assumed to be low. As the CRH secretion by the placenta serves as a constant stimulus to the pituitary, its action may become dominant in the evening hours and may thus explain the pattern of free cortisol concentrations in late pregnancy, which is rapidly reversed after delivery.

In conclusion, our results are consistent with the hypothesis that in late pregnancy, the pituitary-adrenal-axis is under the dual control of the hypothalamus and the placenta.

REFERENCES

1. Stalla GK, Hartwimmer J, v. Werder K, Muller OA (1984) Ovine (o) and human (h) corticotropin releasing factor (CRF) in man: CRF-stimulation and CRF-immunoreactivity. Acta Endocrinol 106:289-297

2. Goland RS, Wardlaw SL, Stark RI, Broiwn LS, Frantz AG (1986) High levels of corticotropin-releasing hormone immunoreactivity in maternal and fetal plasma during pregnancy. J Clin Endocrinol Metab 63:1199-1204
3. Campbell EA, Linton E, Wolfe CDA, Scraggs PR, Jones MT, Lowry PJ (1987) Plasma corticotropin-releasing hormone concentrations during pregnancy and parturition. J Clin Endocirnol Metab 64:1054-1059
4. Sasaki A, Liotta AS, Luckey MM, Margioris AN, Suda T, Krieger DT (1984) Immunoreactive corticotropin-releasing factor is present in human maternal plasma during the third trimester of pregnancy. J Clin Endocrinol Metab 59:812-814
5. Sasaki A, Shinkawa O, Margioris A, Liotta A, Sato S, Murakami O, Go M, Shimizu Y, Hanew K, Yoshinaga K (1987) Immunoreactive corticotropin-releasing hormone in human plasma during pregnancy, labor and delivery. J Clin Edocrinol Metab 64:224-229
6. Allolio B, Winkelmann W, Hipp FX (1981) Effect of medastine, an H_1-antihistamine, on plasma ACTH in adrenal insufficiency. Acta Endocrinol 97:98-102
7. Linton EA, McLean C, Niewenhuyzen Krusemann AC, Tilders FJ, Van der Veen EA, Lowry PJ (1987) Direct measurement of human plasma corticotropin-releasing hormone by a "two-site" immunoradiometric assay. J Clin Endocrinol Metab 64:1047-1053
8. Allolio B, Feltes G, Deuss U, Kaulen D, Winkelmann W (1985) Direkter Readioimmunoassay von freiem Cortisol im Speichel. Lab Med 9:281-284
9. Nolten WE, Lindheimer MD, Rueckert PA, Oparril S, Ehrlich EN (1980) Diurnal patterns and regulation of cortisol secretion in pregnancy. J Clin Endocrinol Metab 51:466-472
10. Usui T, Jingami H, Takahaski H, Fukata J, Nakai Y (1987) Expression of corticotropin releasing factor (CRF) gene in human placenta and other nonhypothalamic tissues. Endocrinology 120 Suppl, 899 (Abstract)
11. Schulte HM, Allolio B, Weisner D, Jofmann J, Winkelmann W, Ohnhaus EE (1988) Regulation of the pituitary-adrenal-axis in late pregnancy: ACTH and cortisol secretion in response to synthetic human corticotropin-releasing hormone. Acta Endocrinol 117 (Supple. 287):98-99

E. GROWTH HORMONE NEUROREGULATION II

THE THERAPEUTIC MODULATION OF GROWTH HORMONE RELEASE

J. A. H. Wass and R. J. M. Ross

Department of Endocrinology
St. Bartholomew's Hospital
London, U.K.

INTRODUCTION

Growth hormone is under the hypothalamic control of two different peptides: growth hormone-releasing hormone (GRF) and somatostatin. It is clear that of these two peptides, GRF is more important because pituitary stalk section results in growth hormone deficiency in man (1). Currently, we are in an exciting phase of our understanding of the neuroregulation of growth hormone. The structure of GRF was simultaneously discovered by Guillemin and Vale in 1982 (2,3). More is known about somatostatin because of its earlier discovery in 1973 (4).

The Chemistry of Growth Hormone

Human growth hormone is synthesized in the acidophylic somatotrophs of the anterior pituitary, where it is stored in membrane-bound vesicles or granules. It is a major synthetic product of the gland, and between 3 and 5 mg of growth hormone can be extracted from the human pituitary. Chemically, it is a globular protein with a molecular weight of 22,000. The final revision of its structure was published in 1971 by Li and Dixon, and it is thought to be a single chain polypeptide containing 191 amino acid residues. There are two disulphide bridges, the first between residues 53 and 165 and the second between residues 182 and 189. Human growth hormone has now been synthesized, and there are important structural similarities between this molecule and human prolactin.

REGIONS OF THE CNS INVOLVED IN THE REGULATION OF GROWTH HORMONE SECRETION

There is abundant evidence that growth hormone secretion is under hypothalamic control. Pituitary stalk section (1) and hypothalamic diseases (5) can impair the release of growth hormone caused by provocative stimuli in humans. Furthermore, lesions in the median eminence impair responsiveness of growth hormone to insulin hypoglycemia in monkeys (6). In rats, lesions placed in the ventromedial nucleus reduce plasma and pituitary growth hormone (7) and impair pulsatile growth hormone release; lesions of the arcuate nucleus have the same effects (8), suggesting that these nuclei contain specific stimulatory areas for the control of growth hormone secretion as no such effect is seen after ablation of other hypothalamic nuclei. Similar results have been seen in humans undergoing stereotactic hypothalatomy. In the human adult hypothalamus, somatostatin-containing cell bodies have been localized in the anterior hypothalamus in the infundibular nucleus, in close proximity to the infundibular recess (9).

The preoptic area and other extrahypothalamic regions are also involved in the control of growth hormone secretion. Thus, stimulation of the preoptic area inhibits growth hormone secretion in man and ablative lesions have the reverse effect, possibly due to interference with somatostatin release, as this has been found in high concentrations in this region of the human brain. Stimulation of the amygdala also causes growth hormone release in humans (10).

It is clear then that GRF and somatostatin regulate growth hormone secretion from the anterior pituitary gland. The secretion of GRF and somatostatin are in turn modulated by neurotransmitters, and a great number of metabolic, hormonal and pharmacological factors can affect human growth hormone release, probably by modifying release of these neurotransmitters.

GRF

The first conclusive evidence for the existence of GRF was provided in porcine hypothalami (11). Deuben and Meites (12) demonstrated that the amount of growth hormone released from the rat pituitary when incubated in vitro was increased by the addition of hypothalamic extracts. Growth hormone releasing material has been isolated from hypothalamic extracts in pigs and from their hypophysial portal blood, and has been shown to be present in man (13).

In November 1982, within a very short space of time, two groups reported the isolation of GRF from two different human pancreatic islet cell tumors causing acromegaly (2,3). Guillemin's group isolated a 44-amino acid peptide, and Vale's group found a peptide that contained the first 40 of these amino acids.

These peptides are synthesized and found to stimulate growth hormone release from rat pituitary cells in vitro.

More recently, both 40 and 44 residue peptides identical to the pancreatic peptides have been extracted from several mouse and human hypothalamic fragments, respectively (14). Furthermore somatostatin-28 and somatostatin-14 inhibit the release of growth hormone in typical non-competitive antagonism. It seems clear, therefore, that human pancreatic and hypothalamic GRF are the same molecule, and we now have further insight into the hypothalamic control of growth hormone release, as well as a very useful tool in the therapy of growth hormone deficiency.

Human GRF (1-40) (1 μg/kg) has been shown to stimulate selectively growth hormone secretion in man (15). It has recently been shown to cause growth hormone release in patients who are growth hormone deficient after experiencing insulin hypoglycemia (16). Esch et al. (17) described three human pancreatic GRFs (1-37, 1-40 and 1-44). It is now known that full biological activity resides in the first 29 amino acids. The entire human GRF gene has been structurally characterized and mapped on chromosome 20. It has been demonstrated in vitro that GRF stimulates transcription of the GH gene as well as growth hormone secretion, and transgenic transfer of the GRF gene has been shown to increase the growth of mice (18).

GRF ANALOGUES

It appears that the biological activity of GRF resides principally in the first 29 residues, as analogues of this length with deletions from the C-terminal end retain most of their GH releasing activity, but the N-terminal amino acid sequence needs to be preserved. Most human studies have been performed with the native peptides, GRF (1-44) and GRF (1-40) or the shorter analogue GRF (1-29). All three, when given as an intravenous bolus, selectively promote growth hormone release in normal male subjects at doses of 1-3 μg/kg (16,19-20), although at high doses such as 10 μg/kg there may also be prolactin release (21). The mean peak serum growth hormone after supramaximal iv bolus of GRF as reported from the above studies on normal adult men, ranged from 32 to 68 mU/l (1 mU/l = 0.5 ng/ml), and the time of the peak varied between 15 and 60 min after administration. The only side effect of GRF reported was facial flushing, which occured in most subjects within one min of an intravenous injection of 100 μg, and lasted no longer than 5 min. When compared in the same subjects, all three analogues elicit an identical pattern of growth hormone release (22,23).

There is considerable individual variation in the response to GRF, and rapid repetitive administration leads to a decreasing GH response. At a time when the pituitary is apparently refractory to GRF, it will still release growth hormone in response to hypoglycemia, suggesting that the lack of response to GRF is not due to depleted pituitary stores of growth hormone (24). In addition, prior administration of GH completely abolishes the growth hormone response to GRF (25). It is likely that changes in somatostatin tone at the pituitary are responsible for the variable response to GRF, with an increase in somatostatin being induced by GH secretion. Recent work suggests that GH feedback is mediated through somatostatin under cholinergic control, as pyridostigmine, an acetylcholinesterase inhibitor, blocks GH feedback (26). A number of different factors affect the growth hormone response to GRF; these are shown in Table 1.

TABLE 1

ALTERATIONS IN GH RESPONSE TO GRF

Decreased GH Response	Increased GH Response	No Effect
Age	Pyridostigmine	Estrogen
	Atenolol	Pubertal status
Repetitive GRF administration		
GH Administration		
IGF-I (Somatomedin-C)		
Obesity		
Cushing's Syndrome		
Hypothyroidism		

GRF TREATMENT (Table 2)

The finding that many GH-deficient children will secrete growth hormone in response to GRF suggests that this might provide an alternative treatment to human growth hormone. At least three studies have now been reported on the use of GRF given as subcutaneous treatment over at least a six month period. Using an increase in growth velocity of greater than 2 cm/year as evidence of an effect of treatment, two children treated with 3 hourly pulses of GRF showed growth acceleration (27), 2 of 5 children treated with 3 hourly pulses overnight grew (28), as did 8 of 18 children treated with twice daily subcutaneous GRF (29). Although one study suggested that there might be a priming effect of GRF, in the largest study there was no conclusive evidence for either a priming or desensitization effect, and an initial good growth hormone response to an iv bolus of GRF predicted the children who would respond to therapy (29).

TABLE 2

LONG TERM STUDIES OF GRF

	Growth Velocity Increased >2cm/yr	Route of Administration
Thorner et al., 1985	2/2	3 hourly pulses sc 1-3 mcg/kg
Smith et al., 1986	2/5	3 hourly pulses sc 1-2 mcg/kg
Ross et al., 1987	8/18	bd sc 250-500 mcg

It is clear from the above studies that GRF will promote growth in at least 50% of children who are GH-deficient on conventional criteria. However, these children are likely to be the most difficult chidren to treat with GRF, as they frequently have prolonged GRF deficiency and, therefore, only a small pituitary pool of growth hormone. The group of patients who make up the largest number who require treatment for short stature are those with idiopathic short stature. These children, who grow slowly, have low physiological release of growth hormone, but a normal growth hormone response to conventional stimuli of growth hormone secretion. They have a structurally normal hypothalamo-pituitary axis and a good growth hormone response to GRF, and are thus likely to grow well on this treatment.

The recent finding that acetylcholinesterase inhibitors, such as pyridostigmine, enhance the growth hormone response to GRF, has important therapeutic implications (30). It may be that the addition of pyridostigmine to any GRF treatment would greatly increase its effect.

Biosynthetic growth hormone is now widely available and, given once daily, is very effective in promoting growth. The present regimes of GRF treatment require at least twice daily injections, and therefore make it less practical than hGH therapy. However, GRF has a number of important advantages over growth hormone. It is a smaller molecule, and therefore can be given intranasally. As already mentioned, it preserves feedback at the pituitary and should therefore avoid any dangers of overtreatment. GRF given as a continuous infusion augments the normal endogenous release of growth hormone, thus preserving the pulsatile pattern. It has already been shown that GRF infusions will augment GH release in partially GH-deficient children (31), suggesting that a depot-preparation of GRF may well promote growth. The development of a depot-preparation of GRF would provide an important alternative treatment to human growth hormone for the treatment of short stature, with the advantage that it could be given infrequently, and would promote physiological growth hormone release without the danger of overtreatment.

SUMMARY

The majority of growth hormone deficient patients show a growth hormone response to GRF, suggesting that they have a hypothalamic defect in GH release. GRF given subcutaneously promotes growth in 50% of GH-deficient children, and it is likely that depot preparations of GRF will provide a practical alternative treatment for growth hormone deficiency.

SOMATOSTATIN

The first evidence for a hypothalamic factor that inhibits growth hormone secretion from the anterior pituitary came from the work of Krulich, Dhariwall and McCann (32) who, when looking for a factor that released growth hormone, discovered a fraction of rat and sheep hypothalami that inhibited growth hormone release from rat anterior pituitaries in vitro. In 1973, Brazeau and Guillemin isolated somatostatin from sheep hypothalami (4). It was sequenced and found to contain 145 amino acid residues, and to have a cyclic structure joined by 2 intramolecular disulphide bonds between the two cysteine residues. The linear sequence was synthesized and purified and its biological activity was found to be the same as the cyclic form in vitro. It was quickly synthesized by other groups. On the basis of its growth hormone release inhibiting activity, the peptide has variously been called growth hormone release-inhibiting hormone (GHRH or GRF), somatotrophin release inhibiting factor (SRIF), or somatostatin.

Exogenously Administered Somatostatin

Actions on the pituitary gland

It is now known that somatostatin is present not only in the human hypothalamus (33), but is also widely distributed throughout the central nervous system. It is present in peripheral nerves, the pancreas, and the gastrointestinal tract. It has further been shown that when administered intravenously, it has a broad spectrum of action other than that of growth hormone inhibition, suggesting a biological importance beyond that of a hypothalamic pituitary regulating hormone (34). After the first report of inhibition of growth hormone release from cultured rat pituitary cells, somatostatin has been demonstrated to be a powerful inhibitor of growth hormone release both in vivo and in vitro and of the growth hormone response to all tested physiological and pharmacological stimuli is inhibited. In man, somatostatin blocks the growth hormone responses to exercise, arginine and insulin hypoglycemia (35-37). In conditions where basal circulating levels of growth hormone are elevated, such as acromegaly and diabetes mellitus, these levels rapidly fall during infusion of somatostatin (35,36,38). Somatostatin also inhibits TSH secretion both basally and in response to TRH. Furthermore, TSH levels in hypothyroidism are suppressed by intravenous somatostatin (35). Basal prolactin is not altered in normal subjects, nor is the increase seen after TRH or insulin-induced hypoglycemia (35). In normal subjects, the secretion of gonadotropins and ACTH do not appear to be altered by somatostatin. The main actions in the hypothalamus and pituitary, therefore, are related to growth hormone and TSH secretion. Somatostatin has a number of other actions outside the hypothalamo-pituitary axis, particularly on gstrointestinal hormones and gastrointestinal function (34).

CLINICAL USES OF SOMATOSTATIN

Native Somatostatin

The major limitation to the clinical use of this compound is its short half-life of less than 3 minutes in circulation. It is not absorbed orally, and therefore has to be given parenterally. Somatostatin has been used as an investigative tool in the hormonal regulation of the metabolism, and has also been used in an attempt to treat brittle diabetes mellitus, secreting pancreatic tumors, upper gastrointestinal bleeding, pancreatitis and acromegaly.

Acromegaly and Somatostatin Analogues

Patients with acromegaly respond to somatostatin infusion with suppressin of growth hormone and improved glucose tolerance (38). Again, the short effect of native somatostatin has precluded long-term trials, hence the development of somatostatin analogues. One such analogue has been developed in Switzerland and is now undergoing extensive clinical trials in the treatment of acromegaly and pancreatic tumors (39). This compound is an octapeptide of somatostatin (SMS 201-995) with an approximate half-life of 80 min in the peripheral circulation. This analogue has now been used in the treatment of many patients with acromegaly (40-42). It appears to control growth hormone secretion with thrice daily administration of 100-200 μg subcutaneously, and with this regime there is often better control of growth hormone levels than that obtained with dopamine agonists such as bromocriptine. Furthermore, addition of bromocriptine to somatostatin does not further suppress growth hormone levels in most patients so far studied. Carbohydrate tolerance does not deteriorate, and may actually improve in some patients, presumably due to growth hormone suppression. Occasionally, patients suffer diarrhea or colicky abdominal pain, and steatorrhea may be seen, presumably because of inhibition of pancreatic exocrine secretion. This is, however, not a long-term effect, and in patients studied for up to two and a half years, to date no abnormalities have been seen in the absorption of fat soluble vitamins. This long acting analogue may prevent adequate emptying of the gall bladder, and gall stones have been found in some patients. Obviously, this point requires further study. Somatostatin octapeptide may well be the most effective medical means to controlling acromegaly in the long-term.

It is important to bear in mind that acromegaly, which is a difficult disease to treat, should be treated first either with surgery or radiotherapy. Often both of these methods of treatment are necessary. Only in cases where these treatments have been ineffective should one resort to medical intervention.

CONCLUSIONS

Somatostatin, with its widespread distribution, has a number of different, but important physiological roles. Analogues of somatostatin are of potential importance therapeutically, particularly in acromegaly. They also have a number of other indications because of the widespread effects of somatostatin, particularly in the gastro-intestinal tract.

REFERENCES

1. Roth J, Glick SM, Yalow RS, Berson SA (1963) Hypoglycemia: a potent stimulus to secretion of growth hormone. Science 140: 987-988
2. Guillemin R, Brazeau P, Bohlen P, Esch F, Ling N, Wehrenberg WB (1982) Growth hormone-releasing factor from a human pancreatic tumor that caused acromegaly. Science 218: 585-587

3. Rivier J, Speiss J, Thorner MO, Vale W (1982) Characterization of a growth hormone-releasing factor from a human pancreatic islet tumor. Nature 300: 1321-1328
4. Brazeau P, Vale W, Burgus R, Ling N, Butcher M, Rivier J, Guillemin R (1973) Hypothalamic polypeptide that inhibits the secretion of immunoreactive pituitary growth hormone. Science 178: 77-79
5. Krieger DT, Glick S, Silverberg A, Krieger HP (1968) A comparative study of endocrine tests in hypothalamic disease. Circadian periodicity of plasma 11-OHCS levels, plasma 11-OHCS and growth hormone response to insulin hypoglycemia and metyrapone responsiveness. J Clin Endocrinol Metab 28: 1589-1598
6. Abrams RL, Parker ML, Blanco S, Reichlin S, Daughaday WH (1966) Hypothalamic regulation of growth hormone secretion. Endocrinology 78: 605-613
7. Frohman LA, Bernadis LL, Burek L, Maran JW, Dhariwal APS (1971) Hypothalamic control of growth hormone secretion in the rat. Program, 2nd International Symposium on Growth Hormone, Milan, 1971. Abstract 21: 12
8. Nemeroff CB, Konkol RJ, Bissette G, Youngblood W, Martin JB, Brazeau P, Rone MS, Prang AJ, Breese GR, Kizer JS (1977) Analysis of the disruption in hypothalamic pituitary regulation in rats treated with monosodium-L-glutamate (MSG): evidence for involvement of the tubero-infundibular cholinergic and dopaminergic systems in neuroendocrine regulation. Endocrinology 101: 613-622
9. Desy L, Pelletier G (1977) Immunohistochemical localization of somatostatin in the human hypothalamus. J Cell and Tissue Research 184: 491-497
10. Luedbeck D, Mueller D, Patino C (1976) Effects of stereotaxis in the human hypothalamus and limbic system on ACTH and GH secretion. Abstract 662 of the 5th International Congress of Endocrinology, Hamburg 1976
11. Frantz J, Haselbch CH, Libert O (1962) Studies of the effect of hypothalamic extracts on somatotrophic pituitary function. Acta Endocrinology 41: 336-350
12. Deuben RR, Meites J (1964) Stimulation of pituitary growth hormone by a hypothalamic extract *in vivo*. Endocrinology 74: 408-414
13. Schally AV, Arimura A, Wakabayashi I, Sawand S, Barret JF, Bowers CY, Redding TW, Mittler JG, Saito M (1970) Chemistry of hypothalmic growth hormone-releasing hormone (GRH). In: Meites J (Ed) Hypophysiotropic Hormones of the Hypothalamus: Assay and Chemistry. Baltimore: Williams and Wilkins: pp 208-222
14. Ling N, Esch F, Bolten P, Brazeau P, Wehrenberg WB, Guillemin R (1984) Isolation, primary structure, and synthesis of hyman hypothalamic somatocrinin: Growth hormone-releasing factor. Proc Natl Acad Sci USA 81: 4302-4306
15. Thorner MO, Rivier J, Speiss J, Borges JLC, Vance ML, Bloom SR, Rogol AD, Cronin MJ, Daiser DL (1983) Human pancreatic growth hormone-releasing factor selectively stimulates growth hormone secretion in man. Lancet i: 24-28
16. Grossman A, Savage MO, Wass JAH, Lytras N, Sueiras-Diaz J, Coy DH, Besser GM (1983) Growth hormone-releasing factor in growth hormone deficiency: the demonstration of a hypothalamic defect in growth hormone release. Lancet ii: 137-138
17. Esch FS, Bohlen P, Ling NC, Brazeau PE, Wehrenberg WB, Guillemin R (1983) Primary structures of three human pancreas peptides with growth hormone-releasing activity. J Biol Chem 258: 1806-1812
18. Hammer RE, Brinster RL, Rosenfeld MG, Evans RM, Mayo KE (1985) Expression of human growth hormone-releasing factor in transgenic mice results in increased somatic growth. Nature 315: 413-416
19. Gelato MC, Pescovitz O, Cassorla F, Loriauz L, Merriam GR (1983) Effects of a growth hormone releasing factor in man. J Clin Endocrinol Metab 57: 674-676
20. Rosenthal SM, Schriock EA, Kaplan SL, Guillemin R, Grumbach MM (1983) Synthetic human pancreas growth hormone-releasing factor (hpCRF1-44-NH2) stimulates growth hormone secretion in normal men. J Clin Endocrinol Metab 57: 677-679
21. Sassolas G, Chatelain P, Cohen R, Boissel JP, Laporte S, Galleyrand J, Claustrat B, Elmcharfl A, Chayvialle JA, Cohen H, Ferry S, Underwood L (1984) Effects of human pancreatic tumor growth hormone-releasing hormone (hpGRH1-44-NH2) on immunoreactive and bioactive plasma growth hormone in normal young men. J Clin Endocrinol Metab 59: 705-709

22. Grossman A, Savage MO, Lytras N, Coy DH, Rees LH, Besser GM (1984) Responses to analogues of growth hormone-releasing hormone in normal subject, and in growth-hormone deficient children and young adults. Clin Endocrinol 21: 321-330
23. Losa M, Schopohl J, Muller OA, von Werder (1984) Stimulation of growth hormone secretion with human growth hormone releasing factor (GRF1-44, GRF1-40, GRF1-29) in normal subject. Klinische Wochen-shrift 62: 1140-1143
24. Shibasaki T, Hotta M, Masuda A, Imaki T, Obara N, Demura H, Ling N, Shizume K (1985) Plasma GH responses to GHRH and insulin-induced hypoglycemia in man. J Clin Endocrin Metab 60: 1265-1267
25. Ross RJM, Tsagarakis S, Grossman A, Preece MA, Rodda C, Davies PSW, Rees LH, Savage MO, Besser GM (1987) Treatment of growth hormone deficiency with growth hormone-releasing hormone. Lancet 1: 5-7
26. Ross RJM, Tsagarakis S, Grossman A, Nhagafoonf L, Touzel RJ, Rees LH, Besser GM (1987) GH feedback occurs through modulation of hypothalamic somatostatin under cholinergic control; studies with pyridostigmine and GRF. Clin Endocrinol
27. Thorner MO, Rescheke J, Chitwood J, Rogol AD, Furlanetto R, Rivier J, Vale W, Blizzard RM (1985) Acceleration of growth in two children treated with human growth hormone releasing factor. N Engl J Med 312: 4-9
28. Smith PJ, Brook CGD, Rivier J, Vale W, Thorner MO (1986) Nocturnal pulsatile growth hormone-releasing hormone treatment in growth hormone deficiency. Clin Endocrinol 25: 35-44
29. Ross RJM, Tsagarakis S, Grossman A, Preece MA, Rodda C, Davies PSW, Rees LH, Savage MO, Besser GM (1987) Treatment of growth hormone deficiency with growth hormone-releasing hormone. Lancet 1: 5-7
30. Ross RJM, Grossman A, Davies PSW, Savage MO, Besser GM (1987) Stilboestrol pretreatment of children with short stature does not effect the GH response to growth hormone-releasing hormone. Clin Endocrinol 27: 155-161
31. Rochiccioli PE, Tauber M, Uboldi F, Coude F, Morre M (1986) Effect of overnight constant infusion of human growth hormone (GH)-releasing hormone-(1-44) on 24-hour GH secretion in children with partial GH deficiency. J Clin Endocrinol Metab 63: 1100-1105
32. Krulich L, Dhariwal APS, McCann SM (1968) Stimulatory and inhibitory effects of purified hypothalamic extracts on growth hormone release from rat pituitary in vitro. Endocrinology 6: 783-790
33. Emson PC, Rossor M, Lee CM (1981) The regional distribution and chromatographic behavior of somatostatin in human brain. Neuroscience Let 22: 319-324
34. Wass JAH (1982) Somatostatin and its physiology in man in health and disease. In: Besser GM and Martini L (Eds) Clinical Neuroendocrinology, Volume II. New York: Academic Press: pp 359-395
35. Hall R, Besser GM, Schally AV, Coy DH, Evered D, Goldie DJ, Dastin AJ, McNeilly AS, Mortimer CH, Turnbridge WHG (1973) Action of growth hormone release inhibitory hormone in healthy men and in acromegaly. Science 318: 518-584
36. Prange Hansen AA, Orskov H, Seyer-Hansen K, Lundbaek D (1973) Some actions of growth hormone release inhibiting factor. Br Med J iii: 523-524
37. Mortimer CH, Carr D, Lind T, Mallinson CN, Schally AV, Turnbridge WMG, Yeomans L, Coy DH, Kastin AJ, Besser GM, Hall R (1974) Effects of growth hormone release inhibiting hormone on circulating glucagon, insulin and growth hormone in normal, diabetic, acromegalic and hypopituitary patients. Lancet i: 697-701
38. Besser GM, Mortimer CH, Carr D, Schally AV, Coy DH, Evered D, Kastin AJ, Turngridge WMG, Thorner MO, Hall R (1974) Growth hormone release inhibiting hormone in acromegaly. Brit Med J i: 352-355
39. Bauer W, Briner U, Dopfner W, Haller R, Huguenin R, Marbach P, Petcher TJ, Pless J (1982) SMS 201-995: a very potent selective octapeptide analogue with prolonged action. Life Sci 31: 1133-1140
40. Plewe G, Beyer J, Krause U, Neufeld M, Del Pozo E (1984 Long-acting and a selective suppression of growth hormone secretion by somatostatin analog SMS 201-995 in acromegaly. Lancet 2: 782-784

41. Lamberts SWJ, Uitterlinden P, Verschoor L, Van Dongen KJ, Del Pozo E (1985) Long-term treatment of acromegaly with somatostatin analog SMS 201-995. N Engl J Med 313: 1576-1580
42. Wass JAH, Lytras N, Besser GM (1986) Somatostatin octapeptide (SMS 201-005) in the medical treatment of acromegaly. Scand J Gastroenterol 21, Supplement 119: 136-140

F. THE FUTURE OF NEUROENDOCRINE RESEARCH

THE FUTURE OF NEUROENDOCRINE RESEARCH: A REALISTIC APPRAISAL

R. Hall

University of of Wales College of Medicine
Heath Park, Cardiff, Wales, U.K.

INTRODUCTION

Progress in neuroendocrinology has accelerated over the last decade, particularly with the application of the techniques of molecular biology. Neuroendocrinology is concerned with the transmission of messages within or between neural and endocrine tissue. This process requires a system of first messengers, receptors, second messengers and their receptors, often at the genomic level, leading to the production of protein effector molecules or hormones. A distinction between neuropeptides, e.g. thyrotropin releasing hormone (TRH), neuromodulators, e.g. acetylcholine, and gut hormones, e.g. somatostatin (SRIF), is no longer appropriate because they may all function in any capacity, depending on their location.

This review of future developments in the field of neuroendocrinology will be divided into progress in the fields of the anterior pituitary and its hormones, neuropeptides and their receptors, and degenerative brain diseases, and to the application of modern brain scanning techniques to psychiatric disorders.

ANTERIOR PITUITARY

Medical Treatment of Pituitary Tumors

The medical treatment of pituitary tumors was revolutionized by the discovery of bromocriptine, a dopamine agonist that reduces prolactin and growth hormone synthesis, and causes a consistent decrease in the size of prolactinomas. A variety of new dopamine agonists with fewer side effects and a longer duration of action have now been identified (1,2). The possibility exists that a permanent medical cure for somatotroph or lactotroph adenomas may be achieved by the use of some of these longer-acting agents.

Somatostatin analogues are now in regular use in the treatment of acromegaly (3), with occasional reports of tumor shrinkage. Other modifications of the SRIF molecule are likely to produce longer acting oral agents, which will be more generally applicable. Thyrotropin-secreting adenomas have also proved to be amenable to SRIF analogue therapy, but an effective medical agent for nonsecretory adenomas and the rare gonadotroph adenomas is still required.

Pituitary Hormones

The list of pituitary hormones continues to expand with the recent identification of cholecystokinin (CCK) and a variety of growth factors in the gland. There is no doubt that the anterior and posterior lobes of the pituitary will continue to be profitable sites for the identification of further hormones, growth factors and neuropeptides. Recent studies have emphasized the importance of the control of glycosylation of pituitary hormones, not only of the traditional glycoprotein hormones, but also of hormones such as prolactin (4). As far as TSH is concerned, glycosylation is a complex process, involving both α- and β-subunits and their combination, which determines the biological activity of the TSH molecule (5). The role of TRH and thyroid hormones in regulating TSH glycosylation, and hence its biological activity, should provide a fruitful area for future study. Similarly, the presence of glycosylated PRL in the circulation (4) may well overturn many previous concepts of PRL action, since in animals glycosylation has variable effects on the biological activity of this hormone.

Etiology of Pituitary Tumors

The role of neuropeptides, growth factors and oncogenes in the genesis of pituitary tumors is under active exploration. Neuropeptides such as growth hormone-releasing hormone can stimulate tritiated thymidine incorporation into somatotrophs in vitro and lead to the expression of the oncogene c-fos (6). A variety of growth factors - including insulin-like growth factors, epidermal growth factor, chondrocyte growth factor, glial growth factor, ovarian growth factor, fibroblast growth factor and platelet-derived growth factor - have been identified in the pituitary. The role of growth factors, their receptors and their modification as a result of proto-oncogene expression should throw some light on the development of pituitary tumors. The distinction between a growth factor, hormone and neurotransmitter has become less distinct: for example EGF has been shown to be identical with the hormone urogastrone.

Transgenic Animal Models

Fusion genes produced by the combination of the mouse metallothionein I promotor region with the rat growth hormone gene injected into fertilized mouse eggs have produced mice with inheritable gigantism and high growth hormone levels (7). Similarly porcine models of acromegaly have also been produced, as well as animals producing excess somatostatin. Production of a variety of models of human disease is thus possible and should throw further light on underlying disease mechanisms.

NEUROPEPTIDES AND THEIR RECEPTORS

Neuropeptides are in general small peptides, often belonging to a family of peptides which themselves arise from preprohormones. Post-translational processing of the preprohormone varies in different tissues such as the brain, gut, gonads and placenta (8). There is considerable parsimony in structure, as seen in the similarity between the subunits of inhibin, Mullerian inhibitory factor and TGF-β. Neurotransmitters can function as neuromodulators, or as growth factors in the brain and also act as circulating hormones in the regulation of a single homeostatic system (9).

For example, angiotensin is involved in the regulation of blood pressure and volume and of thirst; atriopeptin in the regulation of fluid and electrolyte balance; corticotropin-releasing factor (CRF) in cardiovascular regulation; gonadotropin-releasing hormone (GnRH), GnRH associated peptide (GAP) and PRL in sexual behavior and control of the gonads; bombesin in the control of temperature; AVP in memory; and CCK in satiety.

After transcription of the preprohormone, modifications can be produced by alternative splicing, e.g. CGRP/calcitonin (10), and by post-translational processing with glycosylation, amidation, sulphation, aetylation or phosphorylation, leading to a considerable variety of molecules from a single precursor. Most of the genes of the neuropeptides have now been cloned; Table 1 shows the variety of products that have been identified so far.

Cloning the genes for neuropeptide receptors has proved more difficult because of the small amount of the receptor, its liability and complex structure, often being composed of a series of glycosylated subunits. One novel method for identifying neuropeptide receptors has emerged from the "Blalock hypothesis" (11), which states that ligands and their binding sites are encoded on

TABLE 1

CLONING GENES FOR NEUROPEPTIDES

POMC gene	- ACTH; βLPH; α-,β-,γ-MSH; Clip; β-endorphin
PROENKEPHALIN gene	- Leu-enkephalin, met-enkephalin etc.
PRODYNORPHIN gene	- Dynorphin A; dynorphin B; α-, β-neoendorphin
GnRH gene	- GnRH, GAP
AVP gene	- AVP, neurophysin II
CALCITONIN	- Calcitonin, CGRP
SOMATOSTATIN gene	- SRIF-14, SRIF-28 (1-12),
TRH gene	- TRH, various peptides

complementary segments of genomic DNA. To test the hypothesis, mRNA sequences complementary to the mRNA sequences for the receptors for EGF, IL-2 and transferrin were written out and compared for homologies with their ligands when highly significant amino acid and nucleotide homologies were detected. Blalock then generated antibodies to a synthetic peptide (complementary peptide) encoded by an RNA complementary to the mRNA for GnRH (12). This antibody recognized the GnRH receptor. Further developments of this fascinating hypothesis are awaited.

New neuropeptides are continually being identified (Table 2), although a variety of actions still await their causal agents. One example of a neuropeptide with a known action which awaits full structural analysis is ASTI, a presumed nonapeptide with an approximate molecular weight of 1300, present in the circulation. This active sodium transport inhibitor inhibits Na^+/K^+-ATPase, competes with oubain for binding sites, reduces Na^+ efflux from erythrocytes, increases vascular tension and could have a key role in the genesis of essential hypertension (13).

The neuromodulatory role of the neuropeptides has recently been emphasized because they have been colocalized with most of the conventional neurons as serotonin, VIP with acetylacetylcholine, and the enkaphalins and NPY with noradrenaline. Despite their colocalization within neurons, there are still significant differences in the turnover and disposal of the neuropeptides and neurotransmitters. The former show no reuptake by neurons

TABLE 2
NEW PEPTIDES

FSH Inhibition	- Inhibin
ASTI	- ?
SLEEP	- Sleep peptides
AROUSAL	- ?
FEEDING	- ?
SATIETY	- CCK
THIRST	- Angiotensin
REPLETION	- ?
MOTOR ACTIVITY	- ?
MOTOR INACTIVITY	- ?
SEXUAL AROUSAL	- GnRH/PRL
SEXUAL SATIETY	- ?
TEMPERATURE CONTROL	- Bombesin
MEMORY	- AVP

and require new protein synthesis. Transport along the neurons is a slow process, requiring hours, and the range of action is greater. The latter show active reuptake without the need for protein synthesis or transport and act at short range over a period of seconds. Future structural and functional studies of the interaction of neuropeptides and neurotransmitters should provide key information in the field of neurobiology.

Clinical Applications of the Neuropeptides

So far only somatostatin and gonadotropin-releasing hormone have found major clinical applications. A long-acting SRIF analogue, given parenterally, has proven therapeutic effects in acromegaly, islet cell tumors of the pancreas, upper gastro-intestinal bleeding, and in ileostomy diarrhea. It may also have a role in acute pancreatitis, in severe psoriasis, and in combination with insulin in diabetes mellitus. No doubt, further therapeutic applications will emerge for current analogues or their successors.

GnRH and its long-acting analogues have emerged as drugs of considerable therapeutic importance. Pulsatile administration of native GnRH has proved helpful in the treatment of some cases of infertility. Down-regulation of GnRH receptors on the gonadotrophs by continuous administration of GnRH or intermittent use of a long-acting analogue has provided a powerful method of reversible pharmacological castration. This has been applied to the treatment of such diverse conditions as prostatic and breast carcinoma, precocious puberty, endometriosis and fibroids; other potential uses will no doubt emerge.

Disappointingly, thyrotropin-releasing hormone and corticotropin releasing hormone have only found roles as tests of their hypothalamic-pituitary-target organ axes. Growth hormone releasing hormone can be used to induce GH release, and hence stimulate growth in a proportion of children with growth hormone deficiency (15) and may well find a role in the treatment of a subgroup of these patients.

DEGENERATIVE BRAIN DISEASES

Alzheimer's Disease (16,17)

This progressive disease, now known to be responsible for the majority of cases of so-called senile dementia, has become a focus for active study by neuroendocrinologists. A number of clues to its pathogenesis have recently emerged. Premature Alzheimer's disease with extensive deposition of amyloid in the brain is seen in trisomy 21. The gene cloning for amyloid β-protein has also been located on chromosome 21, raising the possibility that reduplication of the gene might be responsible for the production of excessive amyloid. Many problems still face investigators in this field. Only some cases of Alzheimer's appear genetically-determined and many may result from multi-factorial agents, including those in the environment. There are problems in diagnosis of the condition and few, if any, specific tests. Patients and relatives are usually old and DNA is less readily available. The evidence for choline acetyltransferase deficiency is uncertain as is the reduced level of SRIF seen in the frontal and temporal cortex and in the hippocampus.

Huntington's Disease (18)

The gene for Huntington's disease has now been located on the short arm of chromosome 4, and probes closely linked to the gene are available allowing exclusion testing in pregnancy and, in some cases, testing of presymptomatic subjects at risk. Cloning of the gene with isolation of its protein product should not be long delayed.

Motor Neuron Disease (19)

Thyrotropin-releasing hormone, substance P and serotonin are all located in the median raphe nucleus of the brain stem, whose neurons project onto the cranial and spinal lower motor neurons. TRH-positive nerve terminals surround these lower motor neurons and can excite their activity. For some time, hints of transient clinical improvement have been reported after the administration of TRH or its analogues, orally, intravenously or intra-thecally, but these have not yet resulted in any routine form of therapy. It seems likely that TRH may be involved in some way in the pathogenesis of this disease and further studies in the area are awaited.

Parkinson's Disease

Perhaps the best established model of the degenerative brain diseases, Parkinson's disease has proved amenable to therapeutic intervention by dopamine precursors and to dopamine agonists. Two new lines of research have recently focused on the pathogenesis and therapy of this disease, which may also throw light on other degenerative brain diseases. First, studies of Parkinson's disease have directed attention to the "neurotoxin hypothesis" typified by the development of a Parkinson's-like condition after ingestion of methyl-phenyl-tetrahydropyridine (MPTP) (20). MAO type B drugs can protect an individual from the development of the disease, and this raises therapeutic possibilities in the native condition. MPTP can be used to produce an animal model of the disease. Other environmental toxins of relevance that have been identified include B-N-oxalylamino-L-alanine from the chickling pea seed, which can cause cortico-spinal lesions and cramps, and B-N-methylamino-L-alanine, which can cause either motor neuron or Parkinson's disease. It seems likely that endogenous and/or exogenous neurotoxins may play some part in the pathogenesis of the degenerative brain diseases.

Secondly, Parkinson's disease and its animal models have provided useful systems for the application of grafts of neuroendocrine tissue (21). In animals, grafts of intraventricular embryonic substantia nigra, intraventricular adrenal medulla, embryonic ventral mesencephalon to the surface of the neostriatum and intrastriatal nigral cell suspensions have all been used with variable degrees of success. In man, recent reports have been encouraging, but further critical assessment is required.

Application of Modern Brain Scanning Techniques to the Study of Psychiatric Disorders

The techniques of positron-emission tomography (PET) are now being applied to the study of a variety of brain disorders. Appropriate isotopes (^{11}C, ^{18}F, ^{13}N, ^{15}O) produced by a cyclotron are used to label certain drugs or precursors whose location after administration can be detected by the PET technique using a positron camera (22). This method allows the in vivo visualization of hormone or drug receptors and their location, quantitation and characterization both in animals and in man. Drug effects on neuroreceptors can be demonstrated in vivo. The technique has been applied to D_1 and D_2 dopamine receptors and also to receptors for serotonin, benzodiazepines, opiates and muscarinic drugs.

In a high proportion of patients with panic disorders (23), reduced blood flow, volume and oxygen metabolism have been demonstrated in the right parahippocampal gyrus associated with a high whole brain oxygen metabolism.

In obsessive compulsive disorders, increased metabolic rates have been demonstrated in the left orbital gyrus and the caudate nuclei with improvements associated with effective therapy. Similarly, in schizophrenia valuable information on drug action and receptor numbers have been obtained.

Such physical methods of demonstrating clear metabolic changes in many of the classical psychiatric diseases will undoubtedly throw further light on the underlying mechanisms involved in their response to drug and other therapy.

REFERENCES

1. Savino L, Goffi S, Ciccarelli E, Ghigo E, Bertagna A, Gandini G, Avataneo T, Lancranjan I, Camanni F (1987) Long-acting bromocriptine in tumorous hyperprolactinemia treatment. J Endocrinol Invest 10 (Suppl 3): Abst P. 69-82
2. Mattei AM, Ferrari C, Baroldi P, Cavioni V, Parracci A, Galparoli C, Romano C, Spellechia D, Gerevini G, Crosignani PG (1988) Prolactin-lowering effect of acute and once weekly repetitive oral administration of cabergoline at two dose levels in hyperprolactinemic patients. J Clin Endocrinol Metab 66: 193-198
3. Lambers SWJ, Uitterlinden P, Verschoor L, van Dongen KJ, Del Pozo E (1985) Long-term treatment of acromegaly with the somatostatin analogue SMS 201-995. N Engl J Med 313: 1576-1580
4. Markoff E, Lee DW (1987) Glycosylated prolactin is a major circulating variant in human serum. J Clin Endocrinol Metab 65: 1102-1106
5. Weintraub BD, Gesundheit N (1987) Thyroid-stimulating hormone synthesis and glycosylation: Clinical implications. Thyroid Today X: 1-11
6. Billestrup N, Mitchell RM, Verma JM, Vale W (1987) Stimulation of growth and proto-oncogene expression by GRF in primary pituitary cells in vitro. First European Congress of Endocrinology. Copenhagen. Abst 12.392
7. Evans RM, Swanson L, Rosenfeld MG (1985) Creation of transgenic animals to study development and as models for human disease. Rec Prog Horm Res 41: 317-337
8. Brownstein MJ (1982) Post-translational processing of neuropeptide precursors. TINS September: 318-320

9. Standaert DG, Saper CB, Needleman P (1985) Atriopeptin: potent hormone and potential neuromediator TINS 8: 509-511
10. Rosenfled MG, Amara SG, Birnberg NC, Mermod J-J, Murdoch GH, Evans RM (1983) Caltitonin, prolactin, and growth hormone gene expression as model systems for the characterization of neuroendocrine regulation. Rec Prog Horm Res 39: 305-351
11. Blalock JE, Smith EM (1984) Hydropathic anti-complementarity of amino-acids based on the genetic code. Biochem & Biophys Res Comm 121: 203-207
12. Mulchahey JJ, Neill JD, Dion LD, Bost KL, Blalock JE (1986) Antibodies to the binding site of the receptor for luteinizing hormone-releasing hormone (LHRH) P: generation with a synthetic decapeptide encoded by an RNA complementary to LHRH mRNA. Proc Natl Acad Sci USA 83: 9714-9718
13. Mir MA, Chappell SP, Morgan K, Lewis MD, Scanlon MF, Lewis MJ (1986) Hypothalamic sodium transport inhibitor and vascular reactivity. J Hypertension 4 (Suppl 5): S233-S235
14. Lundberg JM, Hokfelt T (1983) Coexistence of peptides and classical neurotransmitters. TINS August: 325-333
15. Low LCK, Wang C, Cheung PT, Ho P, Lam KSL, Young RTT, Yeung CY, Ling N (1988) Long term pulsatile growth hormone (GH)-releasing hormone therapy in children with GH deficiency. J Clin Endocrinol & Metab 66: 611-617
16. Selkoe DJ, Bell DS, Podlisny MB, Price DL, Cork LC (1987) Conservation of brain amyloid proteins in aged mammals and humans with Alzheimer's disease. Science 253: 873-876
17. Bahmanyar S, Higgins GA, Goldgaber D, Lewis DA, Morrison JH, Wilson MC, Shankar SK, Gajdusek DC (1987) Localization of amyloid protein messenger RNA in brains from patients with Alzheimer's disease. Science 237: 77-80
18. Meissen GJ, Myers RH, Mastomauro CA, Koroshetz WJ, Klinger KW, Farrer LA, Watkins PA, Gusella JF, Bird EF, Martin JB (1988) Predictive testing for Huntington's disease with use of a linked DNA marker. New Engl J Med 318: 535-541
19. Bowker RM, Westlund KN, Sullivan MC, Wilber JF (1983) Descending serotonergic, peptidergic and cholinergic pathways from the raphe nuclei: a multiple transmitter complex. Brain Research 288: 33-48
20. Langston JW (1985) MPTP and Parkinson's disease. TINS February: 79-83
21. Dunnett SB, Bjorklund A, Stenevi U (1983) Dopamine-rich transplants in experimental Parkinsonism. TINS July: 266-270
22. Sedvall G, Farde L, Persson A, Wiesel F-A (1986) Imaging of neurotransmitter receptors in the living human brain. Arch Gen Psychiatry 43: 995-1005
23. Reinman EM, Raichle ME, Robins E, Butler FK, Herscovitch P, Fox P, Perlmutter J (1986) The application of positron emission tomography to the study of panic disorders. Am J Psychiatric 143: 469-477
24. Baxter LR, Phelps ME, Mazziotta JC, Guze BH, Schwartz JM, Selin CE (1987) Local cerebral glucose metabolic rates in obsessive-compulsive disorder. Arch Gen Psychiatry 44: 211-218
25. Wong DF, Wagner HN, Tune LE, Dannals RF, Pearlson GD, Links JM, Tamminga CA, Broussolle EP, Ravert HT, Wilson AA, Toung JKT, Malat J, Williams JA, O'Tuama LA, Snyder SH, Kuhar MJ, Gjedde A (1986) Positron emission tomography reveals elevated D_2 dopamine receptors in drug-naive schizophrenics. Science 234: 1558-1563

AUTHOR INDEX

SUBJECT INDEX

MIX
Papier aus verantwortungsvollen Quellen
Paper from responsible sources
FSC® C105338

If you have any concerns about our products,
you can contact us on
ProductSafety@springernature.com

In case Publisher is established outside the EU,
the EU authorized representative is:
Springer Nature Customer Service Center GmbH
Europaplatz 3, 69115 Heidelberg, Germany

Printed by Libri Plureos GmbH
in Hamburg, Germany